Student Solutions Manual to Accompany

COLLEGE ALGEBRA AND TRIGONOMETRY WITH APPLICATIONS

SECOND EDITION

JOSÉ BARROS-NETO
Rutgers University

WEST PUBLISHING COMPANY
St. Paul   New York   Los Angles   San Francisco

WEST'S COMMITMENT TO THE ENVIRONMENT

In 1906, West Publishing Company began recycling materials left over from the production of books. This began a tradition of efficient and responsible use of resources. Today, up to 95% of our legal books and 70% of our college texts are printed on recycled, acid-free stock. West also recycles nearly 22 million pounds of scrap paper annually—the equivalent of 181,717 trees. Since the 1960s, West has devised ways to capture and recycle waste inks, solvents, oils, and vapors created in the printing process. We also recycle plastics of all kinds, wood, glass, corrugated cardboard, and batteries, and have eliminated the use of styrofoam book packaging. We at West are proud of the longevity and the scope of our commitment to our environment.

Production, Prepress, Printing and Binding by West Publishing Company.

ISBN 0–314–65233–7

PREFACE

This Solutions Manual contains worked out solutions for all odd-numbered exercises in College Algebra and Trigonometry with Applications.

At the end of each chapter, we include a Chapter Test containing 30 problems each. These are review problems designed for students who need additional practice. The answers for the Chapter Tests are given at the end of this manual.

Jose Barros-Neto

Table of Contents

Chapter 1
Fundamentals of Algebra

Exercises 1.1

1. $240 = 2 \cdot 120$
$= 2 \cdot 2 \cdot 60$
$= 2 \cdot 2 \cdot 2 \cdot 30$
$= 2 \cdot 2 \cdot 2 \cdot 2 \cdot 15$
$= 2 \cdot 2 \cdot 2 \cdot 2 \cdot 3 \cdot 5$

3. $432 = 2 \cdot 216$
$= 2 \cdot 2 \cdot 108$
$= 2 \cdot 2 \cdot 2 \cdot 54$
$= 2 \cdot 2 \cdot 2 \cdot 2 \cdot 27$
$= 2 \cdot 2 \cdot 2 \cdot 2 \cdot 3 \cdot 9$
$= 2 \cdot 2 \cdot 2 \cdot 2 \cdot 3 \cdot 3 \cdot 3$

5. $2310 = 2 \cdot 1155$
$= 2 \cdot 3 \cdot 385$
$= 2 \cdot 3 \cdot 5 \cdot 77$
$= 2 \cdot 3 \cdot 5 \cdot 7 \cdot 11$

7. $\dfrac{25}{120} + \dfrac{42}{80} = \dfrac{25}{120} + \dfrac{21}{40}$

$= \dfrac{25}{120} + \dfrac{63}{120}$

$= \dfrac{88}{120}$

$= \dfrac{11}{55}$

9. $\dfrac{21}{60} \times \dfrac{24}{84} = \dfrac{1}{5} \times \dfrac{2}{4}$

$= \dfrac{2}{20}$

$= \dfrac{1}{10}$

11. $3/7 = 0.\overline{428571}$

13. $21/40 = 0.525$

15. $3/11 = 0.\overline{27}$

17. $15/120 = 1/8 = 0.125$

19. $0.15 = 15/100 = 3/20$

21. If $r = 0.\overline{123}$, then

$1000r = 123.\overline{123}$
$-r = -0.\overline{123}$
$\overline{ }$
$999r = 123$
$r = 123/999$
$= 41/333$

1

23. If $r = 0.\overline{418}$, then

$$1000r = 418.\overline{18}$$
$$\underline{-10r = \quad -4.\overline{18}}$$
$$990r = \quad 414$$
$$r = \quad 414/990$$
$$= \quad 23/55$$

25. If $r = 1.\overline{106}$, then

$$1000r = 1106.\overline{06}$$
$$\underline{-10r = \quad -11.\overline{06}}$$
$$990r = \quad 1095$$
$$r = \quad 1095/990$$
$$= \quad 73/66$$

27. We have

$$0.444444\ldots$$
$$\underline{+ \; 0.151515\ldots}$$
$$0.595959\ldots$$

Thus $0.\overline{4} + 0.\overline{15} = 0.\overline{59}$.

29. We have

$$0.121212\ldots$$
$$\underline{+ \; 0.203203\ldots}$$
$$0.324415\ldots$$

Thus $0.\overline{12} + 0.\overline{203} = 0.\overline{324415}$

Exercises 1.2

1. Associativity of the addition.

3. Commutativity of the multiplication.

5. Distributivity of the multiplication over the addition.

7. Distributivity property and commutativity of the multiplication.

9. a)

$$\frac{3}{10}\left(\frac{2}{3} + \frac{1}{4}\right) = \frac{3}{10} \cdot \frac{2}{3} + \frac{3}{10} \cdot \frac{1}{4}$$
$$= \frac{1}{5} + \frac{3}{40}$$
$$= \frac{8}{40} + \frac{3}{40}$$
$$= \frac{11}{40}$$

b)

$$\frac{3}{10} \cdot \left(\frac{8}{12} + \frac{3}{12}\right) = \frac{3}{10} \cdot \frac{11}{12}$$
$$= \frac{11}{40}$$

11. a)

$$\frac{5}{12}\left(\frac{-4}{5} + \frac{8}{12}\right) = \frac{5}{12} \cdot \frac{-4}{5} + \frac{5}{12} \cdot \frac{8}{12}$$
$$= \frac{-1}{3} + \frac{10}{36}$$
$$= \frac{-1}{3} + \frac{5}{18}$$
$$= \frac{-6}{18} + \frac{5}{18}$$
$$= \frac{-1}{18}$$

b)

$$\frac{5}{12}\left(\frac{-4}{5} + \frac{8}{12}\right) = \frac{5}{12}\left(\frac{-4}{5} + \frac{2}{3}\right)$$
$$= \frac{5}{12}\left(\frac{-12}{15} + \frac{10}{15}\right)$$
$$= \frac{5}{12}\left(\frac{-2}{15}\right)$$
$$= \frac{-1}{18}$$

13. a)

$$\frac{-4}{9}\left(\frac{1}{3} - \frac{2}{5}\right) = \frac{-4}{27} + \frac{8}{45}$$

$$= \frac{-20}{135} + \frac{24}{135}$$

$$= \frac{4}{135}$$

b)

$$\frac{-4}{9}\left(\frac{1}{3} - \frac{2}{5}\right) = \left(\frac{-4}{9}\right)\left(\frac{5}{15} - \frac{6}{15}\right)$$

$$= \left(\frac{-4}{9}\right)\left(\frac{-1}{15}\right)$$

$$= \frac{4}{135}$$

15.

$$\frac{2 + 1/2}{2 - 1/2} = \frac{(4 + 1)/2}{(4 - 1)/2}$$

$$= \frac{5/2}{3/2} = \frac{5}{3}$$

17.

$$\frac{2/3 + 1/2}{7/9 - 2/3} = \frac{(4 + 3)/6}{(7 - 6)/9}$$

$$= \frac{7/6}{1/9} = \frac{7}{6} \times \frac{9}{1}$$

$$= \frac{21}{2}$$

19.

$$\frac{1/3 - 3/4}{1/2 + 2/3} = \frac{(4 - 9)/12}{(3 + 4)/6}$$

$$= \frac{-5/12}{7/6} = \frac{-5}{12} \times \frac{6}{7}$$

$$= \frac{-5}{14}$$

21.

$$\frac{2/3 + 3/6 - 3/12}{7/8 - 4/3}$$

$$= \frac{8/12 + 6/12 - 3/12}{21/24 - 32/24} = \frac{11/12}{-11/24}$$

$$= \frac{11}{12} \times \frac{-24}{11} = -2$$

23.

$$\frac{4/9 - 1/6 + 5/18}{2/3 + 5/6}$$

$$= \frac{(8 - 3 + 5)/18}{(4 + 5)/6} = \frac{10/18}{9/6}$$

$$= \frac{10}{18} \times \frac{6}{9} = \frac{10}{27}$$

25.

$$\frac{(3/5 \times (2/3 - 3/4)}{4/6 + 2/3 - 2/5}$$

$$= \frac{(3/5) \times (8/12 - 9/12)}{20/30 + 20/30 - 12/30}$$

$$= \frac{(3/5) \times (-1/12)}{28/30} = \frac{-1/20}{28/30}$$

$$= \frac{-1}{20} \times \frac{30}{28} = \frac{-3}{56}$$

27.

$$\frac{3/5 - 3/4 \times 6/8}{1/3 \times 3/7 + 5/4}$$

$$= \frac{3/5 - 9/16}{1/7 + 5/4} = \frac{48/80 - 45/80}{4/28 + 35/28}$$

$$= \frac{3/80}{39/28} = \frac{7}{260}$$

29. No: $\sqrt{2} + (-\sqrt{2}) = 0$ which is a rational number.

31. Let $x = \sqrt{2} + 5/6$. If x is rational, then $x - 5/6$ is also rational. But $x - 5/6 = \sqrt{2}$ which is irrational. Therefore, x cannot be rational.

33. Let $x = a + b$ with a rational and b irrational. If x is rational, then $x - a$ is also rational. But $x - a = b$ is an irrational number. Therefore, x must be irrational.

35. a) rational, b) irrational, c) rational, d) irrational.

37. $\dfrac{22}{7} = 3.\overline{142857}$ and

    $\pi = 3.141592\ldots.$

    By comparing corresponding place values it follows that

    $\dfrac{22}{7} > \pi.$

39. If $\dfrac{a}{b} = \dfrac{c}{d},$ then

    $$\dfrac{a}{b} + 1 = \dfrac{c}{d} + 1$$

    $$\dfrac{a}{b} + \dfrac{b}{b} = \dfrac{c}{d} + \dfrac{d}{d}$$

    $$\dfrac{a + b}{b} = \dfrac{c + d}{d}$$

Exercises 1.3

1. $11^4 = 11 \cdot 11 \cdot 11 \cdot 11 = 14641$

3. $5^{-3} = \dfrac{1}{5^3} = \dfrac{1}{125}$

5. $\left(\dfrac{3}{4}\right)^{-2} = \left(\dfrac{4}{3}\right)^2 = \dfrac{16}{9}$

7. $(-8)^3 = -512$

9. $(-7)^{-3} = \left(\dfrac{1}{-7}\right)^3 = -\dfrac{1}{343}$

11. $3(-4)^{-3} = \dfrac{3}{(-4)^3} = -\dfrac{3}{64}$

13. $3^{-2} + 4^{-1} = \dfrac{1}{9} + \dfrac{1}{4}$

    $$= \dfrac{4}{36} + \dfrac{9}{36}$$

    $$= \dfrac{13}{36}$$

15. $\left(\dfrac{2}{3}\right)^{-1} + 3^{-1} = \dfrac{3}{2} + \dfrac{1}{3}$

    $$= \dfrac{9}{6} + \dfrac{2}{6}$$

    $$= \dfrac{11}{6}$$

17. $(2xy^3)^{-2} = \dfrac{1}{(2xy^3)^2}$

    $$= \dfrac{1}{4x^2y^6}$$

19. $\dfrac{x^9}{x^{21}} = \dfrac{1}{x^{21-9}} = \dfrac{1}{x^{12}}$

21. $x^{-2} + \dfrac{1}{x^{-2}} = \dfrac{1}{x^2} + x^2$

    $$= \dfrac{1}{x^2} + \dfrac{x^4}{x^2}$$

    $$= \dfrac{1 + x^4}{x^2}$$

23. $\dfrac{(x + y)^{10}}{(x + y)^{13}} = \dfrac{1}{(x + y)^3}$

25. $\dfrac{3x^2y^{-4}}{z^{-5}} = \dfrac{3x^2z^5}{y^4}$

27. $(4x^{-8}y^{-3})x^2y^2(3y) = \dfrac{12x^2y^3}{x^8y^3}$

    $$= \dfrac{12}{x^6}$$

29. $(3x^2y^{-1})^2 = 9x^4y^{-2}$

    $$= \dfrac{9x^4}{y^2}$$

31. $\left(\dfrac{4ab^2}{c^2}\right)^{-4} = \left(\dfrac{c^2}{4ab^2}\right)^4$

    $$= \dfrac{c^8}{256a^4b^8}$$

33. $(\dfrac{x^{-2}y^{-3}}{2})^2 = \dfrac{x^{-4}y^{-6}}{4}$

$= \dfrac{1}{4x^4y^6}$

35. $\dfrac{3x^2y^{-1}}{(xy^2)^{-2}} = \dfrac{3x^2y^{-1}}{x^{-2}y^{-4}}$

$= 3x^4y^3$

37. $\dfrac{a^{-2}b^{-3}c^{-4}}{3ab} = \dfrac{1}{3aba^2b^3c^4}$

$= \dfrac{1}{3a^3b^4c^4}$

39. $[(\dfrac{a^{-2}x^2}{y^2})^2]^2 = [(\dfrac{x^2}{a^2y^2})^2]^2$

$= [\dfrac{x^2}{a^4y^4}]^2$

$= \dfrac{x^4}{a^8y^8}$

41. $\dfrac{a^{-n}}{b^{-n}} = \dfrac{\frac{1}{a^n}}{\frac{1}{b^n}} = \dfrac{1}{a^n} \cdot \dfrac{b^n}{1} = \dfrac{b^n}{a^n}$

43. $5{,}150{,}000{,}000 = 5.15 \times 10^9$

45. $0.000\ 000\ 001\ 8 = 1.8 \times 10^{-9}$

47. $0.000\ 000\ 000\ 514\ 2 = 5.142 \times 10^{-10}$

49. $0.213 \times 10^7 = 2.13 \times 10^6$

51. $405.6 \times 10^{-8} = 4.056 \times 10^{-6}$

53. 5.6 ft. has a range of 5.55 to 5.65 ft.

55. 15.76 km has a range of 15.755 to 15.765km.

57. 21.315 kg has a range of 21.3145 to 21.3155 kg.

59. 31.6 has 3 significant digits.

61. 56.012 has 5 significant digits.

63. 0.01506 has 4 significant digits.

65. 3.2×10^{-8} has 2 significant digits.

67. $\dfrac{(3.2 \times 10^{-3})(2.5 \times 10^{-2})}{(0.8 \times 10^4)(0.5 \times 10^{-5})}$

$= \dfrac{(3.2)(2.5)}{(0.8)(0.5)} \times \dfrac{(10^{-3})(10^{-2})}{(10^4)(10^{-5})}$

$= 20 \times 10^{-4}$

$= 2.0 \times 10^{-3}$

69. $\dfrac{(1.2 \times 10^{-8})(4.9 \times 10^6)}{(1.5 \times 10^{-9})(7.0 \times 10^3)}$

$= \dfrac{(1.2)(4.9)}{(1.5)(7.0)} \times \dfrac{(10^{-8})(10^6)}{(10^{-9})(10^3)}$

$= 0.56 \times 10^4$

$= 5.6 \times 10^3$

71. $\dfrac{3.1 \times 10^{-12}}{9.0 \times 10^6} = .34 \times 10^{-18}$

$= 3.4 \times 10^{-19}$

73. $\dfrac{(4.0 \times 10^{12})(3.5 \times 10^{-9})}{(2.0 \times 10^{-4})^2}$

$= \dfrac{(4.0)(3.5) \times (10^{12})(10^{-9})}{4.0 \times 10^{-8}}$

$= 3.5 \times 10^{12-9+8}$

$= 3.5 \times 10^{11}$

75. $\dfrac{(0.000\ 000\ 012)(41000)}{0.000\ 000\ 001\ 5}$

$= \dfrac{(1.2 \times 10^{-8})(4.1 \times 10^4)}{1.5 \times 10^{-9}}$

$= \dfrac{(1.2)(4.1)}{1.5} \times \dfrac{(10^{-8})(10^4)}{10^{-9}}$

$= 3.3 \times 10^5$

77. $\dfrac{(3210)(0.000\ 000\ 000\ 18)}{(321\ 000\ 000)(0.000\ 516)}$

$= \dfrac{(3.21 \times 10^3)(1.8 \times 10^{-10})}{(3.21 \times 10^8)(5.16 \times 10^{-4})}$

$= \dfrac{\cancel{(3.21)}(1.8)}{\cancel{(3.21)}(5.16)} \times \dfrac{10^3 \times 10^{-10}}{10^8 \times 10^{-4}}$

$= 3.5 \times 10^{-12}$

79. 1 hour = 60 minutes = 3600 seconds
    300,000 × 3600

$= (3.0 \times 10^5)(3.6 \times 10^3)$

$= (3.0 \times 3.6)(10^5)(10^3)$

$= 1.08 \times 10^9$ km

81. $5.983 \times 10^{24} \times 10^3 \times 10^{-6}$

$= 5.983 \times 10^{21}$ metric tons

83. $(1.228 \times 10^{-2})(5.983 \times 10^{24})$

$= 7.347 \times 10^{22}$ kg

85. $\dfrac{(6.673 \times 10^{-11})(5.983 \times 10^{24})(7.337 \times 10^{22})}{(3.843 \times 10^8)^2}$

$= \dfrac{292.926 \times 10^{35}}{14.768 \times 10^{16}}$

$= 19.83 \times 10^{19}$

$= 1.983 \times 10^{20}$ N

87. $(2.998 \times 10^8)(6.214 \times 10^{-4})$

$= 18.629 \times 10^4$

$= 1.863 \times 10^5$ miles per second

89. a) in liters;

$\dfrac{22.4}{6.02 \times 10^{23}} = 3.72 \times 10^{-23}$ ℓ

b) m^3:

$\dfrac{22.4 \times 10^{-3}}{6.02 \times 10^{23}} = 3.72 \times 10^{-26}$ m^3

Exercises 1.4

1. $(2x^2 - 3x + 1) + (-x^2 + x - 4)$

$= 2x^2 - x^2 - 3x + x + 1 - 4$

$= x^2 - 2x - 3$

3. $(-\dfrac{1}{2}x^4 + 2x^2 - \dfrac{1}{5}) - (\dfrac{1}{3}x^4 - 3x^2 + \dfrac{1}{2})$

$= -\dfrac{1}{2}x^4 + 2x^2 - \dfrac{1}{5} - \dfrac{1}{3}x^4 + 3x^2 - \dfrac{1}{2}$

$= \dfrac{1}{6}x^4 + 5x^2 - \dfrac{7}{10}$

5. $(x^7 + 5x^5 + 3x^3 + x + 1) +$
    $(4x^4 + 2x^2 + 1)$

$= x^7 + 5x^5 + 4x^4 + 3x^3 + 2x^2 + x + 2$

7. $(t^5 - t^4 + t^3 - t^2 - 1) +$
    $(2 + 3t + 4t^2 + 5t^3)$

$= t^5 - t^4 + 6t^3 + 3t^2 + 3t + 1$

9. $(x^4 - 1) + (x^4 - 3x^2) -$
    $(-x^4 - 2x^3 + x^2 + x - 5)$

$= x^4 - 1 + x^4 - 3x^2 + x^4 + 2x^3 - x^2$
    $- x + 5$

$= 3x^4 + 2x^3 - 4x^2 - x + 4$

11. $5x^3 - 4x^2y + 3xy^2 - 4y^3 + xy + 1$

$\dfrac{- 4x^2 - 3x^2y + 6xy^2 \qquad\ \ + 6xy + 4 - x}{5x^3 - 4x^2 - 7x^2y + 9xy^2 - 4y^3 + 7xy}$
    $- x + 5$

13. $(2x^2 - 3x + 1)(4x - 2)$

$= 8x^3 - 4x^2 - 12x^2 + 6x + 4x - 2$

$= 8x^3 - 16x^2 + 10x - 2$

15. $(3x^2 - 2x + 1)(x^2 - 2x + 3)$

$= 3x^4 - 6x^3 + 9x^2 - 2x^3 + 4x^2 - 6x$
$\quad + x^2 - 2x + 3$

$= 3x^4 - 8x^3 + 14x^2 - 8x + 3$

17. $(x^3 - 4x^2 + x)(x^4 - 2x^2)$

$= x^7 - 4x^6 + x^5 - 2x^5 + 8x^4 - 2x^3$

$= x^7 - 4x^6 - x^5 + 8x^4 - 2x^3$

19. $(x^5 - x^3 + 1)(x^2 + x + 1)$

$= x^7 + x^6 + x^5 - x^5 - x^4 - x^3 + x^2$
$\quad + x + 1$

$= x^7 + x^6 - x^4 - x^3 + x^2 + x + 1$

21. $(a + b)(a^2 - ab + b^2)$

$= a^3 - \cancel{a^2b} + \cancel{ab^2} + \cancel{a^2b} - \cancel{ab^2} + b^3$

$= a^3 + b^3$

23. $(x + y + 1)(x + y - 1)$

$= x^2 + xy - x + xy + y^2 - y$
$\quad + x + y - 1$

$= x^2 + 2xy + y^2 - 1$

25. $(2x - 3y + 4)(2x + 3y - 4)$

$= 4x^2 + \cancel{6xy} - \cancel{8x} - \cancel{6xy} - 9y^2$
$\quad + 12y + \cancel{8x} + 12y - 16$

$= 4x^2 - 9y^2 + 24y - 16$

27. $(x^2 + y^2 + xy)(x^2 - y^2 - xy)$

$= x^4 - x^2y^2 - x^3y + x^2y^2 - y^4 - xy^3$
$\quad + x^3y - xy^3 - x^2y^2$

$= x^4 - y^4 - x^2y^2 - 2xy^3$

29. $(x - 1)(x - 2)(x - 3)$

$= (x - 1)(x^2 - 3x - 2x + 6)$

$= (x - 1)(x^2 - 5x + 6)$

$= x^3 - 5x^2 + 6x - x^2 + 5x - 6$

$= x^3 - 6x^2 + 11x - 6$

31. $(3x - 1)(3x - 2)(3x - 3)$

$= (9x^2 - 9x + 2)(3x - 3)$

$= 27x^3 - 27x^2 + 6x - 27x^2 + 27x - 6$

$= 27x^3 - 54x^2 + 33x - 6$

33. $(2x + 1)(2x + 1)(2x + 1)$

$= (4x^2 + 2x + 2x + 1)(2x + 1)$

$= (4x^2 + 4x + 1)(2x + 1)$

$= 8x^3 + 4x^2 + 8x^2 + 4x + 2x + 1$

$= 8x^3 + 12x^2 + 6x + 1$

35. $(a + b)(a - b)(a^2 + b^2)$

$= (a^2 - b^2)(a^2 + b^2)$

$= a^4 - b^4$

37. $(3x + 2)(3x - 2)(x + 1)(x - 1)$

$= [(3x + 2)(3x - 2)][(x + 1)(x - 1)]$
$\qquad \text{[Associativity]}$

$= (9x^2 - \cancel{6x} + \cancel{6x} - 4)(x^2 - \cancel{x} + \cancel{x} - 1)$

$= (9x^2 - 4)(x^2 - 1)$

$= 9x^4 - 9x^2 - 4x^2 + 4$

$= 9x^4 - 13x^2 + 4$

39. $(6x^4 + 8x^2 - 10x) \div 2x$

$\qquad = 3x^3 + 4x - 5$

41. $(9x^7 - 12x^5 - 3x^4 + 6x^3) \div 3x^2$

$\qquad = 3x^5 - 4x^3 - x^2 + 2x$

43. $\dfrac{16x^9 - 20x^7 + 12x^5}{4x^3}$

$\qquad = 4x^6 - 5x^4 + 3x^2$

45. $(10x^3y^2 - 6x^2y^3) \div 2xy^2$

$\qquad = 5x^2 - 3xy$

47. $\dfrac{3x^2yz - 6x^2yz^3 + 9x^3y^2z^3}{3xyz}$

$\qquad = x - 2xz^2 + 3x^2yz^2$

49. $\dfrac{16a^2u^2v^4 - 8a^3u^3v^3 + 12a^4u^4v^2}{4a^2uv}$

$\qquad = 4uv^3 - 2au^2v^2 + 3a^2u^3v$

Exercises 1.5

1. $(2x + 1)(2x + 3) = 4x^2 + 8x + 3$

3. $(2x + 3a)(2x - a) = 4x^2 + 4ax - 3a^2$

5. $(2ax - 5b)^2 = 4a^2x^2 - 20abx + 25b^2$

7. $(4y - 3b)^2 = 16y^2 - 24by + 9b^2$

9. $(x^2 + 5)(x^2 - 5) = x^4 - 25$

11. $(x + \sqrt{2})(x - \sqrt{2}) = x^2 - 2$

13. $(\sqrt{a} + \sqrt{b})(\sqrt{a} - \sqrt{b}) = a - b$

15. $(x - y + a)^2 = ((x - y) + a)^2$

$\qquad = (x - y)^2 + 2a(x - y) + a^2$

$\qquad = x^2 - 2xy + y^2 + 2ax - 2ay + a^2$

17. $x^2 + 6x + 8 = (x + 2)(x + 4)$

19. $x^2 - 11x + 24 = (x - 3)(x - 8)$

21. $x^2 + 2x - 35 = (x - 5)(x + 7)$

23. $y^2 - 4y - 12 = (y - 6)(y + 2)$

25. $p^2 - 11p + 30 = (p - 5)(p - 6)$

27. $4x^2 + x - 3 = (4x - 3)(x + 1)$

29. $2x^2 + 16x + 30 = 2(x^2 + 8x + 15)$

$\qquad = 2(x + 3)(x + 5)$

31. $5x^2 - 20 = 5(x^2 - 4)$

$\qquad = 5(x + 2)(x - 2)$

33. $3a^2 + 6ax - 9a = 3a(x^2 + 2x - 3)$

$\qquad = 3a(x - 1)(x + 3)$

35. $64a^2 - 9b^2 = (8a + 3b)(8a - 3b)$

37. $16x^4 - 81 = (4x^2 + 9)(4x^2 - 9)$

$\qquad = (4x^2 + 9)(2x + 3)(2x - 3)$

39. $x^3 - 27 = (x - 3)(x^2 + 3x + 9)$

41. $8x^3 + 64 = 8(x^3 + 8)$

$\qquad = 8(x + 2)(x^2 - 2x + 4)$

43. $16ax^3 + 2a = 2a(8x^3 + 1)$

$\qquad = 2a(2x + 1)(4x^2 - 2x + 1)$

45. $2x^2 + 5x - 3 = (2x - 1)(x + 3)$

47. $4x^2 - 4x + 1 = (2x - 1)(2x - 1)$

$\qquad = (2x - 1)^2$

49. $6x^2 - 5x + 1 = (2x - 1)(3x - 1)$

51. $3x^2 + 11x + 6 = (x + 3)(3x + 2)$

53. $2ac - 6ad + bc - 3bd$

$$= 2a\,(c - 3d) + b(c - 3d)$$

$$= (c - 3d)(2a + b)$$

55. $6ax - 9ay - 2bx + 3by$

$$= 3a\,(2x - 3y) - b(2x - 3y)$$

$$= (2x - 3y)(3a - b)$$

57. $x^2y + 2ay - 3bx^2 - 6ab$

$$= y(x^2 + 2a) - 3b(x^2 + 2a)$$

$$= (x^2 + 2a)(y - 3b)$$

59. $(a + b)^3 - 8$

$$= ((a + b)-2)((a + b)^2+2(a + b)+4)$$

$$= (a +b -2)(a^2 +2ab +b^2 +2a +2b +4)$$

61. $x^6 - 64 = (x^3 + 8)(x^3 - 8)$

$$= (x + 2)(x^2 - 2x + 4)\times$$
$$(x - 2)(x^2 + 2x + 4)$$

63. $(x + 2y)^2 - 4z^2$

$$= (x + 2y + 2z)(x + 2y - 2z)$$

65. $(2x - 1)^3 + 8$

$$= (2x -1 +2)((2x - 1)^2-2(2x - 1)+4)$$

$$= (2x +1)(4x^2 - 4x + 1 - 4x + 2 + 4)$$

$$= (2x + 1)(4x^2 - 8x + 7)$$

67. $(a + b)^2 - (a - b)^2$

$$= ((a + b)+(a - b))((a + b)-(a - b))$$

$$= (a + b + a - b)\,(a + b - a + b)$$

$$= (2a)(2b)$$

$$= 4ab$$

69. $\dfrac{u^2}{2} - 8 = \dfrac{1}{2}(u^2 - 16)$

$$= \dfrac{1}{2}(u + 4)(u - 4)$$

71. $x^2 - \dfrac{1}{6}x - \dfrac{1}{6} = \dfrac{1}{6}(3x + 1)(2x - 1)$

73. $\dfrac{x^3}{2} + 4 = \dfrac{1}{2}(x^3 + 8)$

$$= \dfrac{1}{2}(x + 2)(x^2 - 2x + 4)$$

75. $(321.4)^2 - (320.4)^2$

$$= (321.4 + 320.4)(321.4 - 320.4)$$

$$= (641.8)(1)$$

$$= 641.8$$

77. Let n be any natural number. Then $n + 1$ is the next consecutive number. The difference of squares is
$$(n + 1)^2 - n^2$$

$$= (n + 1 + n)(n + 1 - n)$$

$$= (2n + 1)(1)$$

$$= 2n + 1$$

79. Let n be the integer. Then

$$\frac{1}{4} \left[(n + 1)^2 - (n - 1)^2 \right]$$

$$= \frac{1}{4} \left[n^2 + 2n + 1 - n^2 + 2n - 1 \right]$$

$$= \frac{1}{4} (4n)$$

$$= n$$

Exercises 1.6

1. $\dfrac{15a^2x^4}{3ax^2} = 5ax^2$

3. $\dfrac{10x^6 + 15x^4 - 5x^2}{5x} = \dfrac{5x(2x^5 + 3x^3 - x)}{5x}$

$$= 2x^5 + 3x^3 - x$$

5. $\dfrac{6x - 18}{6x - 24} = \dfrac{6(x - 3)}{6(x - 4)}$

$$= \dfrac{x - 3}{x - 4}$$

7. $\dfrac{12z^2 + 6z}{9z^2 - 3z} = \dfrac{6z(2z + 1)}{3z(3z - 1)}$

$$= \dfrac{2(2z + 1)}{(3z - 1)}$$

9. $\dfrac{x + 1}{x^2 + 9x + 8} = \dfrac{x + 1}{(x + 1)(x + 8)}$

$$= \dfrac{1}{x + 8}$$

11. $\dfrac{u^2 - 9}{u^2 + 6u + 9} = \dfrac{(u + 3)(u - 3)}{(u + 3)(u + 3)}$

$$= \dfrac{u - 3}{u + 3}$$

13. $\dfrac{2x^2 + 10x}{2x^2 + 9x - 5} = \dfrac{2x(x + 5)}{(2x - 1)(x + 5)}$

$$= \dfrac{2x}{2x - 1}$$

15. $\dfrac{4x^2 + 7x + 4}{4x^2 + 11x + 6}$ is in simplest form.

17. $3 + \dfrac{x - 1}{x + 2} = \dfrac{3(x + 2)}{x + 2} + \dfrac{x - 1}{x + 2}$

$$= \dfrac{3x + 6 + x - 1}{x + 2}$$

$$= \dfrac{4x + 5}{x + 2}$$

19. $x - 2 - \dfrac{x - 2}{x - 3}$

$$= \dfrac{x(x - 3)}{x - 3} - \dfrac{2(x - 3)}{x - 3} - \dfrac{x - 2}{x - 3}$$

$$= \dfrac{x(x - 3) - 2(x - 3) - (x - 2)}{x - 3}$$

$$= \dfrac{x^2 - 3x - 2x + 6 - x + 2}{x - 3}$$

$$= \dfrac{x^2 - 6x + 8}{x - 3}$$

21. $\dfrac{3}{2x - 1} - \dfrac{x}{x + 4}$

$$= \dfrac{3(x + 4) - x(2x - 1)}{(2x - 1)(x + 4)}$$

$$= \dfrac{3x + 12 - 2x^2 + x}{(2x - 1)(x + 4)}$$

$$= \dfrac{-2x^2 + 4x + 12}{(2x - 1)(x + 4)}$$

23. $\dfrac{x + 4}{x - 4} - \dfrac{x - 1}{x + 4}$

$$= \dfrac{(x + 4)(x + 4)}{(x - 4)(x + 4)} - \dfrac{(x - 1)(x - 4)}{(x - 4)(x + 4)}$$

$$= \dfrac{(x^2 + 8x + 16) - (x^2 - 5x + 4)}{(x - 4)(x + 4)}$$

$$= \dfrac{x^2 + 8x + 16 - x^2 + 5x - 4}{(x - 4)(x + 4)}$$

$$= \dfrac{13x + 12}{(x - 4)(x + 4)}$$

25. $\dfrac{5x + 10}{(x^2 + 6x + 8)} + \dfrac{3}{x + 4}$

$= \dfrac{5x + 10}{(x + 2)(x + 4)} + \dfrac{3}{(x + 4)}$

$= \dfrac{5x + 10 + 3(x + 2)}{(x + 2)(x + 4)}$

$= \dfrac{5x + 10 + 3x + 6}{(x + 2)(x + 4)}$

$= \dfrac{8x + 16}{(x + 2)(x + 4)}$

$= \dfrac{8(\cancel{x + 2})}{(\cancel{x + 2})(x + 4)} = \dfrac{8}{x + 4}$

27. $\dfrac{2x - 1}{x^2 + 6x + 9} + \dfrac{x}{x + 3}$

$= \dfrac{2x - 1}{(x + 3)(x + 3)} + \dfrac{x}{x + 3}$

$= \dfrac{2x - 1 + x(x + 3)}{(x + 3)^2}$

$= \dfrac{2x - 1 + x^2 + 3x}{(x + 3)^2}$

$= \dfrac{x^2 + 5x - 1}{(x + 3)^2}$

29. $\dfrac{x - 2}{2x^2 - x - 1} + \dfrac{x + 2}{2x^2 + 3x + 1}$

$= \dfrac{x - 2}{(2x + 1)(x - 1)} + \dfrac{x + 2}{(2x + 1)(x + 1)}$

$= \dfrac{(x - 2)(x + 1) + (x + 2)(x - 1)}{(2x + 1)(x + 1)(x - 1)}$

$= \dfrac{x^2 - x - 2 + x^2 + x - 2}{(2x + 1)(x + 1)(x - 1)}$

$= \dfrac{2x^2 - 4}{(2x + 1)(x + 1)(x - 1)}$

31. $\dfrac{1}{x - 1} - \dfrac{1}{3x - 1} + \dfrac{1}{4 - 4x}$

$= \dfrac{1}{x - 1} - \dfrac{1}{3x - 1} + \dfrac{1}{4(1 - x)}$

$= \dfrac{1}{x - 1} - \dfrac{1}{3x - 1} + \dfrac{1}{-4(x - 1)}$

$= \dfrac{-4(3x - 1) - 1(-4)(x - 1) + 1(3x - 1)}{-4(x - 1)(3x - 1)}$

$= \dfrac{-12x + 4 + 4x - 4 + 3x - 1}{-4(x - 1)(3x - 1)}$

$= \dfrac{-5x - 1}{-4(x - 1)(3x - 1)}$

$= \dfrac{5x + 1}{4(x - 1)(3x - 1)}$

33. $\dfrac{3}{x^2 + 3x + 2} + \dfrac{2}{x^2 - x - 6}$

$- \dfrac{1}{x^2 - 2x - 3}$

$= \dfrac{3}{(x + 1)(x + 2)}$

$+ \dfrac{2}{(x - 3)(x + 2)} - \dfrac{1}{(x + 1)(x - 3)}$

$= \dfrac{3(x - 3) + 2(x + 1) - 1(x + 2)}{(x + 1)(x + 2)(x - 3)}$

$= \dfrac{3x - 9 + 2x + 2 - x - 2}{(x + 1)(x + 2)(x - 3)}$

$= \dfrac{4x - 9}{(x + 1)(x + 2)(x - 3)}$

35. $\dfrac{x + 1}{x^2 + x} \cdot \dfrac{x}{x - 1}$

$= \dfrac{\cancel{x + 1}}{\cancel{x}(\cancel{x + 1})} \cdot \dfrac{\cancel{x}}{x - 1}$

$= \dfrac{1}{x - 1}$

37. $\dfrac{4y - 8}{4(y + 2)} \cdot \dfrac{y}{2y - 4}$

$$= \dfrac{\cancel{4}(y - 2)}{\cancel{4}(y + 2)} \cdot \dfrac{y}{2(y - 2)}$$

$$= \dfrac{y}{2(y + 2)}$$

39. $\dfrac{5}{8b^2} \cdot \dfrac{4b + 6}{10b + 15}$

$$= \dfrac{\cancel{5}}{\underset{4}{\cancel{8b^2}}} \cdot \dfrac{2(2b + 3)}{5(2b + 3)}$$

$$= \dfrac{1}{4b^2}$$

41. $\dfrac{3x + 2}{x^2 - 1} \cdot \dfrac{4x + 4}{9x + 6}$

$$= \dfrac{\cancel{3x + 2}}{(x - 1)(x + 1)} \cdot \dfrac{4(x + 1)}{3(3x + 2)}$$

$$= \dfrac{4}{3(x - 1)}$$

43. $\dfrac{2x^2 + 5x + 2}{x^2 + 5x - 6} \cdot \dfrac{x^2 + 7x + 6}{4x^2 + 4x + 1}$

$$= \dfrac{(2x + 1)(x + 2)}{(x - 1)(x + 6)} \cdot \dfrac{(x + 1)(x + 6)}{(2x + 1)(2x + 1)}$$

$$= \dfrac{(x + 2)(x + 1)}{(x - 1)(2x + 1)}$$

45. $\dfrac{x^3 - x^2 - 2x}{x^2 - 1} \cdot \dfrac{2x + 1}{x^2 - 2x}$

$$= \dfrac{x(x - 2)(x + 1)}{(x + 1)(x - 1)} \cdot \dfrac{2x + 1}{x(x - 2)}$$

$$= \dfrac{2x + 1}{x - 1}$$

47. $\dfrac{\dfrac{2x}{x^3}}{\dfrac{4x}{x^5}} = \dfrac{2x}{x^3} \cdot \dfrac{x^5}{4x} = \dfrac{x^2}{2}$

49. $\dfrac{\dfrac{2x - 1}{6}}{\dfrac{2x + 1}{12}} = \dfrac{2x - 1}{6} \div \dfrac{2x + 1}{12}$

$$= \dfrac{2x - 1}{\cancel{6}} \cdot \dfrac{\overset{2}{\cancel{12}}}{2x + 1}$$

$$= \dfrac{2(2x - 1)}{2x + 1} = \dfrac{4x - 2}{2x + 1}$$

51. $\dfrac{\dfrac{x + 4}{3x + 12}}{2x} = (x + 4) \div \dfrac{3x + 12}{2x}$

$$= \dfrac{x + 4}{1} \cdot \dfrac{2x}{3(x + 4)}$$

$$= \dfrac{2x}{3}$$

53. $\dfrac{\dfrac{x^2 - 1}{x}}{\dfrac{x + 1}{x^2}} = \dfrac{(x + 1)(x - 1)}{x} \cdot \dfrac{x^2}{x + 1}$

$$= x(x - 1)$$

55. $\dfrac{\dfrac{5x^2}{4x^2 - 1}}{\dfrac{10x}{2x + 1}}$

$$= \dfrac{5x^2}{(2x + 1)(2x - 1)} \cdot \dfrac{2x + 1}{10x}$$

$$= \dfrac{x}{2(2x - 1)}$$

57. $\dfrac{\dfrac{2x^2 - 7x + 3}{2x + 1}}{\dfrac{x^2 - 9}{10x + 5}}$

$= \dfrac{(2x - 1)\cancel{(x - 3)}}{\cancel{2x + 1}} \cdot \dfrac{5\cancel{(2x + 1)}}{(x + 3)\cancel{(x - 3)}}$

$= \dfrac{5(2x - 1)}{x + 3}$

59. $\dfrac{\dfrac{3x^2 - 8x - 3}{4x^2 - 1}}{\dfrac{x^2 + x - 12}{2x^2 + 3x + 1}}$

$= \dfrac{(3x + 1)\cancel{(x - 3)}}{\cancel{(2x + 1)}(2x - 1)} \cdot \dfrac{\cancel{(2x + 1)}(x + 1)}{(x + 4)\cancel{(x - 3)}}$

$= \dfrac{(3x + 1)(x + 1)}{(2x - 1)(x + 4)}$

61. $\dfrac{\dfrac{1}{x + 1} + \dfrac{1}{x - 1}}{\dfrac{2}{x}}$

$= \dfrac{(x - 1) + (x + 1)}{(x + 1)(x - 1)} \cdot \dfrac{x}{2}$

$= \dfrac{2x}{(x + 1)(x - 1)} \cdot \dfrac{x}{2}$

$= \dfrac{x^2}{(x + 1)(x - 1)}$

63. $\dfrac{\dfrac{3}{x - 2} - 1}{3 - \dfrac{1}{x - 2}} = \dfrac{\dfrac{3 - (x - 2)}{x - 2}}{\dfrac{3(x - 2) - 1}{x - 2}}$

$= \dfrac{3 - x + 2}{\cancel{x - 2}} \cdot \dfrac{\cancel{x - 2}}{3x - 6 - 1}$

$= \dfrac{5 - x}{3x - 7}$

65. $\dfrac{\dfrac{2x}{x - 1} + \dfrac{3}{x + 1}}{\dfrac{4x}{x + 1} - \dfrac{2}{x - 1}}$

$= \dfrac{\dfrac{2x(x + 1) + 3(x - 1)}{(x + 1)(x - 1)}}{\dfrac{4x(x - 1) - 2(x + 1)}{(x + 1)(x - 1)}}$

$= \dfrac{2x^2 + 2x + 3x - 3}{\cancel{(x + 1)}\cancel{(x - 1)}} \cdot \dfrac{\cancel{(x + 1)}\cancel{(x - 1)}}{4x^2 - 4x - 2x - 2}$

$= \dfrac{2x^2 + 5x - 3}{4x^2 - 6x - 2}$

67. $\dfrac{1 - \dfrac{5}{x^2 - 4}}{\dfrac{x - 3}{2 - x}}$

$= \dfrac{x^2 - 4 - 5}{x^2 - 4} \cdot \dfrac{2 - x}{x - 3}$

$= \dfrac{x^2 - 9}{x^2 - 4} \cdot \dfrac{2 - x}{x - 3}$

$= \dfrac{(x + 3)\cancel{(x - 3)}}{(x + 2)\cancel{(x - 2)}} \cdot \dfrac{\cancel{2 - x}^{-1}}{\cancel{x - 3}}$

$= -\dfrac{x + 3}{x + 2}$

69. $\dfrac{\dfrac{7}{x^2 - 9} - 1}{\dfrac{1}{x - 3} - 1} = \dfrac{\dfrac{7 - (x^2 - 9)}{x^2 - 9}}{\dfrac{1 - (x - 3)}{x - 3}}$

$= \dfrac{16 - x^2}{x^2 - 9} \cdot \dfrac{x - 3}{4 - x}$

$= \dfrac{(4 + x)\cancel{(4 - x)}}{(x + 3)\cancel{(x - 3)}} \cdot \dfrac{\cancel{x - 3}}{\cancel{4 - x}}$

$= \dfrac{4 + x}{x + 3}$

Exercises 1.7

1. $\sqrt{81} = 9$

3. $\sqrt[5]{-32} = -2$

5. $\sqrt{\dfrac{16}{25}} = \dfrac{4}{5}$

7. $(\sqrt{3} + 2)(\sqrt{3} - 2)$

$\quad = (\sqrt{3})^2 - 2\sqrt{3} + 2\sqrt{3} - 4$

$\quad = 3 - 4 = -1$

9. $(3\sqrt{5} + 4)(3\sqrt{5} - 4)$

$\quad = (3\sqrt{5})^2 - 12\sqrt{5} + 12\sqrt{5} - 16$

$\quad = 9 \cdot 5 - 16$

$\quad = 45 - 16 = 29$

11. $(\sqrt{3} - \sqrt{7})^2 = (\sqrt{3} - \sqrt{7})(\sqrt{3} - \sqrt{7})$

$\quad\quad\quad = 3 - \sqrt{21} - \sqrt{21} + 7$

$\quad\quad\quad = 10 - 2\sqrt{21}$

13. $(2\sqrt{3} + \sqrt{2})(3\sqrt{2} - \sqrt{3})$

$\quad = 2\sqrt{3} \cdot 3\sqrt{2} - 2\sqrt{3} \cdot \sqrt{3}$
$\quad\ + \sqrt{2} \cdot 3\sqrt{2} - \sqrt{2} \cdot \sqrt{3}$

$\quad = 6\sqrt{6} - 2\sqrt{9} + 3\sqrt{4} - \sqrt{6}$

$\quad = 6\sqrt{6} - 2 \cdot 3 + 3 \cdot 2 - \sqrt{6}$

$\quad = 5\sqrt{6}$

15. $(4\sqrt{3} - 2)(6 + \sqrt{3})$

$\quad = 4\sqrt{3} \cdot 6 + 4\sqrt{3} \cdot \sqrt{3} - 2 \cdot 6 - 2\sqrt{3}$

$\quad = 24\sqrt{3} + 12 - 12 - 2\sqrt{3}$

$\quad = 22\sqrt{3}$

17. $(3\sqrt{3} + 1)(2\sqrt{3} - 3)$

$\quad = 3\sqrt{3} \cdot 2\sqrt{3} - 3 \cdot 3\sqrt{3} + 2\sqrt{3} - 3$

$\quad = 18 - 9\sqrt{3} + 2\sqrt{3} - 3$

$\quad = 15 - 7\sqrt{3}$

19. $(\sqrt[3]{5^2} + \sqrt[3]{5} + 1)(\sqrt[3]{5} - 1)$

$\quad = \sqrt[3]{5^2} \cdot \sqrt[3]{5} + \sqrt[3]{5}\,\sqrt[3]{5} + \sqrt[3]{5} - \sqrt[3]{5^2}$
$\quad\ - \sqrt[3]{5} - 1$

$\quad = \sqrt[3]{5^3} + \sqrt[3]{5^2} + \sqrt[3]{5} - \sqrt[3]{5^2} - \sqrt[3]{5} - 1$

$\quad = \sqrt[3]{5^3} - 1$

$\quad = 5 - 1$

$\quad = 4$

21. $(\sqrt[3]{11} + 1)(\sqrt[3]{11^2} - \sqrt[3]{11} + 1)$

$\quad = \sqrt[3]{11} \cdot \sqrt[3]{11^2} - \sqrt[3]{11} \cdot \sqrt[3]{11} + \sqrt[3]{11}$
$\quad\ + \sqrt[3]{11^2} - \sqrt[3]{11} + 1$

$\quad = \sqrt[3]{11^3} - \sqrt[3]{11^2} + \sqrt[3]{11} + \sqrt[3]{11^2}$
$\quad\ - \sqrt[3]{11} + 1$

$\quad = 11 + 1$

$\quad = 12$

23. $\sqrt[3]{8a^6} = 2a^2$

25. $\sqrt[3]{-27a^5} = \sqrt[3]{(-3)^3 a^3 a^2}$

$\quad\quad = -3a \, \sqrt[3]{a^2}$

27. $\sqrt{2x} \, \sqrt{8x^3} = \sqrt{16x^4}$

$\quad\quad = 4x^2$

29. $\sqrt{3ab^3} \, \sqrt{6a^3b} = \sqrt{18a^4b^4}$

$\quad\quad = \sqrt{9 \cdot 2a^4b^4}$

$\quad\quad = 3\sqrt{2}a^2b^2$

31. $\sqrt{2ax^2}\,\sqrt{8a^3x^4}$

$= \sqrt{16a^4x^6}$

$= 4a^2x^3$

33. $\sqrt[3]{2xy^2}\,\sqrt[3]{4x^5y^7}$

$= \sqrt[3]{8x^6y^9}$

$= 2x^2y^3$

35. $\sqrt[6]{4a^2bc^3}\,\sqrt[6]{16a^4b^5c^9}$

$= \sqrt[6]{64a^6b^6c^{12}}$

$= 2abc^2$

37. $\sqrt{\dfrac{16a^2b^4}{c^6}} = \dfrac{4ab^2}{c^3}$

39. $\sqrt[5]{\dfrac{32a^6b^7}{c^5}} = \dfrac{2ab\,\sqrt[5]{ab^2}}{c}$

41. $\sqrt{\dfrac{1}{2a^3b}} = \sqrt{\dfrac{1}{2a^2ab}}$

$= \dfrac{1 \cdot \sqrt{2ab}}{a\sqrt{2ab} \cdot \sqrt{2ab}}$

$= \dfrac{\sqrt{2ab}}{2a^2b}$

43. $\sqrt[3]{\dfrac{4am^2}{9n}} = \dfrac{\sqrt[3]{4am^2} \cdot \sqrt[3]{3n^2}}{\sqrt[3]{9n} \cdot \sqrt[3]{3n^2}}$

$= \dfrac{\sqrt[3]{12am^2n^2}}{\sqrt[3]{27n^3}}$

$= \dfrac{\sqrt[3]{12am^2n^2}}{3n}$

45. $\sqrt{\sqrt[3]{128a^7b^8x^8}}$

$= \sqrt[6]{128a^7b^8x^8}$

$= \sqrt[6]{2^6 \cdot 2a^6ab^6b^2x^6x^2}$

$= 2abx\,\sqrt[6]{2ab^2x^2}$

47. $\sqrt{(m+n)^2} = m + n$

49. $\sqrt{25a\sqrt{25a^3}} = \sqrt{\sqrt{25^2a^2}\,\sqrt{25a^3}}$

$= \sqrt{\sqrt{25^3a^5}}$

$= \sqrt[4]{5^6a^5}$

$= 5a\,\sqrt[4]{25a}$

51. $3\sqrt{2} + 5\sqrt{8} = 3\sqrt{2} + 5\sqrt{4 \cdot 2}$

$= 3\sqrt{2} + 5 \cdot 2\sqrt{2}$

$= 3\sqrt{2} + 10\sqrt{2}$

$= 13\sqrt{2}$

53. $2\sqrt{3} - 4\sqrt{27} = 2\sqrt{3} - 4\sqrt{9 \cdot 3}$

$= 2\sqrt{3} - 4 \cdot 3\sqrt{3}$

$= 2\sqrt{3} - 12\sqrt{3}$

$= -10\sqrt{3}$

55. $\sqrt{2} + \sqrt{5} - 6\sqrt{8} + 3\sqrt{45}$

$= \sqrt{2} + \sqrt{5} - 6\sqrt{4 \cdot 2} + 3\sqrt{9 \cdot 5}$

$= \sqrt{2} + \sqrt{5} - 6 \cdot 2\sqrt{2} + 3 \cdot 3\sqrt{5}$

$= \sqrt{2} + \sqrt{5} - 12\sqrt{2} + 9\sqrt{5}$

$= 10\sqrt{5} - 11\sqrt{2}$

57. $\sqrt[3]{16} + 2\sqrt[3]{54}$

$= \sqrt[3]{8 \cdot 2} + 2\sqrt[3]{27 \cdot 2}$

$= 2\sqrt[3]{2} + 2 \cdot 3\sqrt[3]{2}$

$= 2\sqrt[3]{2} + 6\sqrt[3]{2}$

$= 8\sqrt[3]{2}$

59. $\sqrt{4x} - \sqrt{16x} + \sqrt{25x}$

$= 2\sqrt{x} - 4\sqrt{x} + 5\sqrt{x}$

$= 3\sqrt{x}$

61. $\sqrt{8a^3} + \sqrt{18a^3} + \sqrt{50a^3}$

$= 2a\sqrt{2a} + 3a\sqrt{2a} + 5a\sqrt{2a}$

$= 10a\sqrt{2a}$

63. $\dfrac{1 + \sqrt{3}}{\sqrt{2}} = \dfrac{(1 + \sqrt{3})\sqrt{2}}{\sqrt{2} \cdot \sqrt{2}}$

$= \dfrac{\sqrt{2} + \sqrt{6}}{2}$

65. $\dfrac{a + \sqrt{b}}{\sqrt{c}} = \dfrac{(a + \sqrt{b})\sqrt{c}}{\sqrt{c} \cdot \sqrt{c}}$

$= \dfrac{a\sqrt{c} + \sqrt{bc}}{c}$

67. $\dfrac{3}{1 - \sqrt{2}} = \dfrac{3(1 + \sqrt{2})}{(1 - \sqrt{2})(1 + \sqrt{2})}$

$= \dfrac{3 + 3\sqrt{2}}{1 - 2}$

$= \dfrac{3 + 3\sqrt{2}}{-1}$

$= -3 - 3\sqrt{2}$

69. $\dfrac{\sqrt{5} + \sqrt{2}}{\sqrt{3}} = \dfrac{(\sqrt{5} + \sqrt{2})\sqrt{3}}{\sqrt{3} \cdot \sqrt{3}}$

$= \dfrac{\sqrt{15} + \sqrt{6}}{3}$

71. $\dfrac{\sqrt{a} - \sqrt{c}}{\sqrt{c}} = \dfrac{(\sqrt{a} - \sqrt{c})\sqrt{c}}{\sqrt{c} \cdot \sqrt{c}}$

$= \dfrac{\sqrt{ac} - c}{c}$

73. $\dfrac{\sqrt{x^3} + \sqrt{x^5}}{\sqrt{x^3}} = \dfrac{(\sqrt{x^3} + \sqrt{x^5})\sqrt{x}}{\sqrt{x^3} \cdot \sqrt{x}}$

$= \dfrac{\sqrt{x^4} + \sqrt{x^6}}{\sqrt{x^4}}$

$= \dfrac{x^2 + x^3}{x^2}$

$= \dfrac{\cancel{x^2}(1 + x)}{\cancel{x^2}}$

$= 1 + x$

75. $\dfrac{\sqrt{6} - 4}{2 + \sqrt{5}} = \dfrac{(\sqrt{6} - 4)(2 - \sqrt{5})}{(2 + \sqrt{5})(2 - \sqrt{5})}$

$= \dfrac{2\sqrt{6} - \sqrt{30} - 8 + 4\sqrt{5}}{4 - 5}$

$= \dfrac{2\sqrt{6} - \sqrt{30} + 4\sqrt{5} - 8}{-1}$

$= -2\sqrt{6} + \sqrt{30} - 4\sqrt{5} + 8$

77. $\dfrac{a}{\sqrt{a} + 2} = \dfrac{a(\sqrt{a} - 2)}{(\sqrt{a} + 2)(\sqrt{a} - 2)}$

$= \dfrac{a\sqrt{a} - 2a}{a - 4}$

$= \dfrac{a(\sqrt{a} - 2)}{a - 4}$

79. $\dfrac{\sqrt{x}}{\sqrt{x} + \sqrt{y}} = \dfrac{\sqrt{x}(\sqrt{x} - \sqrt{y})}{(\sqrt{x} + \sqrt{y})(\sqrt{x} - \sqrt{y})}$

$= \dfrac{x - \sqrt{xy}}{x - y}$

81. $\dfrac{\sqrt{x - 1}}{1 - \sqrt{x - 1}}$

$= \dfrac{\sqrt{x - 1}(1 + \sqrt{x - 1})}{(1 - \sqrt{x - 1})(1 + \sqrt{x - 1})}$

$= \dfrac{\sqrt{x - 1}(1 + \sqrt{x - 1})}{1 - (x - 1)}$

$= \dfrac{\sqrt{x - 1} + x - 1}{2 - x}$

83. $\dfrac{1}{a + \sqrt{b}} = \dfrac{(a - \sqrt{b})}{(a + \sqrt{b})(a - \sqrt{b})}$

$= \dfrac{a - \sqrt{b}}{a^2 - b}$

85. $\dfrac{1}{\sqrt{x^2 + 1} + x}$

$= \dfrac{(\sqrt{x^2 + 1} - x)}{(\sqrt{x^2 + 1} + x)(\sqrt{x^2 + 1} - x)}$

$= \dfrac{\sqrt{x^2 + 1} - x}{x^2 + 1 - x^2}$

$= \sqrt{x^2 + 1} - x$

87. $\sqrt{3} + \sqrt{5} = \sqrt{(\sqrt{3} + \sqrt{5})^2}$

$$= \sqrt{(\sqrt{3} + \sqrt{5})(\sqrt{3} + \sqrt{5})}$$

$$= \sqrt{3 + \sqrt{15} + \sqrt{15} + 5}$$

$$= \sqrt{8 + 2\sqrt{15}}$$

89. Let $x = \sqrt[n]{a}$ and $y = \sqrt[n]{b}$. By definition of the nth root, $x^n = a$ and $y^n = b$. Dividing these two expressions, we get

$$\frac{x^n}{y^n} = \frac{a}{b}.$$

By properties of exponents

$$\frac{x^n}{y^n} = \left(\frac{x}{y}\right)^n.$$

Thus

$$\left(\frac{x}{y}\right)^n = \frac{a}{b} \quad \text{and}$$

$$\frac{x}{y} = \sqrt[n]{\frac{a}{b}}.$$

Resubstituting x and y, we obtain

$$\sqrt[n]{\frac{a}{b}} = \frac{\sqrt[n]{a}}{\sqrt[n]{b}}.$$

Exercises 1.8

1. $8^{3/2} = \sqrt{8^3}$

3. $64^{-1/4} = \sqrt[4]{64^{-1}}$

$$= \frac{1}{\sqrt[4]{64}}$$

5. $(a^2 b^3)^{3/5} = \sqrt[5]{(a^2 b^3)^3}$

7. $(x^2 + y^2)^{1/2} = \sqrt{x^2 + y^2}$

9. $\sqrt[3]{5^2} = 5^{2/3}$

11. $\sqrt[4]{x^3} = x^{3/4}$

13. $a^2 \sqrt{a} = a^2 a^{1/2} = a^{5/2}$

15. $a^3 \sqrt[4]{a^5} = a^3 a^{5/4}$

$$= a^{12/4} a^{5/4}$$

$$= a^{17/4}$$

17. $(-8)^{2/3} = (-2)^2 = 4$

19. $81^{-1/2} = 9^{-1} = 1/9$

21. $(-243)^{3/5} = (-3)^3 = -27$

23. $(5u^{1/2})(3u^{3/2}) = 15u^{4/2} = 15u^2$

25. $(-125x^6)^{2/3} = (-5x^2)^2$

$$= 25x^4$$

27. $(m^{-2}n^{-3})^{-1/6} = m^{2/6}n^{3/6}$

$$= m^{1/3}n^{1/2}$$

29. $\left(\frac{5a^{1/4}}{b^{1/2}}\right)^2 = \frac{25a^{2/4}}{b} = \frac{25a^{1/2}}{b}$

31. $\left(\frac{81x^{-8}}{y^4}\right)^{3/4} = \frac{3^3 x^{-6}}{y^3}$

$$= \frac{27}{x^6 y^3}$$

33. $(36a^8 b^2)^{-1/2} = (6a^4 b)^{-1}$

$$= \frac{1}{6a^4 b}$$

35. $(64a^2 b^3)^{-2/3} = 4^{-2} a^{-4/3} b^{-2}$

$$= \frac{1}{16a^{4/3} b^2}$$

37. $\left(\frac{36x^2 y^4}{z^3}\right)^{-1/2} = \left(\frac{6xy^2}{z^{3/2}}\right)^{-1}$

$$= \frac{z^{3/2}}{6xy^2}$$

39. $(a^{-2/3} b^{-1/2})^{-6} = a^4 b^3$

41. $\left(\dfrac{16a^4 x^{-1}}{25a^{-2}x^3}\right)^{1/2} = \left(\dfrac{16a^4 a^2}{25x^3 x}\right)^{1/2}$

$= \left(\dfrac{16a^6}{25x^4}\right)^{1/2}$

$= \dfrac{4a^3}{5x^2}$

43. $\left(\dfrac{4}{9}a^{-1/3}x^4\right)^{-3/2}$

$= \left(\dfrac{4}{9}\right)^{-3/2} a^{1/2} x^{-6}$

$= \left(\dfrac{2}{3}\right)^{-3} \dfrac{a^{1/2}}{x^6}$

$= \dfrac{27a^{1/2}}{8x^6}$

45. $\left(\dfrac{a^2}{b^{1/4}}\right)\left(\dfrac{b^{-1/2}}{a^{3/2}}\right) = \dfrac{a^2}{b^{1/4} b^{1/2} a^{3/2}}$

$= \dfrac{a^{2-3/2}}{b^{1/4+1/2}}$

$= \dfrac{a^{1/2}}{b^{3/4}}$

47. $\sqrt{2} \cdot \sqrt[3]{3} = 2^{1/2} \cdot 3^{1/3}$

$= 2^{3/6} \cdot 3^{2/6}$

$= \sqrt[6]{2^3 \cdot 3^2}$

$= \sqrt[6]{8 \cdot 9}$

$= \sqrt[6]{72}$

49. $\sqrt{a} \cdot \sqrt[4]{2a} = a^{1/2} \cdot (2a)^{1/4}$

$= a^{2/4} \cdot (2a)^{1/4}$

$= \sqrt[4]{2aa^2}$

$= \sqrt[4]{2a^3}$

51. $\dfrac{\sqrt[4]{25}}{\sqrt{5}} = \dfrac{25^{1/4}}{5^{1/2}}$

$= \dfrac{5^{2/4}}{5^{2/4}}$

$= 1$

53. $\dfrac{\sqrt{2a^2}}{\sqrt[4]{2a}} = \dfrac{(2a^2)^{1/2}}{(2a)^{1/4}}$

$= \dfrac{(2a^2)^{2/4}}{(2a)^{1/4}}$

$= \dfrac{2^{2/4} a^{4/4}}{2^{1/4} a^{1/4}}$

$= 2^{1/4} a^{3/4}$

$= \sqrt[4]{2a^3}$

55. $\dfrac{\sqrt[3]{4x^2 y^2}}{\sqrt[6]{2xy}} = \dfrac{(2^2 x^2 y^2)^{1/3}}{(2xy)^{1/6}}$

$= \dfrac{2^{4/6} x^{4/6} y^{4/6}}{2^{1/6} x^{1/6} y^{1/6}}$

$= (2xy)^{3/6}$

$= \sqrt{2xy}$

57. $\dfrac{\sqrt[3]{m^2 v}}{\sqrt[6]{m^3 v}} = \dfrac{(m^2 v)^{1/3}}{(m^3 v)^{1/6}}$

$= \dfrac{(m^2 v)^{2/6}}{(m^3 v)^{1/6}}$

$= \dfrac{m^{4/6} v^{2/6}}{m^{3/6} v^{1/6}}$

$= m^{1/6} v^{1/6}$

$= \sqrt[6]{mv}$

59. $\sqrt[3]{2a\sqrt{2a}} = \sqrt[3]{\sqrt{4a^2} \cdot \sqrt{2a}}$

$= \sqrt[3]{\sqrt{8a^3}}$

$= \sqrt[6]{8a^3}$

Exercises 1.9

1. $(3 + 2i) + (-4 + 3i)$
    $= (3 - 4) + (2i + 3i)$
    $= -1 + 5i$

3. $(5 - 6i) + (3 - 5i)$
    $= (5 + 3) + (-6i - 5i)$
    $= 8 - 11i$

5. $-(3 + 4i) + (5 - 9i)$
    $= -3 - 4i + 5 - 9i$
    $= (-3 + 5) + (-4i - 9i)$
    $= 2 - 13i$

7. $10 - (8 + 5i) = 10 - 8 - 5i$
    $\qquad\qquad\qquad = 2 - 5i$

9. $(7 - 6i) - 3i = 7 - 6i - 3i$
    $\qquad\qquad\qquad = 7 - 9i$

11. $(4 - 3i)(3 + 4i)$
    $= 12 + 16i - 9i - 12i^2$
    $= 12 + 7i - 12(-1)$
    $= 12 + 7i + 12$
    $= 24 + 7i$

13. $(-8 - i)(4 + i)$
    $= -32 - 8i - 4i - i^2$
    $= -32 - 12i - (-1)$
    $= -32 - 12i + 1$
    $= -31 - 12i$

15. $(-\dfrac{1}{2} + \dfrac{3}{2}i)(3 - 4i)$

    $= \dfrac{3}{2} - 2i + \dfrac{9}{2}i - 6i^2$

    $= \dfrac{3}{2} + \dfrac{5}{2}i - 6(-1)$

    $= \dfrac{3}{2} + \dfrac{5}{2}i + \dfrac{12}{2}$

    $= \dfrac{15}{2} + \dfrac{5}{2}i$

17. $(3 + 5i)(3 - 5i)$
    $= 9 - 15i + 15i - 25i^2$
    $= 9 + 25$
    $= 34 + 0i$
    $= 34$

19. $(-\dfrac{2}{3} + \dfrac{3}{4}i)(-\dfrac{2}{3} - \dfrac{3}{4}i)$

    $= \dfrac{4}{9} + \dfrac{1}{2}i - \dfrac{1}{2}i - \dfrac{9}{16}i^2$

    $= \dfrac{4}{9} + \dfrac{9}{16}$

    $= \dfrac{64}{144} + \dfrac{81}{144}$

    $= \dfrac{145}{144} + 0i = \dfrac{145}{144}$

21. $(2 + 5i)(2 + 5i)$
    $= 4 + 10i + 10i + 25i^2$
    $= 4 + 20i - 25$
    $= -21 + 20i$

23. $i(3 - 2i)(3 + 5i)$
    $= (3i - 2i^2)(3 + 5i)$
    $= (2 + 3i)(3 + 5i)$
    $= 6 + 10i + 9i + 15i^2$
    $= -9 + 19i$

25. $3i(1 - 2i)(3 - 4i)$
    $= (3i - 6i^2)(3 - 4i)$
    $= (6 + 3i)(3 - 4i)$
    $= 18 - 24i + 9i - 12i^2$
    $= 18 - 15i + 12$
    $= 30 - 15i$

27. $\dfrac{1}{1 + i} = \dfrac{(1 - i)}{(1 + i)(1 - i)}$

    $= \dfrac{1}{1 - i + i - i^2}$

    $= \dfrac{1 - i}{1 + 1} = \dfrac{1 - i}{2}$

    $= \dfrac{1}{2} - \dfrac{1}{2}i$

29. $\dfrac{3 - 3i}{4 + 4i} = \dfrac{(3 - 3i)(4 - 4i)}{(4 + 4i)(4 - 4i)}$

$$= \dfrac{12 - 12i - 12i + 12i^2}{16 - 16i^2}$$

$$= \dfrac{12 - 24i - 12}{16 + 16}$$

$$= \dfrac{-24i}{32} = 0 - \dfrac{3}{4}i$$

$$= -\dfrac{3}{4}i$$

31. $\dfrac{2 + 5i}{2 - 5i} = \dfrac{(2 + 5i)(2 + 5i)}{(2 - 5i)(2 + 5i)}$

$$= \dfrac{4 + 10i + 10i + 25i^2}{4 - 25i^2}$$

$$= \dfrac{4 + 20i - 25}{4 + 25}$$

$$= \dfrac{-21 + 20i}{29}$$

$$= -\dfrac{21}{29} + \dfrac{20}{29}i$$

33. $\dfrac{3 - i}{2 + 4i} = \dfrac{(3 - i)(2 - 4i)}{(2 + 4i)(2 - 4i)}$

$$= \dfrac{6 - 12i - 2i + 4i^2}{4 - 16i^2}$$

$$= \dfrac{6 - 14i - 4}{4 + 16}$$

$$= \dfrac{2 - 14i}{20} = \dfrac{1}{10} - \dfrac{7}{10}i$$

35. $\dfrac{2 - 3i}{3 + 6i} = \dfrac{(2 - 3i)(3 - 6i)}{(3 + 6i)(3 - 6i)}$

$$= \dfrac{6 - 12i - 9i + 18i^2}{9 - 36i^2}$$

$$= \dfrac{6 - 21i - 18}{9 + 36}$$

$$= \dfrac{-12 - 21i}{45} = -\dfrac{4}{15} - \dfrac{7}{15}i$$

37. $\overline{z + w} = \overline{(a + bi) + (c + di)}$

$$= \overline{(a + c) + (b + d)i}$$

$$= (a + c) - (b + d)i$$

$$= a + c - bi - di$$

$$= (a - bi) + (c - di)$$

$$= \overline{z} + \overline{w}$$

39. We have $z^{-1} = \dfrac{1}{a + bi}$

$$= \dfrac{(a - bi)}{(a + bi)(a - bi)}$$

$$= \dfrac{a - bi}{a^2 - b^2 i^2}$$

$$= \dfrac{a - bi}{a^2 + b^2}$$

$$= \dfrac{a}{a^2 + b^2} - \dfrac{b}{a^2 + b^2}i,$$

so $\overline{z^{-1}} = \dfrac{a}{a^2 + b^2} + \dfrac{b}{a^2 + b^2}i$

$$= \dfrac{a + bi}{a^2 + b^2}$$

$$= \dfrac{a + bi}{a^2 - b^2 i^2}$$

$$= \dfrac{a + bi}{(a + bi)(a - bi)}$$

$$= \dfrac{1}{a - bi} = (\overline{z})^{-1}$$

Review Exercises - Chapter 1

1. $\dfrac{7}{16} = 0.4375$

3. $\dfrac{9}{11} = 0.8181 \ldots = 0.\overline{81}$

5. $\dfrac{15}{64} = 0.234375$

7. $5.18 = 5\dfrac{18}{100} = 5\dfrac{9}{50}$

$$= \dfrac{259}{50}$$

9. Let $r = 0.\overline{141}$

$$1000\ r = 141.\overline{141}$$
$$\underline{-r = -0.\overline{141}}$$
$$999\ r = 141$$

$$r = \frac{141}{999} = \frac{47}{333}$$

11. Let $r = 1.3\overline{18}$

$$1000\ r = 1318.\overline{18}$$
$$\underline{-10\ r = -13.\overline{18}}$$
$$990\ r = 1305$$

$$r = \frac{1305}{990} = \frac{29}{22}$$

13. $0.\overline{6} + 0.\overline{7} = \frac{6}{9} + \frac{7}{9} = \frac{13}{9}$

$$= 1.444\ldots = 1.\overline{4}$$

15. $0.\overline{5} + 0.\overline{13} = \frac{5}{9} + \frac{13}{99}$

$$= \frac{55}{99} + \frac{13}{99} = \frac{68}{99}$$

$$= 0.6868\ldots = 0.\overline{68}$$

Or, $0.5555\ldots$
$$\underline{+\ 0.1313\ldots}$$
$$0.6868\ldots = 0.\overline{68}$$

17. $\dfrac{\dfrac{1}{2} + \dfrac{3}{4}}{\dfrac{5}{6} - \dfrac{2}{3}} = \dfrac{\dfrac{2}{4} + \dfrac{3}{4}}{\dfrac{5}{6} - \dfrac{4}{6}} = \dfrac{\dfrac{5}{4}}{\dfrac{1}{6}}$

$$= \frac{5}{\underset{2}{\cancel{4}}} \cdot \frac{\cancel{6}^{3}}{1} = \frac{15}{2}$$

19. $\dfrac{\dfrac{2}{5}\left(\dfrac{3}{8} - \dfrac{1}{2}\right)}{\dfrac{4}{10} - \dfrac{3}{5}} = \dfrac{\dfrac{2}{5}\left(\dfrac{3}{8} - \dfrac{4}{8}\right)}{\dfrac{4}{10} - \dfrac{6}{10}}$

$$= \frac{\dfrac{\cancel{2}}{5}\left(-\dfrac{1}{\underset{4}{\cancel{8}}}\right)}{-\dfrac{2}{10}}$$

$$= \frac{-\dfrac{1}{20}}{-\dfrac{1}{5}} = \frac{1}{\cancel{20}} \cdot \frac{\cancel{5}}{1}$$

$$= \frac{1}{4}$$

21. $\dfrac{\dfrac{5}{8} \div \left(\dfrac{7}{12} + 1\right)}{\dfrac{3}{14} - \dfrac{3}{7}} = \dfrac{\dfrac{5}{8} \div \left(\dfrac{7}{12} + \dfrac{12}{12}\right)}{\dfrac{3}{14} - \dfrac{6}{14}}$

$$= \frac{\dfrac{5}{8} \div \dfrac{19}{12}}{-\dfrac{3}{14}}$$

$$= \frac{\dfrac{5}{\cancel{8}_{2}} \cdot \dfrac{\cancel{12}^{3}}{19}}{-\dfrac{3}{14}}$$

$$= \frac{15^{5}}{\cancel{38}_{19}} \cdot \left(-\frac{\cancel{14}^{7}}{\cancel{3}_{1}}\right)$$

$$= -\frac{35}{19}$$

23. No; for example $\sqrt{2} - \sqrt{2} = 0$.

25. a) rational: $(3 + 5\sqrt{2})(3 - 5\sqrt{2})$
$= 9 - 50 = -41$

  b) irrational: $\sqrt{3} - 5\sqrt{3}$

  $= \sqrt{3}(1 - 5)$

  $= -4\sqrt{3}$

  c) irrational: nonterminating and nonrepeating decimal

  d) rational: $\dfrac{\sqrt{8}}{\sqrt{2}} = \dfrac{2\sqrt{2}}{\sqrt{2}} = 2.$

27. $(3a^{-5}x^2)(4ay^{-3}) = \dfrac{12ax^2}{a^5y^3}$

  $= \dfrac{12x^2}{a^4y^3}$

29. $\dfrac{6a^2b^{-5}}{4c^{-4}} = \dfrac{6a^2c^4}{4b^5}$

  $= \dfrac{3a^2c^4}{2b^5}$

31. $\dfrac{3my^{-4}}{2y} = \dfrac{3m}{2yy^4}$

  $= \dfrac{3m}{2y^5}$

33. $\left(\dfrac{m^3p^2q^5}{2mp}\right)^3 = \left(\dfrac{m^2pq^5}{2}\right)^3$

  $= \dfrac{m^6p^3q^{15}}{8}$

35. $\left(\dfrac{a^2x^{-3}}{b}\right)^{-2} = \dfrac{a^{-4}x^6}{b^{-2}}$

  $= \dfrac{b^2x^6}{a^4}$

37. $(25a^6x^4)^{-1/2} = \dfrac{1}{(25a^6x^4)^{1/2}}$

  $= \dfrac{1}{5a^3x^2}$

39. $(x^{-1/4}y^{-2/4})^{-2} = x^{2/4}y^{4/4}$

  $= x^{1/2}y$

41. $\left(\dfrac{8a^3x^{-1}}{27a^{-4}x^3}\right)^{1/2} = \left(\dfrac{8a^3a^4}{27x^3x}\right)^{1/2}$

  $= \left(\dfrac{2^3a^7}{3^3x^4}\right)^{1/2}$

  $= \dfrac{2^{3/2}a^{7/2}}{3^{3/2}x^2}$

43. $\dfrac{(b^{-8}y)^{-1/4}}{(b^3y^6)^{-1/6}} = \dfrac{b^2y^{-1/4}}{b^{-1/2}y^{-1}}$

  $= \dfrac{b^2b^{1/2}y}{y^{1/4}}$

  $= b^{5/2}y^{3/4}$

45. $\dfrac{(3.6 \times 10^{-4})(3.5 \times 10^{-2})}{(0.7 \times 10^5)(0.6 \times 10^{-7})}$

  $= \dfrac{30 \times 10^{(-4 - 2)}}{10^{(5 - 7)}}$

  $= \dfrac{30 \times 10^{-6}}{10^{-2}} = 30 \times 10^{-6 + 2}$

  $= 30 \times 10^{-4}$

  $= 3.0 \times 10^{-3}$

47. $\dfrac{(3.1 \times 10^{-3})(4.2 \times 10^{-5})}{(2.0 \times 10^5)(7.0 \times 10^4)}$

  $= \dfrac{.93 \times 10^{(-3-5)}}{10^{5 + 4}} = \dfrac{.93 \times 10^{-8}}{10^9}$

  $= .93 \times 10^{-17} = 9.3 \times 10^{-18}$

49. $\dfrac{(1.313 \times 10^{-6})(2.11 \times 10^9)}{(3.12 \times 10^3)^2(1.4 \times 10^{-8})}$

  $= \dfrac{(1.313)(2.11)}{(9.7344)(1.4)} \times \dfrac{(10^{-6})(10^9)}{(10^6)(10^{-8})}$

  $= \dfrac{2.770}{13.628} \times \dfrac{10^3}{10^{-2}}$

  $= 2.0328716 \times 10^4$

  $= 2.0 \times 10^4$

51. $\dfrac{(2.134 \times 10^{-8})(3.15 \times 10^{-3})}{(5.07 \times 10^{4})(2.312 \times 10^{6})}$

$= \dfrac{(2.134)(3.15)}{(5.07)(2.312)} \times \dfrac{(10^{-8})(10^{-3})}{(10^{4})(10^{6})}$

$= \dfrac{6.72}{11.72} \times \dfrac{10^{-11}}{10^{10}}$

$= .573 \times 10^{-21}$

$= 5.73 \times 10^{-22}$

53. $300,000 = 3.0 \times 10^{5}$
$(3.0 \times 10^{5})(4 \times 10^{5}) = 12.0 \times 10^{10}$
$= 1.2 \times 10^{11}$

55. $(1.516)(1.217)(1.605)$
$\times (10^{3})(10^{4})(10^{5})$
$= 2.961 \times 10^{12}$

57. $(3.0 \times 10^{5})(3.156 \times 10^{7})$
$= 9.468 \times 10^{12}$ km

59. 3.0×10^{5} = speed of light in km/sec.
    8 min. = 8 · 60 sec = 480 sec
$(3.0 \times 10^{5})(4.8 \times 10^{2}) = 14.4 \times 10^{7}$
$= 1.44 \times 10^{8}$ km

61. $(3.785)(170,000)$
$= 3.785 \times 1.7 \times 10^{5}$
$= 6.4345 \times 10^{5}$ liters

63. $(5.983 \times 10^{24})(2.203)$
$= 13.180 \times 10^{24}$
$= 1.318 \times 10^{25}$ lb.

65. $(u - 5)(u + 2) = u^{2} - 3u - 10$

67. $(x^{2} - 2)(x^{2} + 3) = x^{4} + x^{2} - 6$

69. $(3a^{2} + b^{2})(3a^{2} - b^{2}) = 9a^{4} - b^{4}$

71. $(x + \sqrt{a})(x - \sqrt{a}) = x^{2} - a$

73. $(\sqrt{x} + 2)(\sqrt{x} - 2) = x - 4$

75. $v^{2} - 10v + 25 = (v - 5)(v - 5)$
$= (v - 5)^{2}$

77. $3x^{2} + 10x - 8 = (3x - 2)(x + 4)$

79. $2ax + 5bx - 8ay - 20by$
$= x(2a + 5b) - 4y(2a + 5b)$
$= (2a + 5b)(x - 4y)$

81. $2bu - 3bv + 2au - 3av$
$= b(2u - 3v) + a(2u - 3v)$
$= (2u - 3v)(b + a)$

83. $8x^{3} - 18x(a^{2} + 2a + 1)$
$= 2x[4x^{2} - 9(a + 1)^{2}]$
$= 2x(2x + 3(a + 1))(2x - 3(a + 1))$
$= 2x(2x + 3a + 3)(2x - 3a - 3)$

85. $\dfrac{5}{x - 6} + \dfrac{2}{x - 3}$

$= \dfrac{5(x - 3) + 2(x - 6)}{(x - 6)(x - 3)}$

$= \dfrac{5x - 15 + 2x - 12}{(x - 6)(x - 3)}$

$= \dfrac{7x - 27}{(x - 6)(x - 3)}$

87. $\dfrac{5x}{x^{2} - 4} + \dfrac{1}{x - 2}$

$= \dfrac{5x}{(x + 2)(x - 2)} + \dfrac{1}{x - 2}$

$= \dfrac{5x + 1(x + 2)}{(x + 2)(x - 2)}$

$= \dfrac{5x + x + 2}{(x + 2)(x - 2)}$

$= \dfrac{6x + 2}{(x + 2)(x - 2)}$

89. $\dfrac{4}{x^2 - 2x + 1} - \dfrac{3}{x^2 - 1}$

$= \dfrac{4}{(x - 1)(x - 1)} - \dfrac{3}{(x + 1)(x - 1)}$

$= \dfrac{4(x + 1) - 3(x - 1)}{(x - 1)(x - 1)(x + 1)}$

$= \dfrac{4x + 4 - 3x + 3}{(x - 1)(x - 1)(x + 1)}$

$= \dfrac{x + 7}{(x - 1)^2(x + 1)}$

91. $\dfrac{xy}{x^3 - y^3} - \dfrac{y}{x^2 + xy + y^2}$

$= \dfrac{xy}{(x - y)(x^2 + xy + y^2)}$

$- \dfrac{y}{x^2 + xy + y^2}$

$= \dfrac{xy - y(x - y)}{x^3 - y^3}$

$= \dfrac{xy - xy + y^2}{x^3 - y^3} = \dfrac{y^2}{x^3 - y^3}$

93. $\dfrac{1}{3x - 3} + \dfrac{1}{2x + 2} - \dfrac{1}{1 - x^2}$

$= \dfrac{1}{3(x - 1)} + \dfrac{1}{2(x + 1)}$

$- \dfrac{1}{(1 + x)(1 - x)}$

$= \dfrac{1}{3(x - 1)} + \dfrac{1}{2(x + 1)}$

$+ \dfrac{1}{(x + 1)(x - 1)}$

$= \dfrac{2(x + 1) + 3(x - 1) + 6}{6(x + 1)(x - 1)}$

$= \dfrac{2x + 2 + 3x - 3 + 6}{6(x + 1)(x - 1)}$

$= \dfrac{5x + 5}{6(x + 1)(x - 1)}$

$= \dfrac{5(x + 1)}{6(x + 1)(x - 1)} = \dfrac{5}{6(x - 1)}$

95. $\dfrac{x - \dfrac{x^2}{x - 5}}{1 + \dfrac{25}{x^2 - 25}} = \dfrac{\dfrac{x(x - 5) - x^2}{x - 5}}{\dfrac{x^2 - 25 + 25}{(x + 5)(x - 5)}}$

$= \dfrac{x^2 - 5x - x^2}{x - 5} \cdot \dfrac{(x + 5)(x - 5)}{x^2}$

$= \dfrac{-5x(x + 5)}{x^2} = \dfrac{-5(x + 5)}{x}$

97. $\dfrac{x + \dfrac{1}{1 - \dfrac{1}{x}}}{x - \dfrac{1}{1 + \dfrac{1}{x}}} = \dfrac{x + \dfrac{1}{\dfrac{x - 1}{x}}}{x - \dfrac{1}{\dfrac{x + 1}{x}}}$

$= \dfrac{x + \dfrac{x}{x - 1}}{x - \dfrac{x}{x + 1}}$

$= \dfrac{\dfrac{x(x - 1) + x}{x - 1}}{\dfrac{x(x + 1) - x}{x + 1}}$

$= \dfrac{\dfrac{x^2 - x + x}{x - 1}}{\dfrac{x^2 + x - x}{x + 1}}$

$= \dfrac{x^2}{x - 1} \cdot \dfrac{x + 1}{x^2}$

$= \dfrac{x + 1}{x - 1}$

99. $\sqrt{48x^3y^6} = \sqrt{16 \cdot 3 \cdot x^2 \cdot x \cdot y^6}$

$= 4xy^3\sqrt{3x}$

101. $\sqrt[4]{625a^5b^7} = \sqrt[4]{5^4 a^4 ab^4 b^3}$

$= 5ab \sqrt[4]{ab^3}$

103. $\sqrt{3ab^3}\sqrt{2a^2b^5} = \sqrt{6a^3b^8}$

$$= \sqrt{6a^2ab^8}$$

$$= ab^4\sqrt{6a}$$

105. $\sqrt[5]{\dfrac{32ab^3}{c^6}} = \dfrac{2\sqrt[5]{ab^3}}{c\sqrt[5]{c}}$

$$= \dfrac{2\sqrt[5]{ab^3}\cdot\sqrt[5]{c^4}}{c\sqrt[5]{c}\cdot\sqrt[5]{c^4}}$$

$$= \dfrac{2\sqrt[5]{ab^3c^4}}{c^2}$$

107. $\sqrt[3]{\sqrt{64a^6b^8c^9}} = \sqrt[6]{64a^6b^8c^9}$

$$= 2abc\sqrt[6]{b^2c^3}$$

109. $(\sqrt{3ax^2y})^4 = (3ax^2y)^2$

$$= 9a^2x^4y^2$$

111. $\dfrac{4+\sqrt{5}}{\sqrt{3}} = \dfrac{(4+\sqrt{5})\sqrt{3}}{\sqrt{3}\cdot\sqrt{3}}$

$$= \dfrac{4\sqrt{3}+\sqrt{15}}{3}$$

113. $\dfrac{\sqrt{a}+\sqrt{b}}{\sqrt{a}} = \dfrac{(\sqrt{a}+\sqrt{b})\sqrt{a}}{\sqrt{a}\cdot\sqrt{a}}$

$$= \dfrac{a+\sqrt{ab}}{a}$$

115. $\dfrac{\sqrt{2}+\sqrt{5}}{\sqrt{3}-\sqrt{2}} = \dfrac{(\sqrt{2}+\sqrt{5})(\sqrt{3}+\sqrt{2})}{(\sqrt{3}-\sqrt{2})(\sqrt{3}+\sqrt{2})}$

$$= \dfrac{\sqrt{6}+2+\sqrt{15}+\sqrt{10}}{3-2}$$

$$= 2+\sqrt{6}+\sqrt{10}+\sqrt{15}$$

117. $\dfrac{2\sqrt{x}+\sqrt{y}}{\sqrt{x}-\sqrt{y}} = \dfrac{(2\sqrt{x}+\sqrt{y})(\sqrt{x}+\sqrt{y})}{(\sqrt{x}-\sqrt{y})(\sqrt{x}+\sqrt{y})}$

$$= \dfrac{2x+3\sqrt{xy}+y}{x-y}$$

119. $\dfrac{\sqrt{x+1}+\sqrt{x}}{\sqrt{x-1}-\sqrt{x}}$

$$= \dfrac{(\sqrt{x+1}+\sqrt{x})(\sqrt{x-1}+\sqrt{x})}{(\sqrt{x-1}-\sqrt{x})(\sqrt{x-1}+\sqrt{x})}$$

$$= \dfrac{\sqrt{x^2-1}+\sqrt{x(x+1)}+\sqrt{x(x-1)}+x}{(x-1)-x}$$

$$= \dfrac{\sqrt{x^2-1}+\sqrt{x^2+x}+\sqrt{x^2-x}+x}{-1}$$

$$= -\sqrt{x^2-1}-\sqrt{x^2+x}-\sqrt{x^2-x}-x$$

121. $\dfrac{\sqrt{x^2+1}-\sqrt{x^2-1}}{\sqrt{x^2+1}-\sqrt{x^2-1}} = 1$

123. $-(-3+4i)+(5-9i)$

$$= 3-4i+5-9i$$

$$= (3+5)+(-4-9)i$$

$$= 8+(-13)i$$

$$= 8-13i$$

125. $(-8-i)(4+i)$

$$= -32-8i-4i-i^2$$

$$= -32-12i+1$$

$$= -31-12i$$

127. $(-3-2i)(-3+2i)$

$$= 9-6i+6i-4i^2$$

$$= 9+4$$

$$= 13+0i$$

$$= 13$$

129. $i(5-3i)(5+3i)$

$$= (5i-3i^2)(5+3i)$$

$$= (3+5i)(5+3i)$$

$$= 15+9i+25i+15i^2$$

$$= 15+34i-15$$

$$= 0+34i$$

$$= 34i$$

131. $3i(-2-6i)(2+6i)$

$$= (-6i-18i^2)(2+6i)$$

$$= (18-6i)(2+6i)$$

$$= 36+108i-12i-36i^2$$

$$= 36+96i+36$$

$$= 72+96i$$

133. $\dfrac{1}{4 - 5i} = \dfrac{4 + 5i}{(4 - 5i)(4 + 5i)}$

$= \dfrac{4 + 5i}{16 - 25i^2}$

$= \dfrac{4 + 5i}{16 + 25}$

$= \dfrac{4 + 5i}{41} = \dfrac{4}{41} + \dfrac{5}{41}i$

137. If $z = a + bi$, then $\overline{z} = a - bi$,

so $\overline{\overline{z}} = a - (bi)$

$= a + bi$

$= z$

139. First, we have $\dfrac{z}{w} = \dfrac{a + bi}{c + di}$

$= \dfrac{(a + bi)(c - di)}{(c + di)(c - di)}$

$= \dfrac{ac - adi + bci - bdi^2}{c^2 - d^2i^2}$

$= \dfrac{ac + bd + bci - bdi^2}{c^2 + d^2}$

$= \dfrac{(ac + bd)}{c^2 + d^2} + \dfrac{(bc - ad)i}{c^2 + d^2}$

Next,

$\overline{\left(\dfrac{z}{w}\right)} = \dfrac{ac + bd}{c^2 + d^2} - \dfrac{bc - ad}{c^2 + d^2}i$

$= \dfrac{ac + bd - (bc - ad)i}{c^2 + d^2}$

$= \dfrac{ac + bd - bci + adi}{c^2 - d^2i^2}$

$= \dfrac{ac + adi - bdi^2 - bci}{(c + di)(c - di)}$

$= \dfrac{a(c + di) - bi(c + di)}{(c + di)(c - di)}$

$= \dfrac{\cancel{(c + di)}(a - bi)}{\cancel{(c + di)}(c - di)}$

$= \dfrac{a - bi}{c - di} = \dfrac{\overline{z}}{\overline{w}}$

Chapter 1 Test

1. Find the prime factorization of 168.

2. Find the decimal representation of 17/32.
3. Write the following decimal number as a fraction: $0.2\overline{3}$.

4. Write the following sum as a repeating decimal: $0.\overline{24} + 0.\overline{5}$.

5. Evaluate $[-\frac{1}{2} \times (\frac{1}{3} + \frac{1}{4})] \div (\frac{2}{5} + \frac{5}{6})$.

6. Which of the following are rational numbers? Irrational numbers?
    a) $\sqrt{18}/\sqrt{2}$,   b) $(3 + \sqrt{3})^2$,   c) 0.0123,   d) $2\sqrt{7} - \sqrt{7}$.

7. Simplify and express your answer using only positive exponents:
    $(3x^2y^{-3})/(9x^{-1}y^4)$.

8. Simplify and express your answer using only positive exponents:
    $(2a^2b^{-1}/3b^3c^0)^{-2}$.

9. Perform the indicated operations and express your answer in scientific
    notation $\dfrac{(1.45 \times 10^6)(2.3 \times 10^{-6})}{(4.2 \times 10^2)^2}$.

10. Convert to scientific notation and simplify $\dfrac{(0.0004)(25000)}{(1600000)(0.9)}$.

11. Perform the multiplication $(2x + 4)(3x^2 - x + 2)$.

12. Perform the division and simplify $\dfrac{(3ax^2 - 12a^2x + 15a^3x^3)}{3ax}$.

13. Factor $50a^2 - 72$ as completely as possible.

14. Factor the polynomial $x^3 + 2x^2 - x - 2$.

15. Factor $625x^4 - 256$ as completely as possible.

16. Perform the indicated operation and simplify
    $\dfrac{x^2 - 4}{x^2 + 4x + 4} \cdot \dfrac{x + 2}{x^2 + x - 6}$.

17. Perform the division and simplify $\dfrac{x^3 - y^3}{x - y} \div \dfrac{x^2 + xy + y^2}{y^3}$.

18. Perform the indicated operation and simplify $\dfrac{2x + 3}{x^2 + 5x + 6} - \dfrac{3}{x + 2}$.

19. Simplify the expression $\dfrac{(1/x) + (1/y)}{(1/x^2) - (1/y^2)}$.

20. Perform the indicated operations and simplify $\sqrt{3x^3} + \sqrt{12x^3} - \sqrt{27x^3}$.

21. Rationalize the denominator $\dfrac{6 + \sqrt{2}}{6 - \sqrt{2}}$.

22. Rationalize the numerator of the expression $\dfrac{\sqrt{x + h} - \sqrt{x}}{h}$.

23. Write $ab^2 \sqrt{ab}$ using rational exponents.

24. Simplify and express your answer using only positive exponents:
$$\left(\frac{25a^3b^{-5}}{16a^{-3}b}\right)^{-1/2}.$$

25. Simplify and write the expression $\sqrt[3]{4} \cdot \sqrt{3}$ as a radical with least positive index.

26. Write the expression $3(x + 1)^{1/2} + 3x\,[(1/3)(x + 1)^{-1/2}]$ as a single fraction whose numerator is free of radical. Give your answer in simplified form.

27. Show that $2\sqrt{3} - \sqrt{5} = \sqrt{17 - 4\sqrt{15}}$.

28. Evaluate $2i(5 + i)(-2 - 3i)$ and give your answer in the form $a + bi$.

29. Evaluate $\dfrac{2}{6 + i}$ and place the result in the form $a + bi$.

30. Find a real number x so that $x - 2i$ is a square root of $5 - 12i$.

Chapter 2
Linear and Quadratic Equations
and Inequalities

Exercises 2.1

1. $(\frac{3}{2})x - 5 = 0$

$$3x - 10 = 0$$
$$3x = 10$$
$$x = \frac{10}{3}$$

3. $(\frac{2}{3})x + \frac{3}{4} = 0$

$$(12)(\frac{2}{3})x + (12)(\frac{3}{4}) = (12)(0)$$

$$8x + 9 = 0$$
$$8x = -9$$
$$x = -\frac{9}{8}$$

5. $\sqrt{2}x - 3 = 0$
$$\sqrt{2}x = 3$$
$$x = \frac{3}{\sqrt{2}} = \frac{3\sqrt{2}}{2}$$

7. $5(x + 2) = 2(x - 1)$
$$5x + 10 = 2x - 2$$
$$3x + 10 = -2$$
$$3x = -12$$
$$x = -4$$

9. $3(x - 4) + 6 = 2(x + 1) - 7$
$$3x - 12 + 6 = 2x + 2 - 7$$
$$3x - 6 = 2x - 5$$
$$x - 6 = -5$$
$$x = 1$$

11. $0.1(t - 4) + 0.2t = 0.5$
$$0.1t - 0.4 + 0.2t = 0.5$$
$$0.3t - 0.4 = 0.5$$
$$0.3t = 0.9$$
$$t = 3$$

13. $\sqrt{3}x - 4 = 5 - \sqrt{3}x$

$2\sqrt{3}x - 4 = 5$

$2\sqrt{3}x = 9$

$x = \dfrac{9}{2\sqrt{3}}$

$= \dfrac{9\sqrt{3}}{2\sqrt{3}\sqrt{3}}$

$= \dfrac{9\sqrt{3}}{6}$

$= \dfrac{3\sqrt{3}}{2}$

15. $\sqrt{3}y - 2 = \sqrt{2}y + 6$

$(\sqrt{3} - \sqrt{2})y - 2 = 6$

$(\sqrt{3} - \sqrt{2})y = 8$

$y = \dfrac{8}{(\sqrt{3} - \sqrt{2})}$

$y = \dfrac{8(\sqrt{3} + \sqrt{2})}{(\sqrt{3} - \sqrt{2})(\sqrt{3} + \sqrt{2})}$

$y = \dfrac{8(\sqrt{3} + \sqrt{2})}{1}$

$y = 8\sqrt{3} + 8\sqrt{2}$

17. $\dfrac{1}{x + 1} = \dfrac{2}{x + 3}$

$2(x + 1) = 1(x + 3)$

$2x + 2 = x + 3$

$x + 2 = 3$

$x = 1$

19. $\dfrac{2}{x - 3} = \dfrac{3}{x - 4}$

$3(x - 3) = 2(x - 4)$

$3x - 9 = 2x - 8$

$x - 9 = -8$

$x = 1$

21. $\dfrac{1}{t} - \dfrac{1}{4} = \dfrac{2}{3} - \dfrac{1}{2t}$

$\dfrac{12t}{t} - \dfrac{12t}{4} = \dfrac{24t}{3} - \dfrac{12t}{2t}$

$12 - 3t = 8t - 6$

$18 - 3t = 8t$

$18 = 11t$

$t = \dfrac{18}{11}$

23. $\dfrac{x}{5} - \dfrac{x + 1}{4} = \dfrac{3x + 1}{2} - 3$

$\dfrac{20x}{5} - \dfrac{20(x + 1)}{4}$

$\qquad = \dfrac{20(3x + 1)}{2} - (20)(3)$

$4x - 5(x + 1) = 10(3x + 1) - 60$

$4x - 5x - 5 = 30x + 10 - 60$

$-x - 5 = 30x - 50$

$-5 = 31x - 50$

$45 = 31x$

$x = \dfrac{45}{31}$

25. $\dfrac{4x}{x - 3} + \dfrac{12}{3 - x} = \dfrac{1}{5}$

$\dfrac{4x}{x - 3} - \dfrac{12}{x - 3} = \dfrac{1}{5}$

$\dfrac{4x - 12}{x - 3} = \dfrac{1}{5}$

$20x - 60 = x - 3$

$19x - 60 = -3$

$19x = 57$

$x = 3$

No Solution

27. $(x - 1)(x + 3) = x(x - 2) + 4$

$x^2 + 2x - 3 = x^2 - 2x + 4$

$2x - 3 = -2x + 4$

$4x = 7$

$x = \dfrac{7}{4}$

29. $(3x - 1)(x - 2) = 3x(x - 4) + 5$

$3x^2 - 7x + 2 = 3x^2 - 12x + 5$

$-7x + 2 = -12x + 5$

$5x = 3$

$$x = \frac{3}{5}$$

31. $2.105x - 3.14 = 1.276x + 0.317$

$0.829x - 3.14 = 0.317$

$0.829x = 3.457$

$$x = \frac{3.457}{0.829} = 4.17$$

33. $\dfrac{2.516}{x - 3.018} = \dfrac{5.101}{x + 4.021}$

$2.516(x + 4.021) = 5.101(x - 3.018)$

$2.516x + 10.117 = 5.101x - 15.395$

$2.585x = 25.512$

$$x = \frac{25.512}{2.585}$$

$$x = 9.869$$

35. $(5.3 \times 10^8)x - 3.15 \times 10^{-5}$

$\qquad = (1.21 \times 10^7)x + 1.07 \times 10^{-6}$

$(5.3 \times 10^{14})x - 31.5$

$\qquad = (1.21 \times 10^{13})x + 1.07$

$(5.179 \times 10^{14})x = 32.57$

$$x = \frac{32.57}{5.179 \times 10^{14}}$$

$$= 6.29 \times 10^{-14}$$

37. $2xy + 3y = x + 1$

$y(2x + 3) = x + 1$

$$y = \frac{x + 1}{2x + 3}$$

39. $3uv - 5v = 2u - 4$

$3uv - 2u = 5v - 4$

$u(3v - 2) = 5v - 4$

$$u = \frac{5v - 4}{3v - 2}$$

41. $x(2y - 2) = y(4x - 3) + 1$

$2xy - 2x = 4xy - 3y + 1$

$2xy - 4xy = 2x - 3y + 1$

$-2xy + 3y = 2x + 1$

$y(-2x + 3) = 2x + 1$

$$y = \frac{2x + 1}{-2x + 3} = \frac{2x + 1}{3 - 2x}$$

43. $y = \dfrac{x - 5}{3x + 2}$

$y(3x + 2) = x - 5$

$3xy + 2y = x - 5$

$3xy - x = -2y - 5$

$x(3y - 1) = -2y - 5$

$$x = \frac{-2y - 5}{3y - 1} = \frac{2y + 5}{1 - 3y}$$

45. $u = \dfrac{2v + 3}{v - 4}$

$u(v - 4) = 2v + 3$

$uv - 4u = 2v + 3$

$uv - 2v = 4u + 3$

$v(u - 2) = 4u + 3$

$$v = \frac{4u + 3}{u - 2}$$

47. $F = \dfrac{9}{5}C + 32$

$F - 32 = \dfrac{9}{5}C$

$C = \dfrac{5}{9}(F - 32)$

49. a) $\dfrac{PV}{T} = C$

$PV = CT$

$\dfrac{PV}{C} = T$

b) $\dfrac{PV}{T} = C$

$$PV = CT$$

$$V = \dfrac{CT}{P}$$

51. a) $P = 2x + 2y$

$P - 2x = 2y$

$$y = \dfrac{P}{2} - x$$

b) $y = \dfrac{P}{2} - x$

$$y = \dfrac{300cm}{2} - 50cm$$

$$y = 150cm - 50cm$$
$$y = 100cm$$

53. $P = P_0(1 + rt)$

$P_0(1 + rt) = P$
$P_0 + P_0 rt = P$
$P_0 rt = P - P_0$

$$r = \dfrac{P - P_0}{P_0 t}$$

55. $S = 2\pi rh$

$$r = \dfrac{S}{2\pi h}$$

57. $F = \dfrac{Gm_1 m_2}{r^2}$

$r^2 F = Gm_1 m_2$

$$m_1 = \dfrac{r^2 F}{Gm_2}$$

59. $V = RI$

$$\dfrac{V}{I} = \dfrac{R\cancel{I}}{\cancel{I}}$$

$$\dfrac{V}{I} = R$$

Exercises 2.2

1. If x denotes an integer, then x, $x + 1$, $x + 2$ and $x + 3$ are four consecutive integers. Since their sum is 450, it follows that

$$x + (x + 1) + (x + 2) + (x + 3) = 450$$
$$4x + 6 = 450$$
$$4x = 444$$
$$x = 111$$

The integers are 111, 112, 113, 114.

3. If x denotes an even number, then $(x + 2)$ and $(x + 4)$ are the next two consecutive even numbers. Since their sum is 480, then

$$x + (x + 2) + (x + 4) = 480$$
$$3x + 6 = 480$$
$$3x = 474$$
$$x = 158$$

The numbers are 158, 160, 162.

5. If x and $x + 1$ denote two consecutive natural numbers, then

$$(x + 1)^2 - x^2 = 49$$
$$x^2 + 2x + 1 - x^2 = 49$$
$$2x + 1 = 49$$
$$2x = 48$$
$$x = 24$$

The numbers are 24 and 25.

7. If x denotes an even number, then $(x + 2)$ and $(x + 4)$ are the next two consecutive even numbers, then

$$x + (x + 4) = (x + 2) + 64$$
$$2x + 4 = x + 66$$
$$x = 62$$

Then numbers are 62, 64, 66.

9. Let w be the width of the rectangle. Then $\ell = 2w - 5$ is the length. We have

$$2\ell + 2w = 98$$
$$2(2w - 5) + 2w = 98$$
$$4w - 10 + 2w = 98$$
$$6w - 10 = 98$$
$$6w = 108$$
$$w = 18$$

The dimensions are 31 cm by 18 cm.

11. Let x be the second test score.

$$\frac{90 + x + 87}{3} = 85$$

$$90 + x + 87 = 255$$
$$177 + x = 255$$
$$x = 78$$

Her second score was 78.

13. Total Cost = Fixed Cost + Cost of Dolls.
Let x = Cost of dolls
$$4335 = 585 + x$$
$$x = \$3750$$
Each doll costs \$15.

Cost of all dolls = cost of one doll times the number of dolls. Let y = the number of dolls
$$3750 = 15y$$
$$y = 250 \text{ dolls.}$$

15. Let x be the original price. It follows that 12% of x is \$15. Thus

$$\frac{12}{100}x = 15$$

$$12x = 1500$$
$$x = 125$$
The original price was \$125.

17. Let x be the wholesale price of the tape recorder. If we add 45% of x to x, we obtain the retail price of \$116. Thus

$$x + 0.45x = 116$$
$$1.45x = 116$$

$$x = \frac{116}{1.45}$$

$$x = 80$$
The wholesale price is \$80.

19. Let x be the number of people polled.

$$858 + 0.35x = x$$
$$858 = 0.65x$$

$$\frac{858}{0.65} = x$$

1320 people were polled.

21. Let d be the number of dimes. Then $d + 4$ represents the number of quarters and $d + 7$ the number of nickels. Since the total amount is \$4.55, we have

$$0.05(d + 7) + 0.10d + 0.25(d + 4) = 4.55$$
$$5(d + 7) + 10d + 25(d + 4) = 455$$
$$5d + 35 + 10d + 25d + 100 = 455$$
$$40d = 320$$
$$d = 8$$

She has 8 dimes, 12 quarters, 15 nickels.

23. $i = prt$
$i = (2400)(0.08)(1) = \$192$

Amount $= p + i = 2400 + 192 = \$2592.$

25. Let x be the amount he puts into the retirement plan.

    5.5% of \$21,520 is x
    $$(0.055)(21,520) = x$$
    $$1183.60 = x$$

    He puts in \$1183.60.

27. Let x be the amount invested at 7.1%. Then $50,000 - x$ is the amount invested at 14.5%

    $$0.071x + 0.145(50000 - x) = 5622$$
    $$0.071x + 7250 - 0.145x = 5622$$
    $$-0.074x + 7250 = 5622$$
    $$-0.074x = -1628$$
    $$x = 22000$$

    \$22,000 is invested at 7.1%
    \$28,000 is invested at 14.5%.

29. Let x be the number of grams per part.

    $$5x + 3x = 720$$
    $$8x = 720$$
    $$x = 90$$
    $$5x = 450$$
    $$3x = 270$$

    450 grams of chemical A
    270 grams of chemical B.

31. Let x be number of cubic feet per part.

    $$3x + 2x + 2x = 875$$
    $$7x = 875$$
    $$x = 125$$
    $$3x = 375$$
    $$2x = 250$$

    375 cubic ft. cement
    250 cubic ft. sand
    250 cubic ft. stone

33. If 2% of 480 cm^3 is salt, then the volume of salt is

    $$(0.02)(480) = 9.6 \ cm^3.$$

    Let x be the amount of water added to decrease the concentration.

The new volume is $480 + x$. Since a 1.5% concentration is desired, it follows that the volume of salt in the new solution is

$$(0.015)(480 + x).$$

Since the amount of salt remains the same, we get the equation

$$(0.015)(480 + x) = 9.6$$
$$7.2 + 0.015x = 9.6$$
$$0.015x = 2.4$$
$$x = 160 \ cm^3$$

35. Let x be gallons of solution A. Then $160 - x$ is gallons of solution B.

    $$20\%x + 40\%(160 - x) = 35\%(160)$$
    $$20x + 40(160 - x) = 35(160)$$
    $$20x + 6400 - 40x = 5600$$
    $$-20x = -800$$
    $$x = 40$$

    40 gallons of A
    120 gallons of B.

37.

	distance	rate	time
To hill	d	8	x
Back from hill	d	12	$1\frac{1}{4} - x$

distance to hill
= distance back from hill

$$8x = 12(\frac{5}{4} - x)$$

$$8x = 15 - 12x$$
$$20x = 15$$

$$x = \frac{3}{4}$$

It took $\frac{3}{4}$ hour to get to the hill. Therefore the distance to the hill was $8(\frac{3}{4}) = 6$ miles. Total distance was 12 miles.

39.

	distance	rate	time
A to B	d	250	x
B to A	d	300	$5\frac{1}{2} - x$

distance to B = distance to A

$$250x = 300(5\frac{1}{2} - x)$$

$$250x = 1650 - 300x$$
$$550x = 1650$$
$$x = 3 \text{ hours}$$
$$d = 250x = 750 \text{ miles}$$

a) A to B is 750 miles
b) It took 3 hours.

41.

	distance	rate	time
Ferry	225-180	r	x
Car	180	45	$5\frac{1}{4} - x$

$$d = rt$$
$$\text{car: } 180 = 45(5.25 - x)$$
$$180 = 236.25 - 45x$$
$$-56.25 = -45x$$
$$x = 1.25 \text{ hours}$$

$$\text{ferry: } 225 - 180 = r(1.25)$$
$$45 = 1.25r$$
$$36 = r$$
$$36 \text{ mph}$$

43.

	distance	rate	time
1st car	d	50	$x + \frac{1}{2}$
2nd car	d	60	x

distance car 1 = distance car 2

$$50(x + \frac{1}{2}) = 60x$$

$$50x + 25 = 60x$$
$$25 = 10x$$

$$x = 2\frac{1}{2}$$

2.5 h, 12:00 noon

45. Let x be the number of hours it takes both pipes working together. They fill $1/x$ of the pool per hour. Working alone, the first pipe fills $1/6$ of the pool and the second fills $1/4$ of the pool per hour. Thus,

$$\frac{1}{6} + \frac{1}{4} = \frac{1}{x}$$

$$\frac{12x}{6} + \frac{12x}{4} = \frac{12x}{x}$$

$$2x + 3x = 12$$
$$5x = 12$$

$$x = \frac{12}{5}$$

It takes 12/5 hours to fill the pool.

47. Let x be the time it takes to fill the tank if all three pipes are open. Thus, $1/x$ of the tank is filled per hour. Pipes A and B fill $1/2$ and $1/3$ of the tank per hour while pipe C empties $1/4$ of the tank per hour. Thus,

$$\frac{1}{2} + \frac{1}{3} - \frac{1}{4} = \frac{1}{x}$$

$$6x + 4x - 3x = 12$$
$$7x = 12$$

$$x = \frac{12}{7}$$

It takes 12/7 hours to fill the tank.

49. The first machine prints 125 labels per minute and the second prints 200 labels per minute. Together they print 325 labels per minute. If x denotes the number of minutes that takes for printing 6500 labels when working together, then

$$325x = 6500$$
$$x = 20 \text{ minutes}$$

Exercises 2.3

Find the interval sketches for Exercises 11-21 in the answer section of your text.

1. $\sqrt{3} \approx 1.732$
   Thus, $-2 < \sqrt{3} < 2$.

3. $-\dfrac{2}{3} = -0.666\ldots$

   $\dfrac{1}{2} = 0.5$

   $-\sqrt{2} \approx -1.414$

   Thus, $-\sqrt{2} < -\dfrac{2}{3} < \dfrac{1}{2}$

5. $\dfrac{1}{4} = 0.25$

   $\sqrt{5} \approx 2.236$
   $\sqrt{3} \approx 1.732$

   Thus, $\dfrac{1}{4} < \sqrt{3} < \sqrt{5} < 4$

7. $-\sqrt{2} \approx -1.414$
   $-\dfrac{5}{2} = -2.5$

   $\sqrt{6} \approx 2.449$

   Thus, $-\dfrac{5}{2} < -\sqrt{2} < 1 < \sqrt{6}$

9. $-\sqrt{6} \approx -2.449$
   $\sqrt{2} \approx 1.414$

$\dfrac{17}{12} = 1.41666\ldots$

Thus $-2.45 < -\sqrt{6} < \sqrt{2} < \dfrac{17}{12}$

11. $[-2,3] = \{x \in \mathbb{R}: \ -2 \le x \le 3\}$

13. $[2,8] = \{x \in \mathbb{R}: \ 2 \le x < 8\}$

15. $(-\infty,5) = \{x \in \mathbb{R}: \ x < 5\}$

17. $(-2,7)$

19. $(-3,3)$

21. $[-2,\infty)$

23. $|x| = |-5|$
    $|x| = 5$
    $x = \pm 5$

25. $|2x| = |-7|$
    $|2x| = 7$
    $2x = \pm 7$
    $2x = 7$ or $2x = -7$

    $x = \pm \dfrac{7}{2}$

27. $-x = |-2|$
    $-x = 2$
    $x = -2$

29. $|x - 1| = 3$
    $x - 1 = \pm 3$
    $x - 1 = 3$ or $x - 1 = -3$
    $x = 3 + 1$ or $x = -3 + 1$
    $x = 4$ or $x = -2$

31. $|3 - 2x| = 4$
    $3 - 2x = \pm 4$
    $3 - 2x = 4$ or $3 - 2x = -4$
    $-2x = 1$ or $-2x = -7$

    $x = -\dfrac{1}{2}$ or $x = \dfrac{7}{2}$

33. $|3x - 4| = |-2|$
$|3x - 4| = 2$
$3x - 4 = \pm 2$
$3x - 4 = 2$ or $3x - 4 = -2$
$3x = 6$ or $3x = 2$

$x = 2$ or $x = \dfrac{2}{3}$

35. $|3x - 1| + 4 = 7$

Isolate absolute value expression:
$|3x - 1| = 3$.

Therefore
$3x - 1 = 3$ or $3x - 1 = -3$
$3x = 4$ $\qquad$ $3x = -2$

$x = \dfrac{4}{3}$ $\qquad$ $x = \dfrac{-2}{3}$

37. $|2 - 5x| - 1 = 9$

Isolate absolute value expression:
$|2 - 5x| = 10$

Therefore
$2 - 5x = 10$ or $2 - 5x = -10$
$-5x = 8$ $\qquad$ $-5x = -12$

$x = \dfrac{-8}{5}$ $\qquad$ $x = \dfrac{12}{5}$

39. $|3 - 4x| - 10 = -4$

Isolate absolute value expression.
$|3 - 4x| = 6$

Therefore
$3 - 4x = 6$ or $3 - 4x = -6$
$-4x = 3$ $\qquad$ $-4x = -9$

$x = \dfrac{-3}{4}$ $\qquad$ $x = \dfrac{9}{4}$

41. $|3x - 4| = |2x + 5|$
Either $3x - 4 = 2x + 5$
or $3x - 4 = -(2x + 5)$

In the first case,
$3x - 4 = 2x + 5$
$x = 9$

In the second case,
$3x - 4 = -2x - 5$
$5x = -1$

$x = \dfrac{-1}{5}$

43. $|m + 3| = |2 - m|$
$m + 3 = 2 - m$ or $m + 3 = -(2 - m)$
$2m = -1$ $\qquad$ $m + 3 = -2 + m$

$m = -\dfrac{1}{2}$ $\qquad$ $3 = -2$

$\qquad\qquad\qquad$ Contradiction.
$\qquad\qquad\qquad$ No other solution.

45. $|-15| = 15$

47. $|-9 + 4| = |-5| = 5$

49. $|-6 - 8| = |-14| = 14$

51. $\left|\pi - \dfrac{25}{8}\right|$

$\dfrac{25}{8} = 3.125$ and $3.125 < \pi$

Thus $\left|\pi - \dfrac{25}{8}\right| = \pi - \dfrac{25}{8}$

53. Since $\sqrt{3} > \sqrt{2}$, it follows that
$\sqrt{3} - \sqrt{2} > 0$. Thus $|\sqrt{2} - \sqrt{3}|$
$= -(\sqrt{2} - \sqrt{3}) = \sqrt{3} - \sqrt{2}$.

55. If $a < 3$, then $3 - a > 0$. Thus
$|a - 3| = -(a - 3) = 3 - a$.

57. If $x < 5$, then $2x < 10$
and $2x - 10 < 0$. Thus
$|2x - 10| = -(2x - 10) = 10 - 2x$.

59. $A = -6 \qquad B = -4 \qquad C = -1$
$$d(A,B) = |-6 - (-4)|$$
$$= |-6 + 4| = |-2| = 2$$
$$d(A,C) = |-6 - (-1)|$$
$$= |-6 + 1| = |-5| = 5$$
$$d(B,C) = |-4 - (-1)|$$
$$= |-4 + 1| = |-3| = 3$$

61. $A = -5/2 \qquad B = 0 \qquad C = 7/2$
$$d(A,B) = |0 - (-5/2)|$$
$$= |5/2| = 5/2$$
$$d(A,C) = |7/2 - (-5/2)|$$
$$= |7/2 + 5/2| = |12/2| = 6$$
$$d(B,C) = |7/2 - 0|$$
$$= |7/2| = 7/2$$

63. $A = 0.5 \qquad B = 1.26 \qquad C = 0.08$
$$d(A,B) = |1.26 - 0.5|$$
$$= |0.76| = 0.76$$
$$d(A,C) = |0.08 - 0.5|$$
$$= |-0.42| = 0.42$$
$$d(B,C) = |0.08 - 1.26|$$
$$= |-1.18| = 1.18$$

65. $|x| = |-x|$
If $x > 0$, then $|x| = x$
$$|-x| = -(-x) = x$$
If $x < 0$, then $|x| = -x$
$$|-x| = -x$$

67. If $x > 0$, then $1/x > 0$.
$$|1/x| = 1/x = 1/|x|.$$
If $x < 0$, then $1/x < 0$.
$$|1/x| = -(1/x) = 1/-x = 1/|x|.$$

69. Let a be the abscissa of A and let b be the abscissa of B. If $d(A,B) = 0$, then $|b - a| = 0$ and $b = a$ and $A = B$.

Conversely, if $A = B$, then they have the same abscissa a. Thus $d(A,B) = |a - a| = 0$.

Exercises 2.4

Find interval sketches in the answer section of your text.

1. $6x - 3 > 8$
$$6x > 11$$
$$x > 11/6$$

3. $5 - 2x \le 7$
$$-2x \le 2$$
$$x \ge -1$$

5. $\dfrac{3}{4}x - 2 \ge \dfrac{1}{3}$
$$\dfrac{12(3x)}{4} - 12(2) \ge \dfrac{12}{3}$$
$$9x - 24 \ge 4$$
$$9x \ge 28$$
$$x \ge 28/9$$

7. $3x - 2 \ge 4 - 2x$
$$3x + 2x \ge 4 + 2$$
$$5x \ge 6$$
$$x \ge 6/5$$

9. $\dfrac{2x - 3}{4} \ge \dfrac{3x}{2} - \dfrac{1}{3}$
$$\dfrac{12(2x - 3)}{4} \ge \dfrac{12(3x)}{2} - \dfrac{12(1)}{3}$$
[multiply by LCD 12]
$$3(2x - 3) \ge 6(3x) - 4$$
$$6x - 9 \ge 18x - 4$$
$$-12x \ge 5$$
$$x \le -5/12$$

11. $3 \le 4x - 1 \le 5$
[add 1 to all members]
$$4 \le 4x \le 6$$
[divide by 4]
$$1 \le x \le \dfrac{6}{4}$$
$$1 \le x \le \dfrac{3}{2}$$

13. $1 \leq \dfrac{2x - 5}{6} < 3$

$6(1) \leq \dfrac{6(2x - 5)}{6} < 6(3)$

$6 \leq 2x - 5 < 18$
$11 \leq 2x < 23$
$11/2 \leq x < 23/2$

15. $5 > 2 - x/3 \geq 0$
$3(5) > 3(2) - 3(x/3) \geq 3(0)$
$15 > 6 - x \geq 0$
$9 > -x \geq -6$
$-9 < x \leq 6$

17. $(3x + 2)(x - 4) < (x + 2)(3x - 1)$
$3x^2 - 10x - 8 < 3x^2 + 5x - 2$
$-10x - 8 < 5x - 2$
$-15x < 6$
$x > -6/15$
$x > -2/5$

19. $(x + 1)^2 > (x - 2)(x - 1)$
$x^2 + 2x + 1 > x^2 - 3x + 2$
$2x + 1 > -3x + 2$
$5x > 1$
$x > 1/5$

21. $\dfrac{2}{x + 3} > 0$

Since the entire fraction $\dfrac{2}{x + 3}$

is positive and the numerator is positive, it follows that the denominator is also positive. Therefore,

$x + 3 > 0$
$x > -3$

23. $\dfrac{-3}{4 - x} < 0$

Multiply by -1 to get the equivalent inequality

$\dfrac{3}{4 - x} > 0.$

For this fraction to be positive, the denominator must be positive, that is

$4 - x > 0$
$x < 4$

25. $\dfrac{x}{x + 4} > 1$

$\dfrac{x}{x + 4} - 1 > 0$

$\dfrac{x - (x + 4)}{x + 4} > 0$

$\dfrac{-4}{x + 4} > 0$

$\dfrac{4}{x + 4} < 0$

$x + 4 < 0$
$x < -4$

27. $\dfrac{3x}{3x - 1} - 1 < 0$

$\dfrac{3x - (3x - 1)}{3x - 1} < 0$

$\dfrac{3x - 3x + 1}{3x - 1} < 0$

$\dfrac{1}{3x - 1} < 0$

$3x - 1 < 0$
$3x < 1$
$x < 1/3$

29. $\dfrac{2x}{4x-1} > \dfrac{1}{2}$

$\dfrac{2x}{4x-1} - \dfrac{1}{2} > 0$

$\dfrac{4x - (4x-1)}{2(4x-1)} > 0$

$\dfrac{1}{8x-2} > 0$

$8x - 2 > 0$
$8x > 2$
$x > 2/8$
$x > 1/4$

31. $|2x| > 8$
$2x > 8 \text{ or } 2x < -8$
$x > 4 \text{ or } x < -4$

33. $|3x| - 1 \le 5$

Isolate the absolute
value expression

$|3x| \le 6$
$-6 \le 3x \le 6$
$-2 \le x \le 2$

35. $|x - 4| > 1$ is equivalent to
$x - 4 < -1 \text{ or } x - 4 > 1$
$x < 3 \text{ or } x > 5$

37. $|3x + 1| < 9$ is equivalent to
$-9 < 3x + 1 < 9$
$-10 < 3x < 8$
$-10/3 < x < 8/3$

39. $|3 - 6x| > 2$
$3 - 6x < -2 \text{ or } 3 - 6x > 2$
$-6x < -5 -6x > -1$
$x > 5/6 x < 1/6$

41. $|2x - 1| \ge |-3|$
$|2x - 1| \ge 3$
$2x - 1 \ge 3 \text{ or } 2x - 1 \le -3$
$2x \ge 4 2x \le -2$
$x \ge 2 x \le -1$

43. $\left|\dfrac{3-x}{4}\right| < 1$

$-1 < \dfrac{3-x}{4} < 1$

$-4 < 3 - x < 4$
$-7 < -x < 1$
$7 > x > -1$
$-1 < x < 7$

45. $\left|\dfrac{2x-5}{3}\right| < 3$

$-3 < \dfrac{2x-5}{3} < 3$

$-9 < 2x - 5 < 9$
$-4 < 2x < 14$
$-2 < x < 7$

47. $|2x - 1/3| < 4$
$-4 < 2x - 1/3 < 4$
$-12 < 6x - 1 < 12$
$-11 < 6x < 13$
$-11/6 < x < 13/6$

49. $\left|\dfrac{x}{3} - \dfrac{1}{2}\right| \ge 1$

$\dfrac{x}{3} - \dfrac{1}{2} \le -1 \text{ or } \dfrac{x}{3} - \dfrac{1}{2} \ge 1$

$2x - 3 \le -6 2x - 3 \ge 6$
$2x \le -3 2x \ge 9$
$x \le -3/2 \text{ or } x \ge 9/2$

51. $\left|\dfrac{3x}{4} - \dfrac{1}{3}\right| \ge 1$

$\dfrac{3x}{4} - \dfrac{1}{3} \le -1 \text{ or } \dfrac{3x}{4} - \dfrac{1}{3} \ge 1$

$9x - 4 \le -12 9x - 4 \ge 12$
$9x \le -8 9x \ge 16$
$x \le -8/9 \text{ or } x \ge 16/9$

53. $|2x + 6| > 0$

The absolute value of any number $\neq$ 0 is always positive. Thus $2x + 6$ can be any number except 0. To find which value of x makes $2x + 6$ equal zero, set up the equation

$$2x + 6 = 0$$
$$x = -3$$

Therefore, the absolute value inequality is satisfied for all real numbers $x \neq -3$.

55. $F = \dfrac{9}{5}C + 32$

Since $20° < C < 25°$, it follows that

$$\dfrac{9}{5}(20) + 32 < F < \dfrac{9}{5}(25) + 32$$

$$36 + 32 < F < 45 + 32$$
$$68° < F < 77°$$

57. If $V = 200$ volts, then

$$200 = RI \text{ or } R = \dfrac{200}{I}. \quad \text{Now,}$$

$$10 < I < 20$$

$$\dfrac{1}{20} < \dfrac{1}{I} < \dfrac{1}{10}$$

$$\dfrac{200}{20} < \dfrac{200}{I} < \dfrac{200}{10}$$

Thus $10 < R < 20$.

59. Let x be the score on Peter's fourth test. Peter's average is

$$\dfrac{85 + 90 + 95 + x}{4}.$$

If Peter's average is to be between 80 and 85 inclusive, then

$$80 \leq \dfrac{85 + 90 + 95 + x}{4} \leq 85$$

[multiplying by 4]

$$320 \leq 85 + 90 + 95 + x \leq 340$$
$$320 \leq 270 + x \leq 340$$
$$50 \leq x \leq 70$$

61. $\dfrac{PV}{T} = C$

$$\dfrac{PV}{100} = 1$$

$$PV = 100$$

$$V = \dfrac{100}{P}$$

Since $25 \leq P \leq 75$, then

$$\dfrac{100}{75} \leq V \leq \dfrac{100}{25}$$

$$4/3 \leq V \leq 4$$

63. $144 \leq v \leq 208$
Since $v = 32t$
$144 \leq 32t \leq 208$
$4.5\ s \leq t \leq 6.5\ s$

65. $IQ = \dfrac{MA}{CA} \times 100$

$$MA = \dfrac{(CA)(IQ)}{100}$$

$$\dfrac{(15)(70)}{100} \leq MA \leq \dfrac{(15)(140)}{100}$$

$$\dfrac{1050}{100} \leq MA \leq \dfrac{2100}{100}$$

$$10.5 \leq MA \leq 21$$

67. $d + rt$
$r = d/t$
$$\dfrac{150}{3} \leq r \leq \dfrac{150}{2.5}$$

$$50 \text{ mph} \leq r \leq 60 \text{ mph}$$

69. Profit = Revenue − Cost

$$R - C > 0$$
$$15x - (775 + 2.5x) > 0$$
$$15x - 775 - 2.5x > 0$$
$$12.5x > 775$$
$$x > 62$$

62 pairs of jeans

Exercises 2.5

1. $x^2 - 4 = 0$
$$x^2 = 4$$
$$x = \pm 2$$

3. $2x^2 - 32 = 0$
$$2x^2 = 32$$
$$x^2 = 16$$
$$x = \pm 4$$

5. $4x^2 - 18 = 0$
$$4x^2 = 18$$
$$x^2 = 18/4$$
$$x^2 = 9/2$$
$$x = \pm 3/\sqrt{2}$$
$$x = \frac{\pm 3\sqrt{2}}{2}$$

7. $(x + 1)^2 - 2 = 0$
$$(x + 1)^2 = 2$$
$$x + 1 = \pm\sqrt{2}$$
$$x = -1 \pm \sqrt{2}$$

9. $(5x - 2)^2 - 11 = 0$
$$(5x - 2)^2 = 11$$
$$5x - 2 = \pm\sqrt{11}$$
$$5x = 2 \pm \sqrt{11}$$
$$x = \frac{2 \pm \sqrt{11}}{5}$$

11. $x^2 + 15 = 0$
$$x^2 = -15$$
$$x = \pm\sqrt{-15}$$
$$x = \pm i\sqrt{15}$$

13. $(x + 3)^2 + 10 = 0$
$$(x + 3)^2 = -10$$
$$x + 3 = \pm\sqrt{-10}$$
$$x + 3 = \pm i\sqrt{10}$$
$$x = -3 \pm i\sqrt{10}$$

15. $(3x - 2)^2 + 20 = 0$
$$(3x - 2)^2 = -20$$
$$3x - 2 = \pm\sqrt{-20}$$
$$3x - 2 = \pm 2i\sqrt{5}$$
$$3x = 2 \pm 2i\sqrt{5}$$
$$x = \frac{2 \pm 2i\sqrt{5}}{3}$$

17. $x^2 - 6x + 8 = 0$
$$(x - 2)(x - 4) = 0$$
$$x - 2 = 0 \quad x - 4 = 0$$
$$x = 2 \qquad x = 4$$

19. $u^2 + 8u + 15 = 0$
$$(u + 5)(u + 3) = 0$$
$$u + 5 = 0 \quad u + 3 = 0$$
$$u = -5 \qquad u = -3$$

21. $x^2 - 6x + 5 = 0$
$$(x - 5)(x - 1) = 0$$
$$x - 5 = 0 \quad x - 1 = 0$$
$$x = 5 \qquad x = 1$$

23. $v^2 - 3v = 0$
$$v(v - 3) = 0$$
$$v = 0 \quad v - 3 = 0$$
$$v = 0 \qquad v = 3$$

25. $4x^2 - 5x - 6 = 0$
$$(4x + 3)(x - 2) = 0$$
$$4x + 3 = 0 \quad x - 2 = 0$$
$$4x = -3$$
$$x = -3/4 \qquad x = 2$$

27. $12x^2 + 5x = 3$
$$12x^2 + 5x - 3 = 0$$
$$(3x - 1)(4x + 3) = 0$$
$$3x - 1 = 0 \quad 4x + 3 = 0$$
$$3x = 1 \qquad 4x = -3$$
$$x = \frac{1}{3} \qquad x = \frac{-3}{4}$$

29.
$$2y(4y - 3) = 9$$
$$8y^2 - 6y = 9$$
$$8y^2 - 6y - 9 = 0$$
$$(4y + 3)(2y - 3) = 0$$
$$4y + 3 = 0 \quad 2y - 3 = 0$$
$$4y = -3 \qquad 2y = 3$$

$$y = \frac{-3}{4} \quad y = \frac{3}{2}$$

31.
$$x^2 - 4x = 0$$
$$x^2 - 4x + 4 = 0 + 4$$
$$(x - 2)^2 = 4$$
$$x - 2 = \pm 2$$
$$x - 2 = 2 \text{ or } x - 2 = -2$$
$$x = 4 \qquad x = 0$$

33.
$$x^2 + 5x = 0$$

$$x^2 + 5x + \frac{25}{4} = 0 + \frac{25}{4}$$

$$(x + \frac{5}{2})^2 = \frac{25}{4}$$

$$x + \frac{5}{2} = \pm \frac{5}{2}$$

$$x + \frac{5}{2} = \frac{5}{2} \text{ or } x + \frac{5}{2} = \frac{-5}{2}$$

$$x = 0 \quad x = \frac{-10}{2} = -5$$

35.
$$x^2 - 2x - 8 = 0$$
$$x^2 - 2x = 8$$
$$x^2 - 2x + 1 = 8 + 1$$
$$(x - 1)^2 = 9$$
$$x - 1 = \pm 3$$
$$x = 1 \pm 3$$
$$x = -2, 4$$

37.
$$m^2 + 8m + 15 = 0$$
$$m^2 + 8m = -15$$
$$m^2 + 8m + 16 = -15 + 16$$
$$(m + 4)^2 = 1$$
$$m + 4 = \pm 1$$
$$m = -4 \pm 1$$
$$m = -5, -3$$

39.
$$x^2 + 10x + 16 = 0$$
$$x^2 + 10x = -16$$
$$x^2 + 10x + 25 = -16 + 25$$
$$(x + 5)^2 = 9$$
$$x + 5 = \pm 3$$
$$x = -5 \pm 3$$
$$x = -8, -2$$

41.
$$x^2 - 6x + 8 = 0$$
$$x^2 - 6x = -8$$
$$x^2 - 6x + 9 = -8 + 9$$
$$(x - 3)^2 = 1$$
$$x - 3 = \pm 1$$
$$x = 3 \pm 1$$
$$x = 2, 4$$

43.
$$2x^2 - 8x + 4 = 0$$
$$2x^2 - 8x = -4$$

[divide by 2]

$$x^2 - 4x = -2$$
$$x^2 - 4x + 4 = -2 + 4$$
$$(x - 2)^2 = 2$$
$$x - 2 = \pm \sqrt{2}$$
$$x = 2 \pm \sqrt{2}$$

45.
$$y^2 - y + 2/9 = 0$$
$$y^2 - y = -2/9$$
$$y^2 - y + 1/4 = -2/9 + 1/4$$
$$(y - 1/2)^2 = -8/36 + 9/36$$
$$(y - 1/2)^2 = 1/36$$
$$y - 1/2 = \pm 1/6$$
$$y = 1/2 \pm 1/6$$
$$= 3/6 \pm 1/6$$
$$y = 2/6, 4/6$$
$$= 1/3, 2/3$$

47.
$$2x^2 + 3x - 1/2 = 0$$
$$2x^2 + 3x = 1/2$$
$$x^2 + (3/2)x = 1/4$$
$$x^2 + (3/2)x + 9/16 = 1/4 + 9/16$$
$$(x + 3/4)^2 = 13/16$$
$$x + 3/4 = \pm \sqrt{13}/4$$
$$x = -3/4 \pm \sqrt{13}/4$$

$$= \frac{-3 \pm \sqrt{13}}{4}$$

49.
$$2x^2 - 3x + 2 = 0$$
$$2x^2 - 3x = -2$$
$$x^2 - (3/2)x = -1$$
$$x^2 - (3/2)x + 9/16 = -1 + 9/16$$
$$(x - 3/4)^2 = -7/16$$
$$x - 3/4 = \pm\sqrt{-7}/4$$
$$x - 3/4 = \pm\sqrt{7}/4$$
$$x = 3/4 \pm i\sqrt{7}/4$$

$$= \frac{3 \pm i\sqrt{7}}{4}$$

51.
$$3x^2 + 2x + 1/2 = 0$$
$$3x^2 + 2x = -1/2$$
$$x^2 + (2/3)x = -1/6$$
$$x^2 + (2/3)x + 1/9 = -1/6 + 1/9$$
$$(x + 1/3)^2 = -1/18$$
$$x + 1/3 = \pm\sqrt{-1/18}$$
$$x + 1/3 = \pm i/3\sqrt{2}$$
$$x + 1/3 = \pm i\sqrt{2}/6$$
$$x = -1/3 \pm i\sqrt{2}/6$$

$$= \frac{-2 \pm i\sqrt{2}}{6}$$

53. $2x^2 + 4x - 5 = 0$

Here, $a = 2$, $b = 4$, and $c = -5$.

$$x = \frac{-4 \pm\sqrt{4^2 - 4(2)(-5)}}{2(2)}$$

$$= \frac{-4 \pm\sqrt{16 + 40}}{4}$$

$$= \frac{-4 \pm\sqrt{56}}{4} = \frac{-4 \pm 2\sqrt{14}}{4}$$

$$= \frac{-2 \pm\sqrt{14}}{2}$$

55.
$$u^2 = 2 - u$$
$$u^2 + u - 2 = 0$$
$$a = 1,\ b = 1,\ c = -2$$

$$u = \frac{-1 \pm\sqrt{1^2 - 4(1)(-2)}}{2(1)}$$

$$= \frac{-1 \pm\sqrt{1 + 8}}{2} = \frac{-1 \pm\sqrt{9}}{2}$$

$$= \frac{-1 \pm 3}{2}$$

$$u = -2,\ 1$$

57.
$$p + 2p^2 = 3$$
$$2p^2 + p - 3 = 0$$
$$a = 2,\ b = 1,\ c = -3$$

$$p = \frac{-1 \pm\sqrt{1^2 - 4(2)(-3)}}{2(2)}$$

$$= \frac{-1 \pm\sqrt{1 + 24}}{4} = \frac{-1 \pm\sqrt{25}}{4}$$

$$= \frac{-1 \pm 5}{4}$$

$$p = -3/2,\ 1$$

59.
$$x^2 + x = 3/2$$
$$x^2 + x - 3/2 = 0$$
$$a = 1,\ b = 1,\ c = -3/2$$

$$x = \frac{-1 \pm\sqrt{1^2 - 4(1)(-3/2)}}{2(1)}$$

$$= \frac{-1 \pm\sqrt{1 + 6}}{2} = \frac{-1 \pm\sqrt{7}}{2}$$

61. $5m^2 - 2m + 4 = 0$
$$a = 5,\ b = -2,\ c = 4$$

$$m = \frac{2 \pm\sqrt{(-2)^2 - 4(5)(4)}}{2(5)}$$

$$= \frac{2 \pm\sqrt{4 - 80}}{10}$$

$$= \frac{2 \pm\sqrt{-76}}{10} = \frac{2 \pm 2i\sqrt{19}}{10}$$

$$= \frac{1 \pm i\sqrt{19}}{5}$$

63. $x - 2 = x^2$

$x^2 - x + 2 = 0$

$a = 1, \ b = -1, \ c = 2$

$$x = \frac{1 \pm \sqrt{(-1)^2 - 4(1)(2)}}{2(1)}$$

$$= \frac{1 \pm \sqrt{1 - 8}}{2} = \frac{1 \pm \sqrt{-7}}{2}$$

$$= \frac{1 \pm i\sqrt{7}}{2}$$

65. $(1/2)x^2 - (1/3)x + 2 = 0$

$a = 1/2, \ b = -1/3, \ c = 2$

$$x = \frac{1/3 \pm \sqrt{(-1/3)^2 - 4(1/2)(2)}}{2(1/2)}$$

$$= \frac{1/3 \pm \sqrt{1/9 - 4}}{1}$$

$$= 1/3 \pm \sqrt{-35/9}$$

$$= 1/3 \pm i\sqrt{35}/3$$

$$= \frac{1 \pm i\sqrt{35}}{3}$$

67. $x - \dfrac{60}{x} = 11$

Multiply equation by x

$$x^2 - 60 = 11x$$

$$x^2 - 11x - 60 = 0$$

$$(x - 15)(x + 4) = 0$$

$$x - 15 = 0 \text{ or } x + 4 = 0$$

$$x = 15 \qquad x = -4$$

69. $3u = \dfrac{3}{u} - 8$

Multiply equation by u

$3u^2 = 3 - 8u$

$3u^2 + 8u - 3 = 0$

$(3u - 1)(u + 3) = 0$

$3u - 1 = 0 \qquad u + 3 = 0$

$3u = 1 \qquad\qquad u = -3$

$u = 1/3$

71. $\dfrac{5}{x - 4} + \dfrac{4}{x + 2} = 3$

Multiply equation by $(x - 4)(x + 2)$

$5(x + 2) + 4(x - 4) =$
$\qquad\qquad 3(x - 4)(x + 2)$

$5x + 10 + 4x - 16 = 3(x^2 - 2x - 8)$

$9x - 6 = 3x^2 - 6x - 24$

$3x^2 - 15x - 18 = 0$

$(3x + 3)(x - 6) = 0$

$3x + 3 = 0 \qquad x - 6 = 0$

$3x = -3 \qquad\qquad x = 6$

$x = -1$

73. $\dfrac{x - 2}{x + 1} + \dfrac{9}{4} = \dfrac{x + 2}{x - 1}$

Multiply equation by

$4(x + 1)(x - 1)$

$4(x - 1)(x - 2) + 9(x + 1)(x - 1)$
$\qquad\qquad = 4(x + 1)(x + 2)$

$4(x^2 - 3x + 2) + 9(x^2 - 1)$
$\qquad\qquad = 4(x^2 + 3x + 2)$

$4x^2 - 12x + 8 + 9x^2 - 9$
$\qquad\qquad = 4x^2 + 12x + 8$

$13x^2 - 12x - 1 = 4x^2 + 12x + 8$

$9x^2 - 24x - 9 = 0$

$3(3x^2 - 8x - 3) = 0$

$3(3x + 1)(x - 3) = 0$

$3x + 1 = 0 \text{ or } x - 3 = 0$

$3x = -1 \qquad\qquad x = 3$

$$x = \frac{-1}{3}$$

75. $\dfrac{3x - 1}{x + 1} = \dfrac{5x}{x + 4}$

Multiply by $(x + 1)(x + 4)$

$$(3x - 1)(x + 4) = 5x(x + 1)$$
$$3x^2 + 12x - x - 4 = 5x^2 + 5x$$
$$3x^2 + 11x - 4 = 5x^2 + 5x$$
$$2x^2 - 6x + 4 = 0$$
$$2(x^2 - 3x + 2) = 0$$
$$2(x - 1)(x - 2) = 0$$
$$x - 1 = 0 \text{ or } x - 2 = 0$$
$$x = 1 \qquad x = 2$$

77. $0.6x^2 - 0.3x + 0.1 = 0$

$$x = \frac{0.3 \pm \sqrt{(0.3)^2 - 4(0.6)(0.1)}}{2(0.6)}$$

$$= \frac{0.3 \pm \sqrt{0.09 - 0.24}}{1.2}$$

$$= \frac{0.3 \pm \sqrt{-0.15}}{1.2}$$

$$= \frac{0.3 \pm i\sqrt{0.15}}{1.2} = \frac{0.3 \pm 0.39i}{1.2}$$

$$= 0.25 \pm 0.32i$$

79. $1.31x^2 - 3.2x + 5.12 = 0$

$$x = \frac{3.2 \pm \sqrt{(-3.2)^2 - 4(1.31)(5.12)}}{2(1.31)}$$

$$= \frac{3.2 \pm \sqrt{10.24 - 26.8288}}{2.62}$$

$$= \frac{3.2 \pm \sqrt{-16.5888}}{2.62}$$

$$= \frac{3.2 \pm 4.07i}{2.62}$$

$$= 1.22 \pm 1.55i$$

81. $2x^2 - 18 = 0$
$$2x^2 = 18$$
$$x^2 = 9$$
$$x = \pm 3$$

We have
$$2x^2 - 18 = 2(x - 3)(x + 3)$$

83. $\quad 2x^2 + 3x - 2 = 0$
$$(2x - 1)(x + 2) = 0$$
$$2x - 1 = 0 \qquad x + 2 = 0$$
$$2x = 1 \qquad\qquad x = -2$$
$$x = 1/2$$

We have
$$2x^2 + 3x - 2 = 2(x - 1/2)(x + 2)$$
$$\qquad\qquad\quad = (2x - 1)(x + 2)$$

85. $x^2 - 2x - 1 = 0$

$$x = \frac{2 \pm \sqrt{2^2 - 4(1)(-1)}}{2(1)}$$

$$= \frac{2 \pm \sqrt{8}}{2} = \frac{2 \pm 2\sqrt{2}}{2} = 1 \pm \sqrt{2}$$

The solutions of the quadratic equation are $1 + \sqrt{2}$ and $1 - \sqrt{2}$. The factorization is

$$x^2 - 2x - 1$$
$$= [x - (1 - \sqrt{2})][x - (1 + \sqrt{2})]$$
$$= (x - 1 + \sqrt{2})(x - 1 - \sqrt{2})$$

87. According to formula (2.12), we have

$$x_1 = \frac{-b + \sqrt{b^2 - 4ac}}{2a} \text{ and }$$

$$x_2 = \frac{-b - \sqrt{b^2 - 4ac}}{2a}$$

By adding these two expressions we obtain

$$x_1 + x_2 = \frac{-b + \sqrt{b^2 - 4ac}}{2a}$$

$$+ \frac{-b - \sqrt{b^2 - 4ac}}{2a}$$

$$= \frac{-2b + \sqrt{b^2 - 4ac} - \sqrt{b^2 - 4ac}}{2a}$$

$$= \frac{-2b}{2a} = \frac{-b}{a}$$

89. $a(x - x_1)(x - x_2)$
$= a(x^2 - x_1 x - x_2 x + x_1 x_2)$
$= ax^2 - a(x_1 + x_2)x + ax_1 x_2$

$= ax^2 - a(\dfrac{-b}{a})x + a(\dfrac{c}{a})$

$= ax^2 + bx + c$

Exercises 2.6

1. $\sqrt{x} - 3 = 9$

    Isolate the radical.
    $\sqrt{x} = 12$

    Square both sides of the equation.
    $x = 144$

    Check: $\sqrt{144} - 3 = 12 - 3 = 9$

3. $\sqrt{1 - 2x} = 5$

    Square each side of the equation.
    $1 - 2x = 25$
    $-2x = 24$
    $x = -12$

    Check: $\sqrt{1 - 2(-12)} = \sqrt{1 + 24} = \sqrt{25}$
    $= 5$

5. Squaring both sides, we obtain
    $x^2 = 7x - 12$
    $x^2 - 7x + 12 = 0$
    $(x - 4)(x - 3) = 0$
    $x = 4, 3$

    Check: $3 = \sqrt{21 - 12}$ | $4 = \sqrt{28 - 12}$
    $\quad\quad\quad 3 = \sqrt{9}$ | $4 = \sqrt{16}$
    $\quad\quad\quad\quad 3 = 3$ | $4 = 4$

    $x = 3, 4$

7. $3x = \sqrt{9x - 2}$
    $(3x)^2 = 9x - 2$ [square both sides]
    $9x^2 - 9x + 2 = 0$
    $(3x - 2)(3x - 1) = 0$
    $x = 2/3, \ x = 1/3$

Check:
$3(\dfrac{2}{3}) = \sqrt{9(\dfrac{2}{3}) - 2}$

$2 = \sqrt{6 - 2}$
$2 = \sqrt{4}$
$2 = 2$

$3(\dfrac{1}{3}) = \sqrt{9(\dfrac{1}{3}) - 2}$

$1 = \sqrt{3 - 2}$
$1 = \sqrt{1}$
$1 = 1$

$x = 1/3, \ 2/3$

9. $x + 1 = 2\sqrt{x}$
    $(x + 1)^2 = 4x$
    $x^2 + 2x + 1 - 4x = 0$
    $x^2 - 2x + 1 = 0$
    $(x - 1)(x - 1) = 0$
    $x = 1$

    Check: $1 + 1 = 2\sqrt{1}$
    $\quad\quad\quad\quad 2 = 2$

11. $\sqrt{x - 2} = x - 8$
    $x - 2 = (x - 8)^2$
    $x - 2 = x^2 - 16x + 64$
    $x^2 - 17x + 66 = 0$
    $(x - 6)(x - 11) = 0$
    $x = 6, \ x = 11$

    Check:
    $\sqrt{6 - 2} = 6 - 8$ | $\sqrt{11 - 2} = 11 - 8$
    $\quad\sqrt{4} = -2$ | $\quad\sqrt{9} = 3$
    $\quad\quad 2 \neq -2$ | $\quad\quad 3 = 3$
    $x = 11$

13. $3\sqrt{x - 8} = x - 6$
    $9(x - 8) = (x - 6)^2$
    $9x - 72 = x^2 - 12x + 36$
    $x^2 - 21x + 108 = 0$
    $(x - 12)(x - 9) = 0$
    $x = 12 \quad x = 9$

    Check:
    $3\sqrt{12 - 8} = 12 - 6 \quad | \quad 3\sqrt{9 - 8} = 9 - 6$
    $\qquad 3\sqrt{4} = 6 \qquad \qquad 3\sqrt{1} = 3$
    $\qquad 3(2) = 6 \qquad \qquad \quad 3 = 3$
    $\qquad \qquad 6 = 6$
    $x = 12, \; 9$

15. $\sqrt{3x + 7} - 1 = x$
    $\sqrt{3x + 7} = x + 1$
    $3x + 7 = (x + 1)^2$
    $3x + 7 = x^2 + 2x + 1$
    $x^2 - x - 6 = 0$
    $(x - 3)(x + 2) = 0$
    $x = 3 \quad x = -2$

    Check:
    $\sqrt{9 + 7} - 1 = 3 \quad | \quad \sqrt{-6 + 7} - 1 = -2$
    $\quad \sqrt{16} - 1 = 3 \qquad \qquad \sqrt{1} - 1 = -2$
    $\qquad 4 - 1 = 3 \qquad \qquad \quad 1 - 1 = -2$
    $\qquad \qquad 3 = 3 \qquad \qquad \qquad 0 \neq -2$
    $x = 3$

17. $\sqrt{2x + 2} = \sqrt{x + 2} + 1$
    $2x + 2 = (\sqrt{x + 2} + 1)^2$
    $2x + 2 = x + 2 + 2\sqrt{x + 2} + 1$
    $x - 1 = 2\sqrt{x + 2}$
    $(x - 1)^2 = 4(x + 2)$
    $x^2 - 2x + 1 = 4x + 8$
    $x^2 - 6x - 7 = 0$
    $(x - 7)(x + 1) = 0$
    $x = 7 \quad x = -1$

    Check:
    $\sqrt{14 + 2} = \sqrt{7 + 2} + 1$
    $\quad \sqrt{16} = \sqrt{9} + 1$
    $\qquad 4 = 3 + 1$
    $\qquad 4 = 4$
    $\sqrt{-2 + 2} = \sqrt{-1 + 2} + 1$
    $\qquad 0 = 1 + 1$
    $\qquad 0 \neq 2$
    $x = 7$

19. $\sqrt{2x - 3} + \sqrt{x + 2} = 3$
    $\sqrt{2x - 3} = 3 - \sqrt{x + 2}$
    $2x - 3 = (3 - \sqrt{x + 2})^2$
    $2x - 3 = 9 - 6\sqrt{x + 2} + x + 2$
    $x - 14 = 6\sqrt{x + 2}$
    $(x - 14)^2 = 36(x + 2)$
    $x^2 - 28x + 196 = 36x + 72$
    $x^2 - 64x + 124 = 0$
    $(x - 62)(x - 2) = 0$
    $x = 62 \quad x = 2$

    Check:
    $\sqrt{4 - 3} + \sqrt{2 + 2} = 3$
    $\quad \sqrt{1} + \sqrt{4} = 3$
    $\qquad 1 + 2 = 3$
    $\qquad \quad 3 = 3$
    $\sqrt{124 - 3} + \sqrt{62 + 2} = 3$
    $\quad \sqrt{121} + \sqrt{64} = 3$
    $\qquad 11 + 8 = 3$
    $\qquad \quad 19 \neq 3$
    $x = 2$

21. In this exercise, x must be different from zero. Multiplying the equation by x^2, we obtain

    $x^2 + 2x - 3 = 0$
    $(x + 3)(x - 1) = 0$
    $x = -3 \quad x = 1$

    Check:
    $1 + \dfrac{2}{-3} - \dfrac{3}{(-3)^2} = 0 \quad \Big| \quad 1 + 2 - 3 = 0$

    $1 - \dfrac{2}{3} - \dfrac{1}{3} = 0 \qquad \qquad \; 3 - 3 = 0$

    $\qquad \quad 1 - 1 = 0$
    $\qquad \qquad 0 = 0 \qquad \qquad \qquad 0 = 0$
    $x = -3, \; 1$

23. $u = \dfrac{2}{u} - 1$ $\qquad u \neq 0$

$u^2 = 2 - u$ $\quad$ [multiply by u]

$u^2 + u - 2 = 0$

$(u + 2)(u - 1) = 0$

$u = -2 \quad u = 1$

$u = -2, 1$

25. $x^{-1} - 4x^{-2} = 0$

$\dfrac{1}{x} - \dfrac{4}{x^2} = 0$ $\qquad x \neq 0$

$x - 4 = 0$ $\quad$ [multiply by x^2]

$x = 4$

27. $x^{-1} - 2x^{-2} = 1$

$\dfrac{1}{x} - \dfrac{2}{x^2} = 1$ $\qquad x \neq 0$

$x - 2 = x^2$

$x^2 - x + 2 = 0$

$x = \dfrac{1 \pm\sqrt{1 - 4(2)}}{2} = \dfrac{1 \pm\sqrt{-7}}{2}$

$x = \dfrac{1 \pm i\sqrt{7}}{2}$

29. $(x - 1)^2 - 3(x - 1) - 4 = 0$

$x^2 - 2x + 1 - 3x + 3 - 4 = 0$

$x^2 - 5x = 0$

$x(x - 5) = 0$

$x = 0, 5$

31. $\dfrac{3}{(x - 1)^2} - \dfrac{5}{x - 1} - 2 = 0 \quad x \neq 1$

$3 - 5(x - 1) - 2(x - 1)^2 = 0$

$\qquad\qquad$ [multiply by $(x - 1)^2$]

$3 - 5x + 5 - 2(x^2 - 2x + 1) = 0$

$3 - 5x + 5 - 2x^2 + 4x - 2 = 0$

$2x^2 + x - 6 = 0$

$(2x - 3)(x + 2) = 0$

$2x - 3 = 0 \quad\Big|\quad x + 2 = 0$

$\qquad x = 3/2 \quad\Big|\qquad x = -2$

$x = 3/2, -2$

33. $\dfrac{3}{(x^2 - 1)^2} - \dfrac{4}{x^2 - 1} + 1 = 0$

$\qquad\qquad\qquad\qquad x \neq \pm 1$

If we let $u = x^2 - 1$, the equation becomes

$\dfrac{3}{u^2} - \dfrac{4}{u} + 1 = 0$

$3 - 4u + u^2 = 0$

$(u - 1)(u - 3) = 0$

$u = 1 \quad u = 3$

Now replace $x^2 - 1$ for u and solve the corresponding equations

$x^2 - 1 = 1 \quad\Big|\quad x^2 - 1 = 3$

$\quad x^2 = 2 \quad\Big|\qquad x^2 = 4$

$\quad x = \pm\sqrt{2} \quad\Big|\qquad x = \pm 2$

35. $\left(\dfrac{x}{x + 1}\right)^2 - \dfrac{x}{x + 1} - 2 = 0$

$\qquad\qquad\qquad\qquad x \neq -1$

Let $u = \dfrac{x}{x + 1}$

$u^2 - u - 2 = 0$

$(u - 2)(u + 1) = 0$

$u = 2 \quad\Big|\quad u = -1$

$\dfrac{x}{x + 1} = 2 \quad\Big|\quad \dfrac{x}{x + 1} = -1$

$x = 2x + 2 \quad\Big|\quad x = -x - 1$

$-x = 2 \quad\quad\Big|\quad 2x = -1$

$x = -2 \quad\quad\Big|\quad x = -1/2$

$x = -2, -1/2$

37. $u^4 - 6u^2 - 8 = 0$
Let $v = u^2$
$v^2 - 6v - 8 = 0$

$$v = \frac{6 \pm\sqrt{36 + 32}}{2} = \frac{6 \pm\sqrt{68}}{2}$$

$$= \frac{6 \pm 2\sqrt{17}}{2}$$

$v = 3 +\sqrt{17}$	$v = 3 -\sqrt{17}$
$u^2 = 3 +\sqrt{17}$	$u^2 = 3 -\sqrt{17}$
$u = \pm\sqrt{3 + \sqrt{17}}$	$u = \pm\sqrt{3 - \sqrt{17}}$
$u = \pm\sqrt{3 + \sqrt{17}}$	$= \pm i\sqrt{\sqrt{17} - 3}$

39. $4x^4 + 6x^2 - 1 = 0$ Let $u = x^2$
$4u^2 + 6u - 1 = 0$

$$u = \frac{-6 \pm\sqrt{52}}{8} = \frac{-6 \pm 2\sqrt{13}}{8}$$

$$= \frac{-3 \pm\sqrt{13}}{4}$$

Now, $x^2 = \dfrac{-3 \pm\sqrt{13}}{4}$, so

$$x = \frac{\pm\sqrt{-3 + \sqrt{13}}}{2}, \qquad \frac{\pm i\sqrt{3 + \sqrt{13}}}{2}$$

41. $6(m - 2)^4 - 13(m - 2)^2 + 2 = 0$
Let $u = (m - 2)^2$ then $u^2 = (m - 2)^4$
$6u^2 - 13u + 2 = 0$
$(6u - 1)(u - 2) = 0$

$6u - 1 = 0$	$u - 2 = 0$
$6u = 1$	$u = 2$
$u = \dfrac{1}{6}$	

$(m - 2)^2 = \dfrac{1}{6}$	$(m - 2)^2 = 2$
$m - 2 = \pm\sqrt{\dfrac{1}{6}}$	$m - 2 = \pm\sqrt{2}$
$m = 2 \pm\dfrac{\sqrt{6}}{6}$	$m = 2 \pm\sqrt{2}$

$$m = \frac{12 \pm\sqrt{6}}{6}$$

43. $2x^{-4} + 3x^{-2} - 2 = 0$ $x \neq 0$
Let $u = x^{-2}$
$2u^2 + 3u - 2 = 0$
$(2u - 1)(u + 2)$

$u = 1/2$	$u = -2$
$x^{-2} = 1/2$	$x^{-2} = -2$
$x^2 = 2$	$x^2 = 1/-2$
$x = \pm\sqrt{2}$	$x = \pm i\sqrt{2}/2$

$x = \pm\sqrt{2}, \ \pm i\sqrt{2}/2$

45. $x + 6\sqrt{x} - 16 = 0$
$x + 6x^{1/2} - 16 = 0$
Let $u = x^{1/2}$
$u^2 + 6u - 16 = 0$
$(u - 2)(u + 8) = 0$

$u = 2$	$u = -8$
$x^{1/2} = 2$	$x^{1/2} = -8$ not possible
$x = 4$	

47. $p - 15p^{1/2} + 26 = 0$
Let $u = p^{1/2}$
$u^2 - 15u + 26 = 0$
$(u - 13)(u - 2) = 0$

$u = 13$	$u = 2$
$p^{1/2} = 13$	$p^{1/2} = 2$
$p = 169$	$p = 4$
$p = 4, \ 169$	

49. $y^{-1/2} - 5y^{-1/4} + 6 = 0$ $y \neq 0$
Let $u = y^{-1/4}$
$u^2 - 5u + 6 = 0$
$(u - 2)(u - 3) = 0$

$u = 2$	$u = 3$
$y^{-1/4} = 2$	$y^{-1/4} = 3$
$y^{1/4} = 1/2$	$y^{1/4} = 1/3$

$y = 1/16, \ 1/81$

51. $x^{2/3} - 6x^{1/3} + 8 = 0$
Let $u = x^{1/3}$
$u^2 - 6u + 8 = 0$
$(u - 4)(u - 2) = 0$

$u = 4$	$u = 2$
$x^{1/3} = 4$	$x^{1/3} = 2$
$x = 64$	$x = 8$

$x = 8, \ 64$

53. $(x - 2)^{2/3} = x^{1/3}$

   Cube each side of the equation.

   $(x - 2)^2 = x$
   $x^2 - 4x + 4 = x$
   $x^2 - 5x + 4 = 0$
   $(x - 4)(x - 1) = 0$

$x - 4 = 0$	$x - 1 = 0$
$x = 4$	$x = 1$

55. $(x + 1)^{2/3} = (1 - x)^{1/3}$

   Cube each side of the equation.

   $(x + 1)^2 = (1 - x)$
   $x^2 + 2x + 1 = 1 - x$
   $x^2 + 3x = 0$
   $x(x + 3) = 0$
   $x = 0$ or $x + 3 = 0$
   $\qquad\qquad x = -3$

57. $x^6 + 8x^3 + 15 = 0$
   Let $u = x^3$
   $u^2 + 8u + 15 = 0$
   $(u + 3)(u + 5) = 0$

$u = -3$	$u = -5$
$x^3 = -3$	$x^3 = -5$
$x = \sqrt[3]{-3}$	$x = \sqrt[3]{-5}$

   $x = -\sqrt[3]{3}, \ -\sqrt[3]{5}$

59. $(x - 2)^2 - 4(x - 2) = 0$
   Let $u = x - 2$
   $u^2 - 4u = 0$
   $u(u - 4) = 0$

$u = 0$	$u - 4 = 0$
$x - 2 = 0$	$u = 4$
$x = 2$	$x - 2 = 4$
	$x = 6$

   $x = 2, \ 6$

61. $(2x + 1)^2 + 15 = 8(2x + 1)$
   $(2x + 1)^2 - 8(2x + 1) + 15 = 0$
   Let $u = 2x + 1$
   $u^2 - 8u + 15 = 0$
   $(u - 3)(u - 5) = 0$

$u = 3$	$u = 5$
$2x + 1 = 3$	$2x + 1 = 5$
$2x = 2$	$2x = 4$
$x = 1$	$x = 2$

   $x = 1, \ 2$

63. $A = \pi r^2$

   $\dfrac{A}{\pi} = r^2$

   $r = \sqrt{\dfrac{A}{\pi}} = \dfrac{\sqrt{A\pi}}{\pi}$

65. $F = G \dfrac{mM}{d^2}$

   Multiply by d^2

   $Fd^2 = GmM$

   $d^2 = \dfrac{GmM}{F}$

   $d = \sqrt{\dfrac{GmM}{F}} \cdot \sqrt{\dfrac{F}{F}}$

   $d = \dfrac{\sqrt{GmMF}}{F}$

67. $s = s_0 + \dfrac{1}{2}at^2$

   $s - s_0 = \dfrac{1}{2}at^2$

   $2(s - s_0) = at^2$

   $\dfrac{2(s - s_0)}{a} = t^2$

   $t = \sqrt{\dfrac{2(s - s_0)}{a}} \cdot \sqrt{\dfrac{a}{a}}$

   $t = \dfrac{\sqrt{2a(s - s_0)}}{a}$

69. $A = \pi r \sqrt{r^2 + h^2}$

$$\frac{A}{\pi r} = \sqrt{r^2 + h^2}$$

Square each side of the equation.

$$\frac{A^2}{\pi^2 r^2} = r^2 + h^2$$

$$\frac{A^2}{\pi^2 r^2} - r^2 = h^2$$

$$\sqrt{\frac{A^2}{\pi^2 r^2} - r^2} = h$$

$$\sqrt{\frac{A^2 - \pi^2 r^4}{\pi^2 r^2}} = h$$

$$h = \frac{\sqrt{A^2 - \pi^2 r^4}}{\pi r}$$

Exercises 2.7

1. Let x and $x + 1$ denote two consecutive natural numbers. Then

$$x(x + 1) = 272$$
$$x^2 + x = 272$$
$$x^2 + x - 272 = 0$$
$$(x + 17)(x - 16) = 0$$
$$x = -17 \qquad x = 16$$

-17 is not a natural number. The numbers are 16 and 17.

3. Let x and $x + 2$ denote two consecutive even positive integers. Then

$$x(x + 2) = 624$$
$$x^2 + 2x = 624$$
$$x^2 + 2x - 624 = 0$$
$$(x + 26)(x - 24) = 0$$
$$x = -26 \qquad x = 24$$

-26 is not positive. The numbers are 24 and 26.

5. Let x be the first number. Then $27 - x$ is the second number.

$$x(27 - x) = 162$$
$$27x - x^2 = 162$$
$$x^2 - 27x + 162 = 0$$
$$(x - 18)(x - 9) = 0$$
$$x = 18 \qquad \qquad x = 9$$
$$27 - x = 9 \qquad \mid \qquad 27 - x = 18$$

The numbers are 9 and 18.

7. Let x be the number. Its reciprocal is $1/x$.

$$x + 1/x = 26/5$$
$$5x^2 + 5 = 26x$$
$$5x^2 - 26x + 5 = 0$$
$$(5x - 1)(x - 5) = 0$$
$$x = 1/5 \qquad \mid \qquad x = 5$$
$$1/x = 5 \qquad \mid \qquad 1/x = 1/5$$

The numbers are 5 and 1/5.

9. If x denotes one of the integers, then the other is $16 - x$. We have

$$x^2 + (16 - x)^2 = 136$$
$$x^2 + 256 - 32x + x^2 = 136$$
$$2x^2 - 32x + 120 = 0$$
$$x^2 - 16x + 60 = 0$$
$$(x - 6)(x - 10) = 0$$
$$x = 6 \text{ or } x = 10$$

Thus the integers are 6 and 10.

11. $x^2 - 16x - 4m = 0$

Let $x = 6$
$36 - 96 - 4m = 0$
$-60 - 4m = 0$
$-60 = 4m$
$-15 = m$

When $m = -15$,

$x^2 - 16x - 4(-15) = 0$
$x^2 - 16x + 60 = 0$
$(x - 10)(x - 6) = 0$
$x = 10 \quad x = 6$

$m = -15$ and the other solution is 10.

13. Let x be the amount of increase. Then new rectangle has dimensions of $3 + x$ by $6 + x$. Area of original rectangle is $(3)(6) = 18$. Area of new rectangle is $(3 + x)(6 + x)$. Thus,

$(3 + x)(6 + x) = 3(18)$
$18 + 9x + x^2 = 54$
$x^2 + 9x - 36 = 0$
$(x - 3)(x + 12) = 0$
$x = 3 \quad x = -12$

Amount of increase cannot be negative. Therefore increase in 3 inches. New dimensions are 6 in. by 9 in.

15. Let x be the width. Then the length is $x + 4$.

$x(x + 4) = 96$
$x^2 + 4x = 96$
$x^2 + 4x - 96 = 0$
$(x + 12)(x - 8) = 0$
$x = -12 \quad x = 8$

The width cannot be negative. Therefore, the width must be 8 ft. Rectangle is 8 ft. by 12 ft.

17. If ℓ and w denote the length and width of the rectangle, then $\ell w = 28800$, so $w = 28800/\ell$. The perimeter of the rectangle is $2\ell + 2w = 720$ or $\ell + w = 360$.

Substituting $28800/\ell$ for w, get

$$\ell + \frac{28800}{\ell} = 360$$

$\ell^2 + 28800 = 360\ell$
$\ell^2 - 360\ell + 28800 = 0$
$(\ell - 240)(\ell - 120) = 0$
$\ell = 240$ or $\ell = 120$

The dimensions are 120 yards by 240 yards.

19. Let x be the base. Then the height is $2x + 8$. Area of a triangle is $(1/2)bh$.

$1/2(x)(2x + 8) = 45$
$1/2(2x^2 + 8x) = 45$
$x^2 + 4x - 45 = 0$
$(x + 9)(x - 5) = 0$
$x = -9 \quad x = 5$

Length of base cannot be negative. Therefore base is 5 cm.

$2(5) + 8 = 10 + 8 = 18$

Base is 5 cm, height is 18 cm.

21. Let x, y, and h be the lengths of the shorter leg, longer leg, and hypotenuse.
Then $x = y - 2$ and $x = h - 4$, so $y = x + 2$ and $h = x + 4$. Now, by the Pythagorean Therom

$x^2 + y^2 = h^2$
$x^2 + (x + 2)^2 = (x + 4)^2$
$x^2 + x^2 + 4x + 4 = x^2 + 8x + 16$
$x^2 - 4x - 12 = 0$
$(x - 6)(x + 2) = 0$
$x = 6$ or $x = -2$ (impossible).

The dimensions of the triangle are 6, 8, and 10.

23. Let x be the side of the square. When the corners are folded up, the open box will have the following dimensions:
length $x - 4 - 4$ or $x - 8$,
width $x - 4 - 4$ or $x - 8$,
height 4.

Thus,
$$4(x - 8)(x - 8) = 144$$
$$4(x^2 - 16x + 64) = 144$$
$$x^2 - 16x + 64 = 36$$
$$x^2 - 16x + 28 = 0$$
$$(x - 14)(x - 2) = 0$$
$$x = 14 \quad x = 2$$

The square measures 14 in. by 14 in.

25. Let x be the width of the pavement. Dimensions of new rectangle are $20 + 2x$ by $30 + 2x$. Thus

$$(20 + 2x)(30 + 2x) = 704$$
$$4(10 + x)(15 + x) = 704$$
$$(10 + x)(15 + x) = 176$$
$$150 + 25x + x^2 - 176 = 0$$
$$x^2 + 25x - 26 = 0$$
$$(x + 26)(x - 1) = 0$$
$$x = -26 \quad x = 1$$

The width is 1m.

27. Let x be the change in the radius.
Area of a circle is πr^2.
Area of old circle
$\pi(8)^2 = 64\pi$.
Area of new circle
$\pi(8 + x)^2$.
$$\pi(8 + x)^2 = 64\pi + 36\pi$$
$$(8 + x)^2\pi = 100\pi$$
$$(8 + x)^2 = 100$$
$$64 + 16x + x^2 - 100 = 0$$
$$x^2 + 16x - 36 = 0$$
$$(x - 2)(x + 18) = 0$$
$$x = 2 \quad x = -18$$

Increase radius by 2 ft.

29. $P = P_0(1 + r)^t$
$$1323 = 1200(1 + r)^2$$
$$1323 = 1200(1 + 2r + r^2)$$
$$1323 = 1200 + 2400r + 1200r^2$$
$$1200r^2 + 2400r - 123 = 0$$
$$400r^2 + 800r - 41 = 0$$
$$(20r - 1)(20r + 41) = 0$$
$$r = 1/20 \quad r = -41/20$$
$$r = 1/20 = 0.05 = 5\%$$

The rate is 5%.

31. Let x be the number of people chartering the flight. The cost per passenger is

$$\frac{14400}{x}.$$

If 12 more people decide to join the group, then the price per passenger is

$$\frac{14400}{x} - 40.$$

The total cost is the cost per passenger times the number of passengers. So

$$\left(\frac{14400}{x} - 40\right)(x + 12) = 14400$$

$$(14400 - 40x)(x + 12) = 14400x$$
$$40x^2 + 480x - 172800 = 0$$
$$x^2 + 12x - 4320 = 0$$
$$(x - 60)(x + 72) = 0$$
$$x = 60 \qquad x = -72$$
$$\text{(impossible)}$$

There are 60 people chartering the flight.

33. Let x be an increase of 50 posters. Total number of posters $100 + 50x$. Total cost $0.20 - 0.02x$.

$$36 = (100 + 50x)(0.20 - .02x)$$
$$36 = 20 + 10x - 2x - x^2$$
$$x^2 - 8x + 16 = 0$$
$$(x - 4)(x - 4) = 0$$
$$x = 4$$
$$100 + 50(4) = 300$$
300 posters

35. Let x be the rate of the current.

	distance	rate	time
upstream	48	14−x	$\frac{48}{14+x} + 1$
downstream	48	14+x	$\frac{48}{14+x}$

distance up = distance back

$$(14 - x)(\frac{48}{14 + x} + 1)$$

$$= (14 + x)(\frac{48}{14 + x})$$

$(14 - x)(48 + 14 + x)$
$\qquad = (14 + x)(48)$
$(14 - x)(62 + x) = 672 + 48x$
$\quad 868 - 48x - x^2 = 672 + 48x$
$\quad x^2 + 96x - 196 = 0$
$\quad (x + 98)(x - 2) = 0$
$\qquad x = -98 \qquad x = 2$
The current is 2 mph.

37. Let x be Peter's rate.
Then $x - 1$ is Mary's rate.

$$\frac{1}{x} + \frac{1}{x - 1} = \frac{1}{1\frac{1}{5}}$$

$$\frac{1}{x} + \frac{1}{x - 1} = \frac{5}{6}$$

$\quad 6(x - 1) + 6x = 5x(x - 1)$
$\quad 6x - 6 + 6x = 5x^2 - 5x$
$\quad 5x^2 - 17x + 6 = 0$
$\quad (5x - 2)(x - 3) = 0$
$\qquad x = 2/5 \qquad x = 3$

Peter takes 3 days
Mary takes 2 days.

39. Let x be rate of one bus. Then x + 10 is rate of the other bus.

	distance	rate	time
Bus 1	150	x	$\frac{150}{x}$
Bus 2	150	x+10	$\frac{150}{x} - \frac{1}{2}$

distance 1 = distance 2

$$x(\frac{150}{x}) = (x + 10)(\frac{150}{x} - \frac{1}{2})$$

$2x(150) = (x + 10)(300 - x)$
$300x = 300x - x^2 + 3000 - 10x$
$x^2 + 10x - 3000 = 0$
$(x + 60)(x - 50) = 0$
$x = -60 \qquad x = 50$

The buses are going 50 mph and 60 mph.

41. Let x be Mary's running speed.

Distance	Rate	Time
3000	x	3000/x
3000	x − 25	$\frac{3000}{x} + 4$

$$3000 = (x - 25)(\frac{3000}{x} + 4)$$

$3000x = (x - 25)(3000 + 4x)$
$4x^2 - 100x - 75000 = 0$
$\quad x^2 - 25x - 18750 = 0$
$(x + 125)(x - 150) = 0$
$x + 125 = 0 \mid x - 150 = 0$
$x = -125 \mid x = 150$
not possible

Mary's original running speed was 150m/min.

43. $s = v_0 t - 16t^2$ $\qquad$ $v_0 = 960$

    a) $s = 8000$
    $8000 = 960t - 16t^2$
    $16t^2 - 960t + 8000 = 0$
    $t^2 - 60t + 500 = 0$
    $(t - 50)(t - 10) = 0$
    $t = 50 \qquad t = 10$

    At 10 seconds and at 50 seconds.

    b) $s = 0$
    $0 = 960t - 16t^2$
    $16t^2 - 960t = 0$
    $16t(t - 60) = 0$
    $t = 0 \qquad t = 60$

    It will hit ground at 60 seconds.

45.

Square 1	Square 2
Perimeter 16	Perimeter $\ell - 16$
Side $\dfrac{16}{4} = 4$	Side $\dfrac{\ell - 16}{4}$
Area $4^2 = 16$	Area $(\dfrac{\ell - 16}{4})^2$

$$4^2 + (\frac{\ell - 16}{4})^2 = 272$$

$$256 + \ell^2 - 32\ell + 256 = 4352$$
$$\ell^2 - 32\ell - 3840 = 0$$
$$(\ell - 80)(\ell + 48) = 0$$
$$\ell = 80 \qquad \ell = -48$$

The length of the wire is 80 inches.

47. $q = \dfrac{4500}{p}$

    $Q = 3000p - 1500$

    $q = Q$ when $\dfrac{4500}{p} = 3000p - 1500$

    $4500 = 3000p^2 - 1500p$
    $3000p^2 - 1500p - 4500 = 0$
    $2p^2 - p - 3 = 0$
    $(2p - 3)(p + 1) = 0$
    $p = 3/2 \qquad p = -1$
    $p = 3/2 = \$1.50$

49. Let x be the distance from balloon to horizon. Distance from balloon to center of earth is

    $$\begin{array}{r} 6.371 \times 10^6 \\ +2.056 \times 10^3 \\ \hline = 6.373 \times 10^6 \end{array} \quad \Big| \quad \begin{array}{r} 6371. \quad \times 10^3 \\ 2.056 \times 10^3 \\ \hline 6373.056 \times 10^3 \end{array}$$

    Using Pythagorean Theorem we get

    $$(6.371 \times 10^6)^2 + x^2 = (6.373 \times 10^6)^2$$
    $$4.059 \times 10^{13} + x^2 = 4.062 \times 10^{13}$$
    $$x^2 = 2.620 \times 10^{10}$$
    $$x = 1.619 \times 10^5$$

Exercises 2.8

1. $(x + 2)(x - 5) > 0$

	$-\infty$	-2	5
$x + 2$	$-$	$+$	$+$
$x - 5$	$-$	$-$	$+$
$(x + 2)(x - 5)$	$+$	$-$	$+$

The table shows that the product $(x + 2)(x - 5)$ is positive (that is, > 0) when $x < -2$ or $x > 5$. Therefore the solution set is

$(-\infty, -2) \cup (5, +\infty)$.

3. $(x - 2)(x + 2) \geq 0$

	$-\infty$	-2	2	∞
$x - 2$	$-$	$-$	$+$	
$x + 2$	$-$	$+$	$+$	
$(x - 2)(x + 2)$	$+$	$-$	$+$	

The table shows that the inequality holds when $x \leq -2$ or $x \geq 2$. The solution set is

$(-\infty, -2] \cup [2, +\infty)$.

5. $(2x + 1)(x - 1) \geq 0$

	$-\infty$	$-1/2$	1	∞
$(2x + 1)$	$-$	$+$	$+$	
$(x - 1)$	$-$	$-$	$+$	
$(2x + 1)(x - 1)$	$+$	$-$	$+$	

The product $(2x + 1)(x - 1)$ is positive when $x < -1/2$ or $x > 1$; and it is equal to zero when $x = -1/2$ or $x = 1$. Thus, the solution set is

$(-\infty, -1/2] \cup [1, +\infty)$.

7. $x(x - 5) \leq 0$

	$-\infty$	0	5	∞
x	$-$	$+$	$+$	
$x - 5$	$-$	$-$	$+$	
$x(x - 5)$	$+$	$-$	$+$	

The solution set is $[0, 5]$.

9. $4 - x^2 < 0$
$(2 + x)(2 - x) < 0$

	$-\infty$	-2	2	∞
x	$-$	$+$	$+$	
$2 - x$	$+$	$+$	$-$	
$(2 + x)(2 - x)$	$-$	$+$	$-$	

$(-\infty, -2) \cup (2, \infty)$

11. $3x^2 + 1 > 0$
The term x^2 is always ≥ 0.
Therefore $3x^2 \geq 0$ and $3x^2 + 1 > 0$.

Thus, the inequality is true for all real x.

$(-\infty, \infty)$

13. $3x^2 + 5 < 0$

Since $3x^2$ is never negative it follows that $3x^2 + 5$ is never negative. Therefore this problem has no solution.

15. $x^2 + 3x \geq 0$
$x(x + 3) \geq 0$

	$-\infty$	-3	0	∞
x	$-$	$-$	$+$	
$x + 3$	$-$	$+$	$+$	
$x(x + 3)$	$+$	$-$	$+$	

$(-\infty, -3] \cup [0, +\infty)$

17. $x^2 + 3x - 10 > 0$
$(x + 5)(x - 2) > 0$

	$-\infty$	-5	2	∞
$x + 5$	$-$	$+$	$+$	
$x - 2$	$-$	$-$	$+$	
$(x + 5)(x - 2)$	$+$	$-$	$+$	

$(-\infty, -5) \cup (2, +\infty)$

19. $x^2 + 10 > 7x$
$x^2 - 7x + 10 > 0$
$(x - 2)(x - 5) > 0$

	$-\infty$	2	5	∞
$x - 2$	$-$	$+$	$+$	
$x - 5$	$-$	$-$	$+$	
$(x - 2)(x - 5)$	$+$	$-$	$+$	

$(-\infty, 2) \cup (5, +\infty)$

21. $2x^2 - 3 < 0$
$2(x + \sqrt{6}/2)(x - \sqrt{6}/2) < 0$
[Note that $2x^2 - 3 = 0$
$$2x^2 = 3$$
$$x^2 = 3/2$$
$$x = \pm\sqrt{3/2}$$
$$= \pm\sqrt{6}/2 \ ]$$

	$-\infty$	$-\sqrt{6}/2$	$\sqrt{6}/2$	∞
$x + \sqrt{6}/2$	$-$	$+$	$+$	
$x - \sqrt{6}/2$	$-$	$-$	$+$	
$(x+\sqrt{6}/2)(x-\sqrt{6}/2)$	$+$	$-$	$+$	

$(-\sqrt{6}/2, \sqrt{6}/2)$

23. $x^2 - 4x + 1 \geq 0$

$(x - 2 - \sqrt{3})(x - 2 + \sqrt{3}) \geq 0$

[Note that $x^2 - 4x + 1 = 0$

$x = \dfrac{4 \pm \sqrt{16 - 4}}{2}$

$= \dfrac{4 \pm \sqrt{12}}{2}$

$= \dfrac{4 \pm 2\sqrt{3}}{2} = 2 \pm \sqrt{3}$]

	$-\infty$		$2 -\sqrt{3}$		$2 +\sqrt{3}$	∞
$x - 2 -\sqrt{3}$	$-$			$-$		$+$
$x - 2 +\sqrt{3}$	$-$			$+$		$+$
$(x-2-\sqrt{3})(x-2+\sqrt{3})$	$+$			$-$		$+$

$(-\infty, \; 2-\sqrt{3}] \cup [2 +\sqrt{3}, \; +\infty)$

25. $2x^2 + 2x < 7$

$2x^2 + 2x - 7 < 0$

$(x - \dfrac{-1 + \sqrt{15}}{2})(x - \dfrac{-1 - \sqrt{15}}{2}) < 0$

[Note that $2x^2 + 2x - 7 = 0$

$x = \dfrac{-2 \pm \sqrt{4 + 56}}{4}$

$= \dfrac{-2 \pm \sqrt{60}}{4}$

$= \dfrac{-2 \pm 2\sqrt{15}}{4} = \dfrac{-1 \pm \sqrt{15}}{2} .$]

	$-\infty$		$\dfrac{-1-\sqrt{15}}{2}$		$\dfrac{-1+\sqrt{15}}{2}$	∞
$x - \dfrac{-1+\sqrt{15}}{2}$	$-$			$-$		$+$
$x - \dfrac{-1-\sqrt{15}}{2}$	$-$			$+$		$+$
$(x-\dfrac{-1+\sqrt{15}}{2})(x-\dfrac{-1-\sqrt{15}}{2})$	$+$			$-$		$+$

$(\dfrac{-1-\sqrt{15}}{2}, \; \dfrac{-1+\sqrt{15}}{2})$

27. $x^2 - 2x - 15 < 0$

$(x - 5)(x + 3) < 0$

	$-\infty$		-3		5	∞
$x - 5$	$-$			$-$		$+$
$x + 3$	$-$			$+$		$+$
$(x - 5)(x + 3)$	$+$			$-$		$+$

$(-3, \; 5)$

29. $4x^2 - 4x - 1 \geq 0$

First solve the quadratic equation $4x^2 - 4x - 1 = 0$ and find the solutions

$$\frac{1 - \sqrt{2}}{2} \quad \text{and} \quad \frac{1 + \sqrt{2}}{2}$$

Next factor the quadratic polynomial and rewrite the given inequality as follows

$$4(x - \frac{1 - \sqrt{2}}{2})(x - \frac{1 + \sqrt{2}}{2}) \geq 0$$

Now make a sign table.

	$-\infty$		$\frac{1-\sqrt{2}}{2}$		$\frac{1+\sqrt{2}}{2}$		∞
$x - \frac{1-\sqrt{2}}{2}$		$-$		$+$		$+$	
$x - \frac{1+\sqrt{2}}{2}$		$-$		$-$		$+$	
$4(x-\frac{1-\sqrt{2}}{2})(x-\frac{1+\sqrt{2}}{2})$		$+$		$-$		$+$	

The table shows that the solution set is

$$(-\infty, \frac{1 -\sqrt{2}}{2}] \cup [\frac{1 +\sqrt{2}}{2}, \infty)$$

31. $x^2 - 2x + 3 > 0$

Since the discriminant of the quadratic polynomial is $\Delta = (-2)2 - 4(3) = -8$, a negative number, the quadratic equation $x^2 - 2x + 3 = 0$ has no real roots. you can check that $x^2 - 2x + 3 > 0$ is true for all real x.

33. $\frac{x - 1}{x + 3} \leq 0$ $x \neq -3$

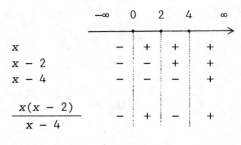

	$-\infty$	-3		1	∞
$x - 1$	$-$		$-$		$+$
$x + 3$	$-$		$+$		$+$
$\frac{x - 1}{x + 3}$	$+$		$-$		$+$

$(-3, 1]$

35. $\frac{x(x - 2)}{x - 4} \geq 0$ $x \neq 4$

	$-\infty$	0	2	4	∞
x		$-$	$+$	$+$	$+$
$x - 2$		$-$	$-$	$+$	$+$
$x - 4$		$-$	$-$	$-$	$+$
$\frac{x(x - 2)}{x - 4}$		$-$	$+$	$-$	$+$

$[0, 2] \cup (4, +\infty)$

37. $\frac{8}{x + 3} < 1$ $x \neq 3$

$$\frac{8}{x + 3} - 1 < 0$$

$$\frac{8 - (x + 3)}{x + 3} < 0$$

$$\frac{5 - x}{x + 3} < 0$$

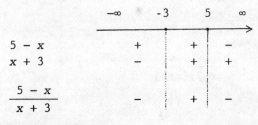

	$-\infty$	-3		5	∞
$5 - x$		$+$		$+$	$-$
$x + 3$		$-$		$+$	$+$
$\frac{5 - x}{x + 3}$		$-$		$+$	$-$

$(-\infty, -3) \cup (5, +\infty)$

39. $\dfrac{3x + 2}{x + 5} \geq 1 \qquad x \neq 5$

$$\dfrac{3x + 2}{x + 5} - 1 \geq 0$$

$$\dfrac{3x + 2 - (x + 5)}{x + 5} \geq 0$$

$$\dfrac{2x - 3}{x + 5} \geq 0$$

	$-\infty$	-5	$3/2$	∞
$2x - 3$	$-$	$-$	$+$	
$x + 5$	$-$	$+$	$+$	
$\dfrac{2x - 3}{x + 5}$	$+$	$-$	$+$	

$(-\infty, -5) \cup [3/2, +\infty)$

41. $\dfrac{2}{x - 2} > \dfrac{1}{x + 1} \qquad x \neq -1, 2$

$$\dfrac{2}{x - 2} - \dfrac{1}{x + 1} > 0$$

$$\dfrac{2(x + 1) - (x - 2)}{(x - 2)(x + 1)} > 0$$

$$\dfrac{x + 4}{(x - 2)(x + 1)} > 0$$

	$-\infty$	-4	-1	2	∞
$x + 4$	$-$	$+$	$+$	$+$	
$x - 2$	$-$	$-$	$-$	$+$	
$x + 1$	$-$	$-$	$+$	$+$	
$\dfrac{x + 4}{(x-2)(x+1)}$	$-$	$+$	$-$	$+$	

$(-4, -1) \cup (2, +\infty)$

43. $\dfrac{3}{x^2} > 1 \qquad x \neq 0$

$$\dfrac{3}{x^2} - 1 > 0$$

$$\dfrac{3 - x^2}{x^2} > 0$$

$$\dfrac{(\sqrt{3} + x)(\sqrt{3} - x)}{x^2} > 0$$

	$-\infty$	$-\sqrt{3}$	0	$\sqrt{3}$	∞
x^2	$+$	$+$	$+$	$+$	
$\sqrt{3} + x$	$-$	$+$	$+$	$+$	
$\sqrt{3} - x$	$+$	$+$	$+$	$-$	
$\dfrac{(\sqrt{3}+x)(\sqrt{3}-x)}{x}$	$-$	$+$	$+$	$-$	

$(-\sqrt{3}, 0 \cup (0, \sqrt{3})$

45. $\dfrac{x^2 - 3x + 2}{x + 3} \leq 0 \qquad x \neq -3$

$$\dfrac{(x - 2)(x - 1)}{x + 3} \leq 0$$

	$-\infty$	-3	1	2	∞
$x - 2$	$-$	$-$	$-$	$+$	
$x - 1$	$-$	$-$	$+$	$+$	
$x + 3$	$-$	$+$	$+$	$+$	
$\dfrac{(x - 2)(x - 1)}{x + 3}$	$-$	$+$	$-$	$+$	

$(-\infty, -3) \cup [1, 2]$

47. $\dfrac{x^2 - 3x - 4}{x^2 + 2x} \le 0$

Factor the numerator and
denominator

$\dfrac{(x - 4)(x + 1)}{x(x + 2)} \le 0$

To determine where this fraction is
negative, look at the signs of
each factor. Note that a fraction
is equal to zero whenever the
numerator is zero

	$-\infty$	-2	-1	0	4	∞
x	$-$	$-$	$-$	$+$		$+$
$x + 1$	$-$	$-$	$+$	$+$		$+$
$x + 2$	$-$	$+$	$+$	$+$		$+$
$x - 4$	$-$	$-$	$-$	$-$		$+$
$\dfrac{(x-4)(x+1)}{x(x + 2)}$	$+$	$-$	$+$	$-$		$+$

$(-2, -1] \cup (0, 4]$

49. $\dfrac{2x^2 + 5x - 3}{x^2 - 3x - 4} > 0$

$\dfrac{(2x - 1)(x + 3)}{(x - 4)(x + 1)} > 0 \qquad x \ne -1, 4$

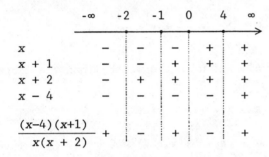

	$-\infty$	-3	-1	$1/2$	4	∞
$2x - 1$	$-$	$-$	$-$	$+$		$+$
$x + 3$	$-$	$+$	$+$	$+$		$+$
$x - 4$	$-$	$-$	$-$	$-$		$+$
$x + 1$	$-$	$-$	$+$	$+$		$+$
$\dfrac{(2x-1)(x+3)}{(x-4)(x+1)}$	$+$	$-$	$+$	$-$		$+$

$(-\infty, -3) \cup (-1, 1/2) \cup (4, +\infty)$

51. $\dfrac{x}{2x - 1} < \dfrac{1}{x + 2} \qquad x \ne -2, 1/2$

$\dfrac{x}{2x - 1} - \dfrac{1}{x + 2} < 0$

$\dfrac{x(x + 2) - (2x - 1)}{(2x - 1)(x + 2)} < 0$

$\dfrac{x^2 + 1}{(2x - 1)(x + 2)} < 0$

	$-\infty$	-2	$1/2$	∞
$x^2 + 1$		$+$	$+$	$+$
$2x - 1$		$-$	$-$	$+$
$x + 2$		$-$	$+$	$+$
$\dfrac{x^2 + 1}{(2x - 1)(x + 2)}$		$+$	$-$	$+$

$(-2, 1/2)$

53. $s = s_0 + v_0 t - 16t^2 \qquad s_0 = 0 \quad v_0 = 960$

a) $960t - 16t^2 > 4400$
    $960t - 16t^2 - 4400 > 0$
    $16t^2 - 960t + 4400 < 0$
    $t^2 - 60t + 275 < 0$
    $(t - 55)(t - 5) < 0$

	$-\infty$	5	55	∞
$t - 55$		$-$	$-$	$+$
$t - 5$		$-$	$+$	$+$
$(t - 55)(t - 5)$	$+$	$-$		$+$

$5 \; s < t < 55 \; s$

b) $960t - 16t^2 > 7200$
    $960t - 16t^2 - 7200 > 0$
    $16t^2 - 960t + 7200 < 0$
    $t^2 - 60t + 450 < 0$
    $(t - 30 - 15\sqrt{2})(t - 30 + 15\sqrt{2}) < 0$

	$-\infty$	$30-15\sqrt{2}$	$30+15\sqrt{2}$	∞
$t-30-15\sqrt{2}$		$-$	$-$	$+$
$t-30+15\sqrt{2}$		$-$	$+$	$+$
product	$+$	$-$		$+$

$30 - 15\sqrt{2} < t < 30 + 15\sqrt{2}$
$8.79 \; s < t < 51.21 \; s$

c) $960t - 16t^2 < 5184$
$16t^2 - 960t + 5184 > 0$
$t^2 - 60t + 324 > 0$
$(t - 6)(t - 54) > 0$

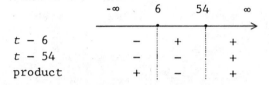

	$-\infty$	6	54	∞
$t - 6$	$-$	$+$	$+$	
$t - 54$	$-$	$-$	$+$	
product	$+$	$-$	$+$	

$0 < t < 6\ s$ or $54\ s < t < 60\ s$

Note that the projectile hits the ground at time $t = 60\ s$.

55. $600p - 4p^2 > 0$
$150p - p^2 > 0$
$p(150 - p) > 0$

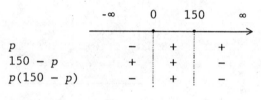

	$-\infty$	0	150	∞
p	$-$	$+$	$+$	
$150 - p$	$+$	$+$	$-$	
$p(150 - p)$	$-$	$+$	$-$	

$0 < p < 150$

57. $1000 + 50t - 5t^2 < 1000$
$50t - 5t^2 < 0$
$5t(10 - t) < 0$
$t(10 - t) < 0$

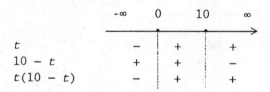

	$-\infty$	0	10	∞
t	$-$	$+$	$+$	
$10 - t$	$+$	$+$	$-$	
$t(10 - t)$	$-$	$+$	$+$	

Since t is time, t must be > 0, so $t > 10$ hours.

59. A profit is made when

$P = -4x^2 + 24x - 27 > 0$
or $4x^2 - 24x + 27 < 0$
$(2x - 9)(2x - 3) < 0$

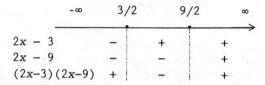

	$-\infty$	3/2	9/2	∞
$2x - 3$	$-$	$+$	$+$	
$2x - 9$	$-$	$-$	$+$	
$(2x-3)(2x-9)$	$+$	$-$	$+$	

Thus a profit is made when

$$1.5 = \frac{3}{2} < x < \frac{9}{2} = 4.5$$

61. If x is the width, then the length is $x + 4$.

$A = x(x + 4) > 45$
$x^2 + 4x - 45 > 0$
$(x - 5)(x + 9) > 0$

Make a table and check that the last product is > 0 when $x > 5$ or $x < -9$ (impossible). Thus the width must be greater than 5 ft.

63. $\Delta = b^2 - 4ac > 0$
$a = 1,\ b = k,\ c = 4$
$k^2 - 4(1)(4) > 0$
$k^2 - 16 > 0$
$(k - 4)(k + 4) > 0$

Now check that the product is positive when $k < -4$ or $k > 4$.

65. The cost C is given in ten-thousands of dollars, so we must solve the inequality

$$2 \leq \frac{8x}{100 - x} \leq 8$$

First, we solve

$$2 \leq \frac{8x}{100 - x}$$

Since x represents percentage, $100 - x \geq 0$. Multiplying both sides of the last inequality by $100 - x$, we obtain

$2(100 - x) \leq 8x$

$200 - 2x \leq 8x$

$200 \leq 10x$

$20 \leq x$

Next, solving

$$\frac{8x}{100 - x} \leq 8$$

we obtain $x \leq 50$. Thus, the solution is

$$20 \leq x \leq 50.$$

67. The inequality
$x^2 - (a + b)x + ab \leq 0$
is equivalent to
$(x - a)(x - b) \leq 0$.

Since $a < b$, you may check that the last product is nonpositive for all x such that

$$a \leq x \leq b.$$

69. $ax^2 + bx + c = 0$

The roots are found by using the quadratic formula

$$x = \frac{-b \pm \sqrt{b^2 - 4ac}}{2a}$$

If one root is rational and the other irrational, then $\sqrt{b^2 - 4ac}$

must be both rational and irrational, which is impossible.

Review Exercises - Chapter 2

1. $2x - 3(x - 1) = 4$
    $2x - 3x + 3 = 4$
    $-x + 3 = 4$
    $-x = 1$
    $x = -1$

3. $(2x - 1)(x + 3) = 2x(x - 4) + 1$
    $2x^2 + 5x - 3 = 2x^2 - 8x + 1$
    $5x - 3 = -8x + 1$
    $5x + 8x = 1 + 3$
    $13x = 4$
    $x = 4/13$

5. $$\frac{4}{x - 1} = \frac{2}{x - 3}$$

    $4(x - 3) = 2(x - 1)$
    $4x - 12 = 2x - 2$
    $4x - 2x = -2 + 12$
    $2x = 10$
    $x = 5$

7. $$\frac{3}{2(x + 2)} = \frac{1}{4(x + 3)} \quad x \neq -3, -2$$

    $12(x + 3) = 2(x + 2)$
    $12x + 36 = 2x + 4$
    $12x - 2x = 4 - 36$
    $10x = -32$
    $x = -32/10$
    $x = -16/5$

9. $$\frac{2}{x} - \frac{1}{3} = \frac{1}{2x} - \frac{1}{4} \quad (x \neq 0)$$

    $$12x(\frac{2}{x}) - 12x(\frac{1}{3}) = 12x(\frac{1}{2x}) - 12x(\frac{1}{4})$$

    $24 - 4x = 6 - 3x$
    $-4x + 3x = 6 - 24$
    $-x = -18$
    $x = 18$

11. $\dfrac{x - 7}{x^2 - 4x + 4} = \dfrac{6}{x - 2}$ $(x \neq 2)$

$x - 7 = 6(x - 2)$
$x - 7 = 6x - 12$
$x - 6x = -12 + 7$
$-5x = -5$
$x = 1$

13. $x^2 - 7x + 12 = 0$
$(x - 4)(x - 3) = 0$
$x = 4 \qquad x = 3$
$x = 3, 4$

15. $6x^2 - x - 2 = 0$
$(3x - 2)(2x + 1) = 0$
$x = 2/3 \qquad x = -1/2$
$x = -1/2, 2/3$

17. $3x + 1 = \sqrt{12x}$
$(3x + 1)^2 = 12x$
$9x^2 + 6x + 1 = 12x$
$9x^2 - 6x + 1 = 0$
$(3x - 1)(3x - 1) = 0$
$x = 1/3$

Check: $3(1/3) + 1 = \sqrt{12(1/3)}$
$1 + 1 = \sqrt{4}$
$1 + 1 = 2$
$2 = 2$

19. $\sqrt{3x - 5} + 3 = x$
$\sqrt{3x - 5} = x - 3$
$3x - 5 = (x - 3)^2$
$3x - 5 = x^2 - 6x + 9$
$x^2 - 9x + 14 = 0$
$(x - 2)(x - 7) = 0$
$x = 2 \qquad x = 7$

Check: $\sqrt{6 - 5} + 3 = 2$
$\sqrt{1} + 3 = 2$
$1 + 3 = 2$
$4 \neq 2$

$\sqrt{21 - 5} + 3 = 7$
$\sqrt{16} + 3 = 7$
$4 + 3 = 7$
$7 = 7$

$x = 7$

21. $\dfrac{x - 2}{3x + 4} = \dfrac{x + 1}{2x - 5}$ $x \neq -4/3, 5/2$

$(x - 2)(2x - 5) = (x + 1)(3x + 4)$
$2x^2 - 9x + 10 = 3x^2 + 7x + 4$
$x^2 + 16x - 6 = 0$

$x = \dfrac{-16 \pm \sqrt{256 + 24}}{2} = \dfrac{-16 \pm \sqrt{280}}{2}$

$= \dfrac{-16 \pm 2\sqrt{70}}{2}$

$x = -8 \pm \sqrt{70}$

23. $3 - \dfrac{6}{x} + \dfrac{2}{x^2} = 0$ $x \neq 0$

$3x^2 - 6x + 2 = 0$

$x = \dfrac{6 \pm \sqrt{36 - 24}}{6} = \dfrac{6 \pm \sqrt{12}}{6}$

$= \dfrac{6 \pm 2\sqrt{3}}{6}$

$x = \dfrac{3 \pm \sqrt{3}}{3}$

25. $\dfrac{4}{x} = \dfrac{3}{x^2}$ $(x \neq 0)$

$4x = 3$
$x = 3/4$

27. $\dfrac{4}{x-5} + \dfrac{5}{2} = \dfrac{9}{x+5}$

Multiply by $2(x-5)(x+5)$

$4(2)(x+5) + 5(x-5)(x+5)$
$\qquad = 9(2)(x-5)$
$8(x+5) + 5(x^2-25) = 18(x-5)$
$8x + 40 + 5x^2 - 125 = 18x - 90$
$5x^2 + 8x - 85 = 18x - 90$
$5x^2 - 10x + 5 = 0$
$5(x^2 - 2x + 1) = 0$
$5(x-1)(x-1) = 0$
$x - 1 = 0$
$x = 1$

29. $2x = \dfrac{2}{x} - 3 \qquad (x \neq 0)$

$2x^2 = 2 - 3x$
$2x^2 + 3x - 2 = 0$
$(2x-1)(x+2) = 0$
$x = 1/2 \qquad x = -2$
$x = -2,\ 1/2$

31. $x^{1/2} - 5x^{1/4} + 6 = 0$
Let $u = x^{1/4}$

$u^2 - 5u + 6 = 0$
$(u-3)(u-2) = 0$
$\quad u = 3 \quad \big| \quad u = 2$
$x^{1/4} = 3 \quad \big| \quad x^{1/4} = 2$
$\quad x = 81 \quad \big| \quad x = 16$
$x = 16,\ 81$

33. $\dfrac{2}{x-1} + \dfrac{2}{x-7} = \dfrac{2}{4-x}$

$(x \neq 1,\ 4,\ 7)$
$2(x-7)(4-x) + 5(x-1)(4-x)$
$\qquad = 2(x-1)(x-7)$
$2(-x^2 + 11x - 28) + 5(-x^2 + 5x - 4)$
$\qquad = 2(x^2 - 8x + 7)$
$-2x^2 + 22x - 56 - 5x^2 + 25x - 20$
$\qquad = 2x^2 - 16x + 14$

$9x^2 - 63x + 90 = 0$
$x^2 - 7x + 10 = 0$
$(x-5)(x-2) = 0$
$\quad x = 5 \qquad x = 2$
$x = 2,\ 5$

35. $3\sqrt{x^2 - 9} = 4(x-2)$
$9(x^2 - 9) = 16(x-2)^2$
$9x^2 - 81 = 16(x^2 - 4x + 4)$
$9x^2 - 81 = 16x^2 - 64x + 64$
$7x^2 - 64x + 145 = 0$
$(7x - 29)(x - 5) = 0$
$\quad x = 29/7 \qquad x = 5$
$x = 29/7,\ 5$

37. $A = \dfrac{1}{2}(b_1 + b_2)h$

$2A = (b_1 + b_2)h$

$\dfrac{2A}{h} = b_1 + b_2$

$\dfrac{2A}{h} - b_2 = b_1$

$b_1 = \dfrac{2A}{h} - b_2$

39. $L = L_0(1 + \alpha t)$

$\dfrac{L}{L_0} = 1 + \alpha t$

$\dfrac{L}{L_0} - 1 = \alpha t$

$\dfrac{L - L_0}{L_0} = \alpha t$

$\dfrac{L - L_0}{\alpha L_0} = t$

$t = \dfrac{L - L_0}{\alpha L_0}$

41. $s = \dfrac{a}{2}t^2 + v_0 t$

$\dfrac{a}{2}t^2 + v_0 t - s = 0$

$t = \dfrac{-v_0 \pm \sqrt{v_0^{\,2} - 4\left(\frac{a}{2}\right)(-s)}}{2\left(\frac{a}{2}\right)}$

$= \dfrac{-v_0 \pm \sqrt{v_0^{\,2} + 2as}}{a}$

$t = \dfrac{-v_0 \pm \sqrt{v_0^{\,2} + 2as}}{a}$

43. $S = 2\pi r(r + h)$
$S = 2\pi r^2 + 2\pi rh$
$2\pi r^2 + 2\pi rh - S = 0$

$r = \dfrac{-2\pi h \pm \sqrt{(2\pi h)^2 - 4(2\pi)(-S)}}{2(2\pi)}$

$= \dfrac{-2\pi h \pm \sqrt{4\pi^2 h^2 - 8\pi S}}{4\pi}$

$= \dfrac{-2\pi h \pm 2\sqrt{\pi^2 h^2 + 2\pi S}}{4\pi}$

$r = \dfrac{-\pi h \pm \sqrt{\pi^2 h^2 + 2\pi S}}{2\pi}$

$= \dfrac{-h \pm \sqrt{h^2 + 2S/\pi}}{2}$

45. $\dfrac{x}{5} - 2 > \dfrac{x - 3}{4} + 2$

$4x - 40 > 5(x - 3) + 40$
$4x - 40 > 5x - 15 + 40$
$4x - 5x > 25 + 40$
$-x > 65$
$x < -65$

47. $\dfrac{x - 5}{9} - \dfrac{3}{4} \leq \dfrac{7x}{12} + \dfrac{x - 1}{6}$

$4(x - 5) - 9(3) \leq 3(7x) + 6(x - 1)$
$4x - 20 - 27 \leq 21x + 6x - 6$
$-21x + 4x - 6x \leq -6 + 47$
$-23x \leq 41$
$x \geq -41/23$

49. $\dfrac{x - 1}{x + 3} < 1 \qquad (x \neq -3)$

$\dfrac{x - 1}{x + 3} - 1 < 0$

$\dfrac{x - 1 - (x + 3)}{x + 3} < 0$

$\dfrac{-4}{x + 3} < 0$

$\dfrac{4}{x + 3} > 0$

$x + 3 > 0$
$x > -3$

51. $\dfrac{x}{x - 5} \geq 3 \qquad (x \neq 5)$

$\dfrac{x}{x - 5} - 3 \geq 0$

$\dfrac{x - 3(x - 5)}{x - 5} \geq 0$

$\dfrac{15 - 2x}{x - 5} \geq 0$

	$-\infty$		5		15/2		∞
$15 - 2x$		+		+		−	
$x - 5$		−		+		+	
$\dfrac{15 - 2x}{x - 5}$		−		+		−	

$5 < x \leq 15/2$

53. $\dfrac{(2x - 1)(x - 3)}{3x - 5} > 0 \qquad x \neq 5/3$

	$-\infty$		1/2		5/3		3		∞
$2x - 1$		−		+		+		+	
$x - 3$		−		−		−		+	
$3x - 5$		−		−		+		+	
$\dfrac{(2x-1)(x-3)}{3x - 5}$		−		+		−		+	

$(1/2,\ 5/3) \cup (3,\ +\infty)$

55. $\dfrac{3x^2 + 5x - 2}{x^2 - 10x + 21} < 0$

$\dfrac{(3x - 1)(x + 2)}{(x - 7)(x - 3)} < 0 \qquad x \neq 3, 7$

	$-\infty$	-2	$1/3$	3	7	∞
$3x - 1$		$-$	$-$	$+$	$+$	$+$
$x + 2$		$-$	$+$	$+$	$+$	$+$
$x - 7$		$-$	$-$	$-$	$-$	$+$
$x - 3$		$-$	$-$	$-$	$+$	$+$
$\dfrac{(3x-1)(x+2)}{(x-7)(x-3)}$		$+$	$-$	$+$	$-$	$+$

$-2 < x < 1/3$ or $3 < x < 7$

57. Let x be the first integer. The next 3 consecutive integers are $x + 1$, $x + 2$, $x + 3$.

$2(x + x + 1 + x + 2)$
$\qquad = 102 + 3(x + 3)$
$2(3x + 3) = 102 + 3(x + 3)$
$6x + 6 = 102 + 3x + 9$
$3x = 105$
$x = 35$

The numbers are 35, 36, 37 and 38.

59. Let x be total number of people.

$990 + 90 + 0.45x = x$
$1080 = 0.55x$

$\dfrac{1080}{0.55} = x$

$x = 1964$
1964 people voted.

61. Let x be the number of \$8.50 tickets sold. Then $x + 70$ is the number of \$6 tickets.

$8.5x + 6(x + 70) = 2537$
$85x + 60(x + 70) = 25370$
$85x + 60x + 4200 = 25370$
$\qquad\qquad 145x = 21170$
$\qquad\qquad x = 146$

146 at \$8.50
216 at \$6.00

63. Let x be number of gallons of antifreee.

$0\%(4) + 90\%x = 30\%(4 + x)$
$9x = 3(4 + x)$
$9x = 12 + 3x$
$6x = 12$
$x = 2$

2 gallons

65. Let x be the amount that each person will pay. Then $8x$ is the total cost of the boat.

$8x = 10(x - 10)$
$8x = 10x - 100$
$-2x = -100$
$x = 50$
$8x = 400$

Boat rents for \$400.

67. Let x be the width. Then $2x - 3$ is the length.

$x(2x - 3) = 44$
$2x^2 - 3x - 44 = 0$
$(2x - 11)(x - 4) = 0$
$x = 11/2 \qquad x = -4$
$2x - 3 = 8$

Dimensions are 5.5m by 8m.

69. Let x be one side.

Area is $\dfrac{x^2\sqrt{3}}{4}$

$\dfrac{x^2\sqrt{3}}{4} = 16\sqrt{3}$

$\dfrac{x^2}{4} = 16$

$x^2 = 64$
$x = \pm 8$
8m.

71. Let x be number of glasses. Then $x - 10$ is number of glasses sold.

Cost per glass $\dfrac{2400}{x}$

Selling price per glass

$\dfrac{2400}{x} + 1.50$

Profit = Revenue − Cost

$1765 = (x - 10)(\dfrac{2400}{x} + 1.5)$

$\quad - 2400$

$1765 = (x - 10)(\dfrac{2400 + 1.5x}{x})$

$\quad - 2400$

$1765x = (x - 10)(2400 + 1.5x)$
$\quad - 2400x$
$0 = 2400x + 1.5x^2 - 24000 - 15x$
$\quad - 4165x$
$1.5x^2 - 1780x - 24000 = 0$
$15x^2 - 17800x - 240000 = 0$
$3x^2 - 3560x - 48000 = 0$
$(3x + 40)(x - 1200) = 0$
$x = -40/3 \qquad x = 1200$

He bought 1200 glasses.

73. Let x be the length of one leg. Then $x + 7$ is other leg and $x + 8$ is the hypotenuse.

$x^2 + (x + 7)^2 = (x + 8)^2$
$x^2 + x^2 + 14x + 49 = x^2 + 16x + 64$
$x^2 - 2x - 15 = 0$
$(x - 5)(x + 3) = 0$
$x = 5 \qquad x = -3$

The triangle measures 5 by 12 by 13.

75. $I = \dfrac{c}{d^2}$

$\dfrac{90}{d^2} = \dfrac{40}{20^2}$

$\dfrac{90}{d^2} = \dfrac{40}{400}$

$\dfrac{90}{d^2} = \dfrac{1}{10}$

$900 = d^2$
$d = \pm 30$
30 ft.

77. Let x be the change in the radius.

Area of sphere $4\pi r^2 =$
$4\pi(3.015)^2 = 114.23$

Area of new sphere
$4\pi(3.015 + x)^2$
$4\pi(3.015 + x)^2 = 3(114.23)$
$(3.015 + x)^2 = 27.27$
$3.015 + x = \pm 5.22$
$x = 2.21$ cm.
Increase radius by 2.21 cm.

79. $I = \dfrac{c}{d^2}$

Let x be intensity at 1 mile.

$C = Id^2$
$(1.5 \times 10^6)(55)^2 = x(5280)^2$
$(1.5 \times 10^6)(3025) = 27878400x$
$0.0001627 \times 10^6 = x$

$x = 162.7$ ft. candles

Chapter 2 Test

1. Solve for x the equation $3x - 2\sqrt{5} = 1$.
2. Write the given numbers in numerical ascending order: $3/2$, $\sqrt{2}$, $7/5$, $17/12$.
3. Find four consecutive even numbers whose sum is 468.
4. Find and sketch the solution set of the inequality $2(x - 3) > x + 4$.
5. Solve the equation $3x^2 - 243 = 0$ by moving the constant term to the right-hand side and taking square roots.
6. Solve the equation $x = \sqrt{9x - 20}$ and check for extraneous solutions.
7. Find two consecutive integers whose product is 156.
8. Solve the inequality $(x + 4)(x - 5) > 0$.
9. Solve for x the equation $3(x - 4)/4 + 2/3 = 2(x - 1)/3 - 1/6$.
10. A sugar factory has monthly maintenance costs of \$1200. In addition, it costs \$2 to make 5 pounds of sugar. If during a month the total production costs are \$1684, how many pounds of sugar has the factory produced?
11. Solve for x the absolute value equation $|x - 1| = 4$.
12. Solve for x the inequality $x/(x - 2) > 1$.
13. Solve by factoring the quadratic equation $x^2 - 11x - 12 = 0$.
14. Reduce $1 + 2/x + 1/x^2 = 0$ to a quadratic equation and solve it for x. Check for extraneous solutions.
15. The dimensions of a framed picture are 20 inches by 26 inches. The picture itself has area 432 square inches. What is the distance from the edge of the picture to the edge of the frame if the frame has uniform width?
16. Find the solution set of the inequality $1/x^2 > 1$.
17. Solve the equation $x/(x - 3) = 2x/(x - 3) + 4$.
18. A non-prescription drug contains 6 parts acetylsalicylic acid, 3 parts phenacetin, and 1 part caffeine. If each tablet contains 500 mg, how much of each chemical is needed to make one tablet?
19. Write the expression $|-3x^3|$, $x < 0$, without absolute value bars.
20. What is the solution set of the inequality $|2 - 3x| > 4$?
21. Complete-the square and solve the equation $x^2 - 12x + 27 = 0$.
22. Solve the equation $(x + 3)^2 + 9 = 6(2x + 3)$.
23. A projectile is fired straight upward at a speed of 128 ft/s. How long will it take before the projectile returns to the ground?
24. The volume of a frustum of a right circular cone is given by $V = \pi(r^2 + rR + R^2)h/3$. Solve this equation for h.
25. Pipes A and B are used to fill a swimming pool, while pipe C is used to empty it. Pipe A alone can fill the pool in 1 hour. Pipe B alone can fill the pool in 2 hours. if the three pipes are opened together, it takes 48 minutes to fill the pool. How long does it take for pipe C alone to empty the pool?
26. The abscissas of A, B, and C are 3, 6, and $\sqrt{2}$, respectively. Find d(A, B), d(A, C), and d(B, C).
27. Leadfoot Louie wants to drive 375 miles in a time interval of 2.5 to 3 hours. What is his corresponding speed range?
28. Solve by using the quadratic formula $2x^2 - 3x + 17 = 0$.
29. Solve the inequality $2x(3x - 1)/(x - 1)(2x + 3) \geq 0$.
30. In a genetics lab, the number of fruit flies at time t is given by $N = 500 + 50t - t^2$. When will the population be less than 125?

Chapter 3
Coordinate Geometry

Exercises 3.1

Find the graph for Exercise 1 in the answer section of your text.

3. $M(\dfrac{-1 + (-2)}{2}, \dfrac{4 + 5}{2})$,

   that is $M(-\dfrac{3}{2}, \dfrac{9}{2})$

5. $M(\dfrac{-1 + 1/4}{2}, \dfrac{-3/2 + 3}{2})$,

   $M(\dfrac{-3/4}{2}, \dfrac{3/2}{2})$,

   thus $M(-\dfrac{3}{8}, \dfrac{3}{4})$

7. $M(\dfrac{2 + a}{2}, \dfrac{5 + (-5)}{2})$,

   so $M(\dfrac{2 + a}{2}, 0)$

9. $d(P, Q) = \sqrt{(1 - 5)^2 + (1 - 5)^2}$

   $= \sqrt{16 + 16} = \sqrt{32} = 4\sqrt{2}$

11. $d(P, Q) = \sqrt{(6 - 5)^2 + (-2 + 6)^2}$

   $= \sqrt{1 + 16} = \sqrt{17}$

13. $d(P, Q) = \sqrt{(0 - 0)^2 + (-\dfrac{2}{3} - \dfrac{5}{3})^2}$

   $= \sqrt{(-\dfrac{7}{3})^2} = 7/3$

15. $d(P, Q) = \sqrt{(2 - 3)^2 + (2 - (-3))^2}$

   $= \sqrt{1 + 25} = \sqrt{26}$

17. $d(P, Q) = \sqrt{(1 - 3)^2 + (b - (-b))^2}$

   $= \sqrt{4 + (2b)^2} = \sqrt{4 + 4b^2}$

   $= \sqrt{4(1 + b^2)} = 2\sqrt{1 + b^2}$

19. $d(P, Q)$

   $= \sqrt{(t - (t+3))^2 + ((t+2) - (t-2))^2}$

   $= \sqrt{(t - t - 3)^2 + (t + 2 - t + 2)^2}$

   $= \sqrt{(-3)^2 + 4^2} = \sqrt{9 + 16} = \sqrt{25}$

   $= 5$

21. $d(P, Q)$

$= \sqrt{(21.05 - (-3.114))^2 + (1.314 - 21.26)^2}$

$= \sqrt{(24.164)^2 + (-19.946)^2}$

$= \sqrt{583.8989 + 397.84292}$

$= \sqrt{981.74182}$

$= 31.332759 \approx 31.333$

23. We have
$d(A, B)^2 = (7 - 4)^2 + (2 - 6)^2$
$= 9 + 16 = 25$
$d(A, C)^2 = (7 - (-4))^2 + (2 - 0)^2$
$= 121 + 4 = 125$
$d(B, C)^2 = (4 - (-4))^2 + (6 - 0)^2$
$= 64 + 36 = 100$

Since $125 = 100 + 25$, it follows by the converse of the Pythagorean Theorem that ABC is a right triangle with right angle at B.

25. ABC is a right triangle whose legs AB and BC measure 5 and 10 units of length, respectively. Thus, the area of ABC is

$$\frac{5 \times 10}{2} = 25$$

27. We have
$d(A, B) = \sqrt{(1 - (-4))^2 + (-2 - 2)^2}$

$= \sqrt{25 + 16} = \sqrt{41}$
and
$d(B, C) = \sqrt{(-4 - 1)^2 + (2 - 6)^2}$

$= \sqrt{25 + 16} = \sqrt{41}$
Since $d(A, B) = d(B, C)$, it follows that the triangle ABC is isoceles.

29. We need to show that all sides have the same length and that there is at least one right angle.

$d(A, B) = \sqrt{(-1 - 4)^2 + (-2 - 3)^2}$

$= \sqrt{25 + 25} = \sqrt{50} = 5\sqrt{2}$

$d(A, D) = \sqrt{(-1 - (-6))^2 + (-2 - 3)^2}$

$= \sqrt{25 + 25} = 5\sqrt{2}$

$d(C, D) = \sqrt{(-1 - (-6))^2 + (8 - 3)^2}$

$= 5\sqrt{2}$

$d(B, C) = \sqrt{(4 - (-1))^2 + (3 - 8)^2}$

$= 5\sqrt{2}$

Next, check that $d(D, B) = 10$ and that
$d(D, B)^2 = d(C, D)^2 + d(B, C)^2$.
Thus the angle at C is a right angle.

31. By the inverse of the Pythagorean Theorem we need to have

$d(0, A)^2 + d(0, B)^2 = d(A, B)^2$.

Since

$d(0, A)^2 = (0 - 2)^2 + (0 - 3)^2$
$= 4 + 9 = 13$
$d(0, B)^2 = (0 - 3)^2 + (0 - b)^2$
$= 9 + b^2$
$d(A, B)^2 = (2 - 3)^2 + (3 - b)^2$
$= 1 + (3 - b)^2$,

it follows that

$13 + 9 + b^2 = 1 + (3 - b)^2$
$22 + b^2 = 1 + 9 - 6b + b^2$
$6b = -12$
$b = -2$

33. $(\frac{x - 5}{2}, \frac{y + 8}{2})$

are the coordinates of the midpoint M(4, -7). Thus
$4 = \frac{x - 5}{2}$ and $-7 = \frac{y + 8}{2}$

$8 = x - 5 \qquad -14 = y + 8$
$x = 13 \qquad y = -22$

35. $\dfrac{2 - 3}{2} = a$ $\dfrac{-6 + b}{2} = 5$

    $-1 = 2a$ $-6 + b = 10$
    $a = -1/2$ $b = 16$

37. Let $M(x, y)$. The relation

    $d(A, M) = \dfrac{1}{4}d(A, B)$

    implies that

    $x - 3 = \dfrac{1}{4}(7 - 3)$

    and $y - 1 = \dfrac{1}{4}(5 - 1)$.

    Solving for x and y, obtain

    $x = 4$ and $y = 2$

39. Every point P on the perpendicular bisector of the segment AB is such that

    $d(A, P) = d(B, P)$.

    Since $P(4, 2)$, $A(-1, 3)$, and $B(3, -3)$, obtain

    $d(A, P) = \sqrt{26}$ and $d(B, P) = \sqrt{26}$.

    Thus $P(4, 2)$ lies on the perpendicular bisector of the segment with endpoints

    $A(-1, 3)$ and $B(3, -3)$.

41. If $P(x, y)$ is on the perpendicular bisector of AB, where $A(-1, 3)$ and $B(3, -3)$, then

    $d(A, P) = d(B, P)$

    that is

$\sqrt{(-1 - x)^2 + (3 - y)^2}$

$\qquad = \sqrt{(3 - x)^2 + (-3 - y)^2}.$

Squaring and simplifying give us

$(-1 - x)^2 + (3 - y)^2$
$\qquad = (3 - x)^2 + (-3 - y)^2$
$1 + 2x + x^2 + 9 - 6y + y^2$
$\qquad = 9 - 6x + x^2 + 9 + 6y + y^2$
$8x - 12y - 8 = 0$
$2x - 3y - 2 = 0$

43. $P(x, x)$ is $\sqrt{5}$ units from $(2, 3)$.

$\sqrt{(x - 2)^2 + (x - 3)^2} = \sqrt{5}$
$(x - 2)^2 + (x - 3)^2 = 5$
$x^2 - 4x + 4 + x^2 - 6x + 9 = 5$
$2x^2 - 10x + 8 = 0$
$x^2 - 5x + 4 = 0$
$(x - 1)(x - 4) = 0$
$x = 1, \ x = 4$

The points are $(4, 4)$ and $(1, 1)$.

45. $\sqrt{(x - 2)^2 + (y - (-4))^2} = 3$
$(x - 2)^2 + (y + 4)^2 = 9$

47. We have

$d(P, Q) = \sqrt{(-2 - 4)^2 + (-3 - 3)^2}$

$\qquad = \sqrt{36 + 36} = \sqrt{72} = 6\sqrt{2}$

$d(P, R) = \sqrt{(-2 - 2)^2 + (-3 - 1)^2}$

$\qquad = \sqrt{16 + 16} = 4\sqrt{2}$

$d(R, Q) = \sqrt{(2 - 4)^2 + (1 - 3)^2}$

$\qquad = \sqrt{4 + 4} = 2\sqrt{2}$

Since $6\sqrt{2} = 4\sqrt{2} + 2\sqrt{2}$, then R lies on the line segment PQ.

49. We have

$$d(P, Q) = \sqrt{(-3 - \frac{1}{3})^2 + (\frac{1}{2} - 2)^2}$$

$$= \sqrt{481}/6$$

$$d(P, R) = \sqrt{(-3 - 1)^2 + (\frac{1}{2} - 1)^2}$$

$$= \sqrt{65}/2$$

$$d(R, Q) = \sqrt{(1 - \frac{1}{3})^2 + (1 - 2)^2}$$

$$= \sqrt{13}/3$$

Since $d(P, Q) \neq d(P, R) + d(R, Q)$, we see that R does not lie on the segment PQ.

Exercises 3.2

Find corresponding art in the answer section of your text.

1. A(2, 2), B(-1/2, -1/2)
   The slope is

   $$m = \frac{-1/2 - 2}{-1/2 - 2} = 1.$$

3. P(-2, -1/4), Q(3, -2/3)
   The slope of the line is

   $$m = \frac{-\frac{2}{3} - (-\frac{1}{4})}{3 - (-2)} = \frac{\frac{-5}{12}}{5} = -\frac{1}{12}$$

5. P(1, 2), Q(3, 5), R(5, 8)
   Slope of line through P and Q:

   $$m = \frac{5 - 2}{3 - 1} = \frac{3}{2}$$

   Slope of line through Q and R:

   $$m' = \frac{8 - 5}{5 - 3} = \frac{3}{2}$$

Since $m = m'$, it follows that the three points lie on the same line.

7. P(2, 2), Q(-2, 4), R(0, 3)
   Slope of line through P and Q:

   $$m = \frac{4 - 2}{-2 - 2} = \frac{-2}{4} = -\frac{1}{2}$$

   Slope of line through Q and R:

   $$m' = \frac{3 - 4}{0 - (-2)} = \frac{-1}{2}$$

   Since $m = m'$ it follows that the three points lie on the same line.

9. Since the line is vertical, it follows that all its points have the same first coordinate. Thus, $x = -2$ is an equation of the line.

11. If the slope is 3 and the y-intercept is -2, then the equation of the line in slope-intercept form is $y = 3x - 2$.

13. Since the line passes through (0, 0) and has slope $m = -2/5$, it follows that the point-slope equation of the line is

    $$y - 0 = -\frac{2}{5}(x - 0)$$

    $$y = -\frac{2}{5}x.$$

15. $y = mx + b$
    $m = -4$, (2, 0) lies on the line.

    $0 = -4(2) + b$
    $0 = -8 + b$
    $b = 8$
    $y = -4x + 8$

17. A line with slope $m = 0$ is horizontal. If the line passes through the point (-6, -2) then an equation of the line is $y = -2$.

19. $y = mx + b$
    $b = 2$, $(-3, 0)$ lies on the line.

    $0 = m(-3) + 2$
    $0 = -3m + 2$
    $3m = 2$
    $m = 2/3$
    $y = (2/3)x + 2$
    $2x - 3y + 6 = 0$

21. To find the slope-intercept form of the line $3x - 4y + 5 = 0$, solve this equation for y:

    $4y = 3x + 5$

    $y = \dfrac{3}{4}x + \dfrac{5}{4}$

    Thus, $m = 3/4$, $(0, 5/4)$, $(-5/3, 0)$

23. We have

    $x = (\dfrac{3}{5})y + 3$

    $5x = 3y + 15$
    $3y = 5x - 15$

    $y = \dfrac{5}{3}x - 5$

    $m = 5/3$, $(0, -5)$, $(3, 0)$.

25. $2y + 3 = 0$
    $2y = -3$
    $y = -3/2$

    $m = 0$, $(0, -3/2)$, no x-intercept.

27. $\dfrac{x}{2} + \dfrac{y}{3} = 1$

    $\dfrac{6x}{2} + \dfrac{6y}{3} = 6$

    $3x + 2y = 6$

$y = -\dfrac{3}{2}x + 3$

$m = -3/2$, $(0, 3)$, $(2, 0)$.

29. The slope of the line $2x + 7y - 2 = 0$ can by obtained by solving the equation for y:

    $y = -2/7x + 2/7$

    Thus, $m = -2/7$. Similarly, the slope of the line $7x - 2y + 1 = 0$ is $m' = 7/2$. Since $m' = -1/m$; it follows that the two lines are perpendicular.

31. By rewriting the equation $x + y + 1 = 0$ in slope-intercept form we obtain

    $y = -x - 1$.

    This line has slope $m = -1$. By rewriting the equation $x - y = 0$ in slope-intercept form we obtain

    $y = x$.

    This line has slope $m' = 1$. Since $m' = -1/m$, the two lines are perpendicular.

33. Every line parallel to the line $y = 3x - 2$ has slope $m = 3$. Since the line passes through $(0, 0)$, we see that its point-slope equation is

    $y = 3x$
    or
    $3x - y = 0$,

    in standard form.

35. Slope of line $y = 2x - 5$ is $m = 2$.
Slope of perpendicular line is
$m' = -1/2$. Since line passes
through $(-1, -2)$, we have

$$y - (-2) = -\frac{1}{2}(x - (-1))$$

$$y + 2 = -\frac{1}{2}(x + 1)$$

$$2y + 4 = -x - 1$$
$$x + 2y + 5 = 0$$

37. Slope of main diagonal $y = x$ is
$m = 1$. Slope of perpendicular
line is $m' = -1$.
Given point: $(1, -4)$. Equation:

$$y - (-4) = -1(x - 1)$$
$$y + 4 = -x + 1$$
$$x + y + 3 = 0$$

39. Main diagonal $y = x$
$m = 1$
$$y - y_1 = m(x - x_1)$$
$$y - 7 = 1(x - (-3))$$
$$y - 7 = x + 3$$
$$x - y + 10 = 0$$

41. The equation
$(1/2)x + (2/3)y - 3/4 = 0$ written
in slope-intercept form is

$$y = -\frac{3}{4}x + \frac{9}{8}.$$

Thus the line has slope $m' = -\frac{3}{4}$.

Since the line we are looking for
is parallel to the given line, its
slope is

$$m = -\frac{3}{4}.$$

But it passes through $(3, -3)$, so

$$y + 3 = -\frac{3}{4}(x - 3)$$

$$4y + 12 = -3x + 9$$
$$4y + 3x + 3 = 0$$

43. The slope of the line
$x - 4y + 1 = 0$ is $m = 1/4$, while
the slope of the line
$ax + 3y - 5 = 0$ is $m' = -a/3$. If
the two lines are parallel, then

$$-\frac{a}{3} = \frac{1}{4}, \text{ so } a = -\frac{3}{4}.$$

45. The slope of the line
$2x + 2y - 5 = 0$ is $m = -1$, while
the slope of the line
$5x - by - 3 = 0$ is

$$m' = \frac{5}{b} \ (b \neq 0).$$

If the two lines are
perpendicular, then

$$-1 = -\frac{b}{5}$$

$$-5 = -b$$
$$b = 5$$

47. The slope of the line passing
through $Q(-2, 9)$ and $R(3, -10)$ is

$$m = \frac{-10 - 9}{3 - (-2)} = \frac{-19}{5}.$$

Any line perpendicular to the line
through Q and R has slope

$$m' = -\frac{1}{m} = \frac{5}{19}.$$

Thus, an equation of the
perpendicular line passing through
$P(-2, -5)$ is

$$y - 5 = \frac{5}{19}(x + 2).$$

In standard form the equation is
$5x - 19y - 85 = 0$.

49. $m_{AB} = \dfrac{9 - 5}{4 - 2} = \dfrac{4}{2} = 2$

$m_{BC} = \dfrac{9 - 8}{4 - 6} = \dfrac{1}{-2} = \dfrac{-1}{2}$

$m_{AC} = \dfrac{8 - 5}{6 - 2} = \dfrac{3}{4}$

Since m_{AB} and m_{BC} are opposite reciprocals then AB ⊥ BC and triangle ABC is a right triangle.

51. $O(0, 0)$, $A(2, 3)$, $B(3, b)$

Slope of the side OA: $m = \dfrac{3}{2}$

Slope of the side OB: $m' = \dfrac{b}{3}$

Slope of the side AB:

$m'' = \dfrac{b - 3}{3 - 2} = b - 3.$

Since OAB is a right triangle with right angle at A, we must have

$m'' = -\dfrac{1}{m}$, that is

$b - 3 = -\dfrac{2}{3}$ and so

$b = \dfrac{7}{3}.$

53. $m_{PQ} = \dfrac{-1 - (-3)}{6 - 2} = \dfrac{2}{4} = \dfrac{1}{2}$

A line perpendicular to PQ must have slope -2. The midpoint of PQ is

$(\dfrac{2 + 6}{2}, \dfrac{-3 - 1}{2}) = (\dfrac{8}{2}, \dfrac{-4}{2})$

$= (4, -2)$

$y - y_1 = m(x - x_1)$
$y - (-2) = -2(x - 4)$
$y + 2 = -2x + 8$
$2x + y - 6 = 0$

55. Make a table of values as follows

u	v
5	10
8	15

Since u and v are linearly related, it follows that they satisfy a linear equation in u and v that represents the line through the points (5, 10) and (8, 15). The slope of this line is

$m = \dfrac{15 - 10}{8 - 5} = \dfrac{5}{3}$

and its slope-intercept equation is

$v = \dfrac{5}{3}u + \dfrac{5}{3}.$

This is a linear equation relating u and v. Now, if we set $u = 2$, we obtain

$v = \dfrac{5}{3}(2) + \dfrac{5}{3} = \dfrac{15}{3} = 5.$

57. Since $A(1, 2)$ lies on the line $ax - 3y + 4 = 0$, we must have

$a(1) - 3(2) + 4 = 0$
$a - 6 + 4 = 0$
$a = 2$

59. Making a table of values for w and s we get

w	s
5	1
0	0

Thus we are looking for the equation of the line passing through the points (5, 1) and (0, 0).

a) The slope of the line is

$$m = \frac{1 - 0}{5 - 0} = \frac{1}{5}.$$

The equation is

$$s - 0 = \frac{1}{5}(w - 0)$$

$$s = \frac{1}{5}w$$

b) When $w = 2$,

$$s = \frac{1}{5}(2)$$

$$s = \frac{2}{5} = 0.4$$

A 2 - lb weight stretches the spring 0.4 inches.

61. To say that a typewriter is being depreciated linearly over a certain number of years means that its price p is linearly related to the number of years t:

$$p = at + b,$$

where a and b are constants. Since $p = 1000$, when $t = 0$, it follows that $b = 1000$, and we rewrite the last equation as

$$p = at + 1000.$$

Now, after $t = 5$ years, $p = 0$ so that

$$0 = a(5) + 1000$$
$$a = -200.$$

Thus
$$p = -200t + 1000$$

gives the depreciated value of the typewriter in t years. After 3 years its value is

$$p = -200(3) + 1000 = \$400.$$

63. We have that $v = -32t + v_0$.

Since $v_0 = 192$, then
$v = 192 - 32t$.
When $t = 3$ seconds, then

$$v = 192 - 32(3)$$
$$= 192 - 96$$
$$= 96 \text{ ft/s.}$$

Now, $v = 0$ implies $192 - 32t = 0$,

so $t = \dfrac{192}{32} = 6$,

i.e., the velocity will be zero after 6 seconds.

65. Case 1: $a = 0$ and $b \neq 0$.

In this case the equations of the two lines are

$by + c = 0$ or $y = -c/b$
and
$bx + d = 0$ or $x = -d/b$.

We see that the first line is horizontal while the second is vertical. Thus the lines are perpendicular to each other.

Case 2: $a \neq 0$ and $b = 0$.

In this case the first line $ax + c = 0$ is vertical, while the second $-ay + d = 0$ is horizontal. Thus the lines are perpendicular to each other.

Case 3: $a \neq 0$ and $b \neq 0$.

In this case the slope of the line $ax + by + c = 0$ is $m = -a/b$ and the slope of the line $bx - ay + d = 0$ is $m' = b/a$. Since $m' = -1/m$, it follows that the two lines are perpendicular to each other.

67. Case 1: $b = 0$

In this case the line is $ax + c = 0$ ($a \neq 0$), a vertical line. The line perpendicular to $ax + c = 0$ is horizontal. If it is to pass through the point (x_0, y_0), then $y = y_0$ must be true, or $y - y_0 = 0$, or $-a(y - y_0) = 0$, as required.

Case 2: $b \neq 0$
The slope of the line

$ax + by + c = 0$ is $m' = -\dfrac{a}{b}$.

The line perpendicular to $ax + by + c$ passing through (x_0, y_0), has slope

$m = -\dfrac{1}{m'} = \dfrac{b}{a}$ and is given by

$y - y_0 = \dfrac{b}{a}(x - x_0)$ or

$b(x - x_0) - a(y - y_0) = 0$
as required.

69. If the point (x_1, y_1) lies on the line $y = mx + b$, then $y_1 = mx_1 + b$ is true.
If the point (x_2, y_2) lies on the line $y = mx + b$, then $y_2 = mx_2 + b$ is also true.
Thus

$y_2 - y_1 = (mx_2 + b) - (mx_1 + b)$
$y_2 - y_1 = mx_2 + b - mx_1 - b$
$y_2 - y_1 = mx_2 - mx_1$
$y_2 - y_1 = m(x_2 - x_1)$

$\dfrac{y_2 - y_1}{x_2 - x_1} = m; \ x_2 \neq x_1$

Solving the equation $y_2 = mx_2 + b$ for b, we get

$b = y_2 - mx_2$.

Replacing m with

$\dfrac{y_2 - y_1}{x_2 - x_1}$, we get

$b = y_2 - \left(\dfrac{y_2 - y_1}{x_2 - x_1}\right)x_2$

$b = \dfrac{y_2(x_2 - x_1) - x_2(y_2 - y_1)}{x_2 - x_1}$

$b = \dfrac{y_2 x_2 - y_2 x_1 - x_2 y_2 + x_2 y_1}{x_2 - x_1}$

$b = \dfrac{x_2 y_1 - x_1 y_2}{x_2 - x_1}$

Exercises 3.3

Find corresponding art in the answer section of your text.

11. Parabola symmetric with respect to the y-axis.

13. Parabola symmetric with respect to the y-axis.

15. Parabola symmetric with respect to the x-axis.

17. Symmetry with respect to the origin.

19. Symmetry with respect to the point $(0, 2)$.

21. Symmetry with respect to the origin.

23. $(x - 1)^2 + (y - 1)^2 = 5$

25. The equation of a circle centered at $(0, 0)$ and with radius r is

$$x^2 + y^2 = r^2.$$

If the point $(2, 2)$ lies on the circle, then

$$2^2 + 2^2 = r^2$$
$$8 = r^2$$

Thus, the equation is $x^2 + y^2 = 8$.

27. $C(2, -1)$
$$(x - 2)^2 + (y - (-1))^2 = r^2$$

r = distance from C to the origin, so

$$r = \sqrt{2^2 + (-1)^2} = \sqrt{4 + 1} = \sqrt{5}$$
So,
$$(x - 2)^2 + (y + 1)^2 = 5.$$

29. Since the center $C(3, 5)$ of the circle has y-coordinate 5 and the circle is tangent to the x-axis, it follows that its radius is 5. Thus,

$$(x - 3)^2 + (y - 5)^2 = 25.$$

31. The radius of the circle is half the distance d(P, Q):

$$r = \frac{d(P, Q)}{2}$$

$$= \frac{\sqrt{(-1 - 2)^2 + (3 - 5)^2}}{2}$$

$$= \frac{\sqrt{9 + 4}}{2} = \frac{\sqrt{13}}{2}$$

The center is the midpoint of the segment PQ:

$$C(\frac{-1 + 2}{2}, \frac{3 + 5}{2}), \text{ that is,}$$

$$C(\frac{1}{2}, 4) \quad \text{Thus,}$$

$$(x - \frac{1}{2})^2 + (y - 4)^2 = \frac{13}{4}.$$

33. Since the circle with radius 3 is tangent to both coordinate axes and since the center is in the first quadrant, its center must be $C(3,3)$. Thus, the equation of the circle is

$$(x - 3)^2 + (y - 3)^2 = 9$$

35. $x^2 + y^2 - 2y - 8 = 0$
$x^2 + y^2 - 2y = 8$
$x^2 + y^2 - 2y + 1 = 8 + 1$
$x^2 + (y - 1)^2 = 9$

Circle with center $C(0, 1)$ and radius $r = 3$.

37. $2x^2 + 2y^2 = 18$
$x^2 + y^2 = 9$
$(x - 0)^2 + (y - 0)^2 = 3^2$

$C(0, 0), r = 3$

39. $x^2 + y^2 - 6x - 4y = -9$
$(x^2 - 6x) + (y^2 - 4y) = -9$

By completing the squares, obtain

$$(x^2 - 6x + 9) + (y^2 - 4y + 4)$$
$$= -9 + 9 + 4$$
$$(x - 3)^2 + (y - 2)^2 = 4$$

$C(3, 2), r = 2$

41. $x^2 + y^2 - 4y = 5$
$x^2 + (y^2 - 4y + 4) = 5 + 4$
$x^2 + (y - 2)^2 = 9$

$C(0, 2), \ r = 3$

43. $x^2 + y^2 + 6x + 4y + 13 = 0$
$(x^2 + 6x) + (y^2 + 4y) = -13$
$(x^2 + 6x + 9) + (y^2 + 4y + 4)$
$\qquad = 9 + 4 - 13$
$(x + 3)^2 + (y + 2)^2 = 0$

The only ordered pair that satisfies this equation is $(-3, -2)$. We may say that the equation represents a point.

45. $x^2 + y^2 + x - 3y = 0$
$x^2 + x + y^2 - 3y = 0$

$x^2 + x + \dfrac{1}{4} + y^2 - 3y + \dfrac{9}{4} = \dfrac{1}{4} + \dfrac{9}{4}$

$(x + \dfrac{1}{2})^2 + (y - \dfrac{3}{2})^2 = \dfrac{10}{4}$

$C(-\dfrac{1}{2}, \dfrac{3}{2}), \ r = \dfrac{\sqrt{10}}{2}$

47. $x^2 + y^2 - 2x + 4y + 6 = 0$
$x^2 - 2x + y^2 + 4y = -6$
$x^2 - 2x + 1 + y^2 + 4y + 4$
$\qquad = -6 + 1 + 4$
$(x - 1)^2 + (y + 2)^2 = -1$

This is not a circle since r^2 cannot be negative.

49. $3x^2 + 3y^2 + 9x + 12y - 6 = 0$
$x^2 + y^2 + 3x + 4y - 2 = 0$
$(x^2 + 3x) + (y^2 + 4y) = 2$

$(x^2 + 3x + \dfrac{9}{4}) + (y^2 + 4y + 4)$

$\qquad\qquad = \dfrac{9}{4} + 4 + 2$

$(x + \dfrac{3}{2})^2 + (y + 2)^2 = \dfrac{33}{4}$

$C(-\dfrac{3}{2}, -2), \ r = \dfrac{\sqrt{33}}{2}$

Exercises 3.4

Find the corresponding graphs in the answer section of your text.

1. Write the equation $y = 4x^2$

as $x^2 = \dfrac{1}{4}y$.

This is an equation of the form $x^2 = 4py$,

with $4p = \dfrac{1}{4}$,

that is, $p = 1/16$. It follows that the parabola has vertex at the origin and focus $F(0, 1/16)$. The principal axis is the y-axis and the directrix is the horizontal line $y = -1/16$.

3. $y = 3x^2 + 1$ can be written as $y - 1 = 3(x - 0)^2$ or

$(x - 0)^2 = 4(\dfrac{1}{12})(y - 1)$

which is an equation of the form $(x - h)^2 = 4p(y - k)$. Therefore, the vertex is $V(0, 1)$, the principal axis is $x = 0$ (y-axis), the focus is

$F(0, 1 + \dfrac{1}{12})$ or

$F(0, \dfrac{13}{12})$,

and the directrix is $y = 11/12$.

5. $x^2 - 4y + 8 = 0$
$x^2 = 4y - 8$
$x^2 = 4(y - 2) \quad (p = 1)$
Vertex: $(0, 2)$
Directrix $y = 1$
Principal axis $x = 0$ (y-axis)
Focus: $(0, 3)$

7. Write $(x - 1)^2 = y + 1$ as follows

$$(x - 1)^2 = 4(\frac{1}{4})(y - (-1))$$

$$(p = \frac{1}{4})$$

So $V(1, -1)$, $F(1, -3/4)$,
PA: $x = 1$, D: $y = -5/4$

9. Note that $V(0, 0)$ and $F(3/2, 0)$ lie on the x-axis, so the line $y = 0$ is the principal axis. It follows that $p = 3/2$, thus the equation of the parabola is

$$(y - 0)^2 = 4(3/2)(x - 0)$$
$$y^2 = 6x$$

11. Both $V(1, 2)$ and $F(2, 2)$ lie on the line $y = 2$ which is the principal axis. We have

$$p = 2 - 1 = 1,$$

and the directrix is the y-axis. Thus, the equation is

$$(y - 2)^2 = 4(1)(x - 1)$$
$$(y - 2)^2 = 4(x - 1)$$

13. The principal axis is the x-axis,

$$p = 3 - 1 = 2, \text{ and } F(5, 0). \text{ Thus,}$$

$$(y - 0)^2 = 4(2)(x - 3)$$
$$y^2 = 8(x - 3)$$

15. PA: $x = -1$, $p = 1/2$
F $= (-1, -3/2)$
$$(x + 1)^2 = -4(1/2)(y + 1)$$
$$(x + 1)^2 = -2(y + 1)$$

17. $V(1, 2)$, PA: $x = 1$ contains the point $(2, 6)$. Thus, the parabola opens up and has an equation of the form

$$(x - 1)^2 = 4p(y - 2).$$

Since $(2, 6)$ lies on the curve, it follows that

$$(2 - 1)^2 = 4p(6 - 2)$$
$$1 = 16p$$
$$p = 1/16$$

Equation: $(x - 1)^2 = 1/4(y - 2)$.

19. Rewrite the given equation as follows

$$2x^2 - 8y + 6 = 0$$
$$x^2 - 4y + 3 = 0$$
$$x^2 = 4y - 3$$
$$x^2 = 4(y - 3/4)$$

Parabola opening upwards,

$V(0, 3/4)$, $F(0, 7/4)$,
PA: $x = 0$, D: $y = -1/4$

21. $y^2 - 2y - x = 0$

Complete the square:

$$y^2 - 2y = x$$
$$y^2 - 2y + 1 = x + 1$$

$$(y - 1)^2 = 4(\frac{1}{4})(x + 1)$$

Parabola opening to the right,
$V(1, -1)$, $F(-3/4, 1)$, PA: $y = 1$,
D: $x = -5/4$.

23. In order for the parabola

$$y = x^2 + 4x + c$$

to have two x-intercepts, the quadratic equation

$$x^2 + 4x + c = 0$$

must have two real solutions. This happens when the discriminant

$b^2 - 4ac > 0$. So,

$$4^2 - 4(1)c > 0$$
$$16 - 4c > 0$$
$$4c < 16$$
$$c < 4$$

25. To find the points of intersection of the line $y = 2x + 1$ and the parabola $y = 3x^2 + 5x - 5$, set $2x + 1 = 3x^2 + 5x - 5$ and solve this equation for x:

$$3x^2 + 3x + 6 = 0$$
$$x^2 + x - 2 = 0$$
$$(x - 1)(x + 2) = 0,$$

so $x = 1$ and $x = -2$. Substituting 1 for x into the equation $y = 2x + 1$, obtain

$$y = 2(1) + 1 = 3.$$

Thus, $(1,3)$ is one of the points of intersection. Substituting -2 for x into the same equation gives us the other point: $(-2,-3)$.

27. The equation $\dfrac{x^2}{16} + \dfrac{y^2}{9} = 1$

represents an ellipse with center $C(0,0)$. The major axis is along the x-axis and $a = 4$. The minor axis is along the y-axis and $b = 3$.

$$c^2 = a^2 - b^2$$
$$= 4^2 - 3^2 = 7$$
$$c = \sqrt{7}$$
$$V(\pm 4,0), \quad F(\pm\sqrt{7},0)$$

29. $4x^2 + 9y^2 = 36$

$$\frac{x^2}{9} + \frac{y^2}{4} = 1$$

Major axis: x-axis, $a = 3$
Minor axis: y-axis, $b = 2$
$$c = \sqrt{5}$$
$$C(0,0), \quad V(\pm 3,0), \quad F(\pm\sqrt{5},0)$$

31. $\dfrac{(x - 1)^2}{25} + \dfrac{(y + 1)^2}{4} = 1$

Ellipse centered at $C(1,-1)$.
Major axis: $y = -1$, $a = 5$
Minor axis: $x = 1$, $b = 2$
$$c = \sqrt{21}$$
$$V_1(-4,-1), \quad V_2(6,-1), \quad F(1\pm\sqrt{21},-1)$$

33. If $C(0,0)$ and $F_1(-2,0)$, then $c = 2$. Since $a = 4$, it follows from $c^2 = a^2 - b^2$ that $b^2 = 12$. The equation of the ellipse is

$$\frac{x^2}{16} + \frac{y^2}{12} = 1$$

35. $C(1,2)$, $F_1(6,2)$, so $a = 6 - 1 = 5$. Since the center is $C(1,2)$ and one endpoint of the minor axis is $(1,6)$, it follows that $b = 6 - 2 = 4$. Thus the equation is

$$\frac{(x - 1)^2}{25} + \frac{(y - 2)^2}{16} = 1.$$

37. $F_1(0,0)$, $F_2(6,0)$, so $c = 6$. $V_1(-2,0)$, $V_2(8,0)$, thus

$$a = \frac{8 - (-2)}{2} = 5$$

From $c^2 = a^2 - b^2$, obtain $b = 4$. The center is the midpoint of the segment V_1,V_2, so $C(3,0)$.

$$\frac{(x - 3)^2}{25} + \frac{y^2}{16} = 1$$

39. $C(0,0)$, $F_1(-5,0)$, so $c = 5$. Now

$$e = \frac{1}{3} = \frac{5}{a}, \text{ so } a = 15.$$

Next check that $b^2 = 200$. Thus the equation is

$$\frac{x^2}{225} + \frac{y^2}{200} = 1.$$

41. $C(0,0)$, $F_1(2,0)$, $c = 2$.
$C(0,0)$, $V_1(3,0)$, $a = 3$.
$b^2 = a^2 - c^2 = 9 - 4 = 5$

$$\frac{x^2}{9} + \frac{y^2}{5} = 1.$$

43. $V(\pm 5,0)$, so $C(0,0)$ and $a = 5$. The equation is of the form

$$\frac{x^2}{25} + \frac{y^2}{b^2} = 1.$$

To find b notice that $(4, 12/5)$ lies on the ellipse, so

$$\frac{4^2}{25} + \frac{(12/5)^2}{b^2} = 1$$

$$\frac{16}{25} + \frac{144}{25b^2} = 1$$

$$16b^2 + 144 = 25b^2$$
$$144 = 9b^2$$

$$\frac{144}{9} = b^2$$

$$\frac{12}{3} = b$$

$$4 = b$$

Equation: $\dfrac{x^2}{25} + \dfrac{y^2}{16} = 1$

45. Complete the squares by arranging your work as follows

$$9x^2 + 4y^2 - 18x + 8y - 23 = 0$$
$$9(x^2 - 2x) + 4(y^2 + 2x) = 23$$

$$9(x^2 - 2x + 1) + 4(y^2 + 2x + 1)$$
$$= 23 + 9 + 4$$
$$9(x - 1)^2 + 4(y + 1)^2 = 36$$

$$\frac{(x - 1)^2}{4} + \frac{(y + 1)^2}{9} = 1$$

$C(1,-1)$, $a = 3$, $b = 2$

47. Complete the squares:

$$4x^2 + 9y^2 - 16x + 18y - 11 = 0$$
$$4(x^2 - 4x) + 9(y^2 + 2y) = 11$$
$$4(x^2 - 4x + 4) + 9(y^2 + 2y + 1)$$
$$= 11 + 16 + 9$$
$$4(x - 2)^2 + 9(y + 1)^2 = 36$$

$$\frac{(x - 2)^2}{9} + \frac{(y + 1)^2}{4} = 1$$

$C(2,-1)$, $a = 3$, $b = 2$

49. We are given $e = \dfrac{c}{a} = 1.670 \times 10^{-2}$

and $2a = 1.858 \times 10^8$, that is $a = 9.29 \times 10^7$. It follows that

$$c = a \cdot e$$
$$= (1.670 \times 10^{-2})(9.29 \times 10^7)$$
$$= 1.55143 \times 10^6.$$

At perihelion, the distance from the Earth to the sun is

$$a - c = 9.29 \times 10^7 - 1.55143 \times 10^6$$
$$\simeq 9.135 \times 10^7 \text{ miles.}$$

At aphelion, the distance is

$$a + c = 9.29 \times 10^7 + 1.55143 \times 10^6$$
$$= 9.445 \times 10^7 \text{ miles.}$$

51. $\dfrac{x^2}{64} - \dfrac{y^2}{36} = 1$

$C(0,0)$, $a = 8$, $b = 6$.

From $b^2 = c^2 - a^2$ obtain $c = 10$.
$V(\pm 8,0)$, $F(\pm 10,0)$.

53. $\dfrac{(x-2)^2}{81} - \dfrac{(y+3)^2}{144} = 1$

    $C(2,-3)$, $a = 9$, $b = 12$, $c = 15$.
    $V_1(-7,-3)$, $V_2(11,-3)$
    $F_1(-13,-3)$, $F_2(17,-3)$.

55. $C(0,0)$, $F_1(4,0)$, so $c = 4$.
    Now, use the relation
    $b^2 = c^2 - a^2$ and $a = 2$
    to obtain $b^2 = 12$.

    $\dfrac{x^2}{4} - \dfrac{y^2}{12} = 1$

57. $V_1(-2,1)$, $V_2(6,1)$, so

    $a = \dfrac{6-(-2)}{2} = 4$

    and $C(2,1)$ (midpoint of V_1V_2).
    $F_1(-3,1)$, $F_2(7,1)$ so

    $c = \dfrac{7-(-3)}{2} = 5$

    Check that $b = 3$.

    $\dfrac{(x-2)^2}{16} - \dfrac{(y-1)^2}{9} = 1$

59. Since $V(\pm 1,0)$, it follows that
    $C(0,0)$ and $a = 1$. Now, the lines
    $y = \pm 2x$ are asymptotes, so

    $\dfrac{b}{a} = 2$

    $b = 2a = 2(1) = 2$

    $x^2 - \dfrac{y^2}{4} = 1$

61. $F_1(-4,1)$, $F_2(6,1)$, so

    $c = \dfrac{6-(-4)}{2} = 5$

and $C(1,1)$ (midpoint of F_1F_2).

From $b^2 = c^2 - a^2$ and $a = 4$ obtain
$b^2 = 9$.

$\dfrac{(x-1)^2}{16} - \dfrac{(y-1)^2}{9} = 1.$

63. $V_1(0,-4)$, $V_2(0,4)$, $a = 8$ and
    $C(0,0)$. The equation is of the
    form

    $\dfrac{y^2}{16} - \dfrac{x^2}{b^2} = 1$

    Let $(x,y) = (9/4,5)$.

    $\dfrac{5^2}{16} - \dfrac{(9/4)^2}{b^2} = 1$

    $\dfrac{25}{16} - \dfrac{81}{16b^2} = 1$

    $25b^2 - 81 = 16b^2$
    $9b^2 = 81$
    $b^2 = 9$
    $b = 3$

    $\dfrac{y^2}{16} - \dfrac{x^2}{9} = 1$

65. $2y^2 + 4y - x^2 = 0$

    $2(y^2 + 2y) - x^2 = 0$

    $2(y^2 + 2y + 1) - x^2 = 2$

    $2(y+1)^2 - x^2 = 2$

    $(y+1)^2 - \dfrac{x^2}{2} = 1$

    $C(0,-1)$, $a = 1$

    $V_1(0,0)$, $V_2(0,-2)$

67. $x^2 - y^2 + 4x + 2y + 2 = 0$

$x^2 + 4x - (y^2 - 2y) = -2$

$x^2 + 4x + 4 - (y^2 - 2y + 1)$
$\quad = -2 + 4 - 1$

$(x + 2)^2 - (y - 1)^2 = 1$

$C(-2,1), \quad a = b = 1$

$V_1(-3,1), \quad V_2(-1,1)$

69. $4x^2 - 9y^2 - 16x - 18y - 29 = 0$

$4(x^2 - 4x) - 9(y^2 + 2y) = 29$

$4(x^2 - 4x + 4) - 9(y^2 + 2y + 1)$
$\quad = 29 + 16 - 9$

$4(x - 2)^2 - 9(y + 1)^2 = 36$

$\dfrac{(x - 2)^2}{9} - \dfrac{(y + 1)^2}{4} = 1$

$C(2,-1), \quad a = 3, \quad b = 2$

$V_1(-1,-1), \quad V_2(5,-1)$

Review Exercises - Chapter 3

1. $P(2,8), \quad Q(3,-5)$

$M\left(\dfrac{2 + 3}{2}, \dfrac{8 - 5}{2}\right)$

$M\left(\dfrac{5}{2}, \dfrac{3}{2}\right)$

3. $P(1/3, -1/2), \quad Q(3/4, 1/5)$

$M\left(\dfrac{1/3 + 3/4}{2}, \dfrac{-1/2 + 1/5}{2}\right)$

$= \left(\dfrac{13/12}{2}, \dfrac{-3/10}{2}\right)$

$M(13/24, -3/20)$

5. $A(2,-3), \quad M(4,-1)$
If $B(x,y)$, then

$\dfrac{2 + x}{2} = 4$ and $\dfrac{-3 + y}{2} = -1$

$x = 8 - 2 = 6$ and $y = -2 + 3 = 1$.

So $B(6,1)$.

7. A point Q on the x-axis has coordinates $(x,0)$. If $P(2,-5)$ and $d(Q,P) = 6$, then

$\sqrt{(x - 2)^2 + (0 - (-5))^2} = 6$

$\sqrt{x^2 - 4x + 4 + 25} = 6$

$x^2 - 4x + 29 = 36$

$x^2 - 4x - 7 = 0$

$x = \dfrac{4 \pm \sqrt{16 + 28}}{2} = \dfrac{4 \pm 2\sqrt{11}}{2}$

$= 2 \pm \sqrt{11}$

Thus the points are $(2 + \sqrt{11}, 0)$ and $(2 - \sqrt{11}, 0)$.

9. $A(-4,0), \quad B(1,10), \quad C(4,6)$

$d(A,B)^2 = (-4 - 1)^2 + (0 - 10)^2$
$\quad = 25 + 100 = 125$

$d(A,C)^2 = (-4 - 4)^2 + (0 - 6)^2$
$\quad = 64 + 36 = 100$

$d(B,C)^2 = (1 - 4)^2 + (10 - 6)^2$
$\quad = 9 + 16 = 25$

Since $d(A,C)^2 + d(B,C)^2 = d(A,B)^2$, it follows by the converse of the Pythagorean theorem that ABC is a right triangle with right angle at C. Its area is

$\dfrac{d(A,C) \times d(B,C)}{2} = \dfrac{10 \times 5}{2} = 25$

11. $A(-3,-4)$, $B(1,-8)$, $C(7,-2)$, $D(3,2)$

$d(A,B)^2 = (-3-1)^2 + (-4+8)^2 = 32$
$d(C,D)^2 = (7-3)^2 + (-2-2)^2 = 32$
$d(A,D)^2 = (-3-3)^2 + (-4-2)^2$
$\qquad = 72$
$d(B,C)^2 = (1-7)^2 + (-8+2)^2 = 72$

Since $d(A,B) = d(C,D) = 4\sqrt{2}$ and $d(A,D) = d(B,C) = 6\sqrt{2}$, it follows that ABCD is a parallelogram.

13. $O(0,0)$, $A(2,3)$, $B(3,b)$

If OAB is a right triangle with right angle at A, then

$\qquad d(O,A)^2 + d(A,B)^2 = d(O,B)^2.$

Now
$\qquad d(O,A)^2 = 2^2 + 3^2 = 13$

$\qquad d(A,B)^2 = (3-2)^2 + (b-3)^2$
$\qquad\qquad = b^2 - 6b + 10$

$\qquad d(O,B)^2 = 9 + b^2,$ so

$\qquad 13 + \cancel{b^2} - 6b + 10 = 9 + \cancel{b^2}$
$\qquad\qquad\qquad -6b = -14$
$\qquad\qquad\qquad\quad b = 7/3$

15. Symmetry with respect to the origin

17. Symmetry with respect to the line $x = 2$.

19. $A(-3,-5)$, $B(4,7)$

$\qquad m = \dfrac{7-(-5)}{4-(-3)} = \dfrac{12}{7}$

21. $M(2.157,-3.019)$, $N(0.026,-3.107)$

$\qquad m = \dfrac{-3.107 - (-3.019)}{0.026 - 2.157} = \dfrac{-0.088}{-2.131}$

$\qquad \approx 0.041$

23. $y - (-6) = -\dfrac{1}{3}(x - (-2))$

$\qquad y + 6 = -\dfrac{1}{3}(x + 2)$

$\qquad 3y + 18 = -x - 2$

$\qquad x + 3y + 20 = 0$

25. $y = \dfrac{5}{3}x + \dfrac{3}{2}$

$\qquad 6y = 10x + 9$

$\qquad 10x - 6y + 9 = 0$

27. The equation is of the form $y = mx + 6.$

If 3/5 is the x-intercept, then (3/5,0) lies on the line:

$\qquad 0 = m(3/5) + 6$
$\qquad 0 = 3m + 30$
$\qquad m = -10$

Thus, the equation of the line is $y = -10x + 6$ or $10x + y - 6 = 0$.

29. The slope of the line is

$\qquad m = \dfrac{-2.15 + 7.21}{1.07 - 2.35}$

$\qquad = -\dfrac{5.06}{1.28} \approx -3.95.$

Thus $y + 7.21 = -3.95(x - 2.35)$ or $3.95x + y - 2.08 = 0$.

31. $\dfrac{x}{2} + \dfrac{y}{3} = 1$

    Solving for y, we get

    $y = -\dfrac{3}{2}x + 3$, thus the slope of

    the line is -3/2. The point-slope equation of the line through (1,6) and parallel to the given line is

    $$y - 6 = -\dfrac{3}{2}(x - 1)$$

    or, in standard form,
    $3x + 2y - 15 = 0$

33. Since (-1,3) lies on the line $y = 5x + b$, it follows that

    $$3 = 5(-1) + b$$
    $$b = 3 + 5 = 8$$

35. $5.17x - 3.21y + 1.72 = 0$
    $3.21y = 5.17x + 1.72$

    $$y = \dfrac{5.17}{3.21}x + \dfrac{1.72}{3.21}$$

    Thus the slope is

    $$m = \dfrac{5.17}{3.21} \approx 1.61$$

    and the y-intercept is

    $$b = \dfrac{1.72}{3.21} \approx 0.54.$$

37. If a denotes the x- and y-intercepts, the the points $(a,0)$ and $(0,a)$ lie on the line. It follows that the slope of the line is

    $$m = \dfrac{0 - a}{a - 0} = \dfrac{-a}{a} = -1.$$

    Since the point (3,5) lies on the line, its point-slope equation is

    $$y - 5 = -(x - 3).$$

Or
$x + y - 8 = 0.$

39. Making a table for v and t, we get

t	v
0	15
1.5	45

    We are looking for an equation of a line which contains the points (0,15) and (1.5,45). The slope is

    $$m = \dfrac{45 - 15}{1.5 - 0} = 20.$$

    The equation is
    $$v - 15 = 20(t - 0)$$
    $$v = 20t + 15$$

    If $v = 85$, then
    $$85 = 20t + 15$$
    $$20t = 70$$
    $$t = 3.5$$

41. When $t = 0$, V = \$20,000. When $t = 1$, V = 20,000 + 0.06 × 20,000 = \$21,200. We want an equation of a line passing through the points (0,20000) and (1,21200). The slope of the line is $m = 21200 - 20000 = 1200$, and an equation is

    $$V - 20000 = 1200(t - 0)$$
    $$V - 20000 = 1200t$$
    $$V = 1200t + 20000$$

43. Since the center is the point (0,0) and the circle passes through (-1,3), we get the equation

    $$(-1 - 0)^2 + (3 - 0)^2 = r^2$$
    $$1 + 9 = r^2$$
    $$10 = r^2.$$

    Thus, the equation of the circle is $x^2 + y^2 = 10$.

45. Since the center is the point $(-1,1)$ and the circle passes through $(2,-3)$, we get the equation

$$(2 + 1)^2 + (-3 - 1)^2 = r^2$$
$$9 + 16 = r^2$$
$$25 = r^2.$$

Thus the equation of the circle is

$$(x + 1)^2 + (y - 1)^2 = 25.$$

47. If the center $C(a,b)$ is in the third quadrant, then $a < 0$ and $b < 0$. Since the circle is tangent to both axes and its radius is $\sqrt{10}$, it follows that $a = b = -\sqrt{10}$. Thus the center-radius equation of the circle is

$$(x + \sqrt{10})^2 + (y + \sqrt{10})^2 = 10.$$

49. $x^2/2 + y^2/2 = 8$
$x^2 + y^2 = 16$
$(x - 0)^2 + (y - 0)^2 = 4^2$
Thus $C(0,0)$ and $r = 4$.

51. $x^2 + y^2 - 3y - 1 = 0$
$x^2 + (y^2 - 3y) = 1$

$$x^2 + (y^2 - 3y + \frac{9}{4}) = 1 + \frac{9}{4}$$

$$x^2 + (y - \frac{3}{2})^2 = \frac{13}{4}$$

$$(x - 0)^2 + (y - \frac{3}{2})^2 = (\frac{\sqrt{13}}{2})^2.$$

Therefore

$$C(0,\frac{3}{2}) \text{ and } r = \frac{\sqrt{13}}{2}.$$

53. $x^2 + y^2 + 3x + 5y - \frac{1}{2} = 0$

$$(x^2 + 3x) + (y^2 + 5y) = \frac{1}{2}$$

$$(x^2 + 3x + \frac{9}{4}) + (y^2 + 5y + \frac{25}{4})$$

$$= \frac{1}{2} + \frac{9}{4} + \frac{25}{4}$$

$$(x + \frac{3}{2})^2 + (y + \frac{5}{2})^2 = \frac{36}{4} = 9$$

$$\text{or } (x + \frac{3}{2})^2 + (y + \frac{5}{2})^2 = 3^2.$$

Thus $C(-\frac{3}{2},-\frac{5}{2})$ and $r = 3$.

55. If the focus is $F(0,3)$ and the directrix is the horizontal line $y = -3$, then the principal axis of the parabola is the y-axis. The vertex has coordinates

$$(0,\frac{3 + (-3)}{2}) = (0,0)$$

and $p = 3$. Thus the equation is

$$(x - 0)^2 = 4(3)(y - 0)$$
$$x^2 = 12y.$$

57. Since the vertex is $(0,0)$ and the principal axis is horizontal, it follows that the principal axis is the x-axis. If $(2,4)$ is on the parabola, then

$$(4 - 0)^2 = 4p(2 - 0)$$
$$16 = 8p$$
$$p = 2.$$

Thus the parabola has equation

$$(y - 0)^2 = 4(2)(x - 0)$$
$$y^2 = 8x.$$

59. If the vertex is (2,1) and the pricipal axis is vertical, then the parabola opens either upward or downward. Since (0,4) is on the parabola, we get the equation

$$(0 - 2)^2 = 4p(4 - 1)$$
$$4 = 4p(3)$$
$$4 = 12p$$
$$1/3 = p.$$

Thus the parabola has equation

$$(x - 2)^2 = 4(1/3)(y - 1)$$
$$(x - 2)^2 = (4/3)(y - 1).$$

61. Solving the equation $3x - 2y - 1 = 0$ for y we get $y = 3/2x - 1/2$. To find the intersection of the parabola $y = x^2$ and the line $y = 3/2x - 1/2$, set $x^2 = 3/2x - 1/2$ and solve for x:

$$2x^2 = 3x - 1$$
$$2x^2 - 3x + 1 = 0$$
$$(2x - 1)(x - 1) = 0$$
$$x = 1/2 \quad x = 1$$

If $x = 1/2$, $y = 1/4$ and if $x = 1$, $y = 1$. The points of intersection are (1,1) and (1/2,1/4).

63. Since C(-1,0) and (-1,2) is a focus, it follows that $c = 2 - 0 = 2$ and the major axis lies on the vertical line $x = -1$. Now $a = 3$ and $b^2 = a^2 - c^2 = 9 - 4 = 5$. Thus the equation is

$$\frac{(x + 1)^2}{5} + \frac{y^2}{9} = 1$$

65. The fact that the foci are (-2,3) and (4,3) indicate that the major axis lies on the horizontal line $y = 3$. The coordinates of the center (midpoint of the segment joining the foci) are (1,3). We have $c = 4 - 1 = 3$ and $b^2 = a^2 - c^2 = 5^2 - 3^2 = 16$. Thus the equation of the ellipse is

$$\frac{(x - 1)^2}{25} + \frac{(y - 3)^2}{16} = 1$$

67. Since the vertices are V($\pm 5/3$,0), it follows that C(0,0), $a = 5/3$, and the major axis lies on the line $y = 0$. Since the point (1,1) is on the ellipse, we get the equation

$$\frac{(1 - 0)^2}{(5/3)^2} + \frac{(1 - 0)^2}{b^2} = 1$$

$$\frac{9}{25} + \frac{1}{b^2} = 1$$

$$9b^2 + 25 = 25b^2$$
$$25 = 16b^2$$

$$\frac{25}{16} = b^2.$$

The ellipse has the equation

$$\frac{(x - 0)^2}{\frac{25}{9}} + \frac{(y - 0)^2}{\frac{25}{16}} = 1 \text{ or}$$

$$9x^2 + 16y^2 = 25.$$

69. If the foci are F(0,± 5), then C(0,0), and

$$c = \frac{d(F_1, F_2)}{2} = \frac{10}{2} = 5.$$

Since the vertices are V(0,± 3), it follows that

$$a = \frac{d(V_1, V_2)}{2} = 3$$

So $b^2 = c^2 - a^2 = 25 - 9 = 16$ and the hyperbola has equation

$$\frac{y^2}{9} - \frac{x^2}{16} = 1$$

71. If $(-5,0)$ is a focus, $(-2,0)$ is a vertex, and $(0,0)$ is the center of the hyperbola, then $c = 5$, $a = 2$, and $b^2 = c^2 - a^2 = 25 - 4 = 21$. Thus the equation is

$$\frac{x^2}{4} - \frac{y^2}{21} = 1$$

73. Since $(10,0)$ is a focus and $(0,0)$ the center, it follows that $c = 10$. Now

$$e = \frac{c}{a} = \frac{10}{a} = \frac{5}{4}$$

so $40 = 5a$
$8 = a$.

Next, $b^2 = c^2 - a^2 = 100 - 64 = 36$.

$$\frac{x^2}{64} - \frac{y^2}{36} = 1$$

75. $x^2 + y^2 - 4x - 6y + 4 = 0$
$(x^2 - 4x + 4) + (y^2 - 6y + 9)$
$\quad = -4 + 4 + 9$
$(x - 2)^2 + (y - 3)^2 = 9$
$C(2,3)$, $r = 3$

77. $9x^2 + 4y^2 + 18x + 8y - 12 = 0$
$9(x^2 + 2x + 1) + 4(y^2 + 2y + 1)$
$\quad = 12 + 9 + 4$
$9(x + 1)^2 + 4(y + 1)^2 = 25$

$$\frac{(x + 1)^2}{25/9} + \frac{(y + 1)^2}{25/4} = 1$$

$C(-1,-1)$, $a = 5/2$, $b = 5/3$

79. $y^2 - 6y - 3x = 0$
$y^2 - 6y = 3x$
$y^2 - 6y + 9 = 3x + 9$
$(y - 3)^2 = 3(x + 3)$
$(y - 3)^2 = 4(3/4)(x + 3)$

Parabola opening to the right,
$p = 3/4$, $V(-3,3)$.

81. $x^2 - 4y^2 - 2x - 8y - 12 = 0$
$(x^2 - 2x) - 4(y^2 + 2y) = 12$
$(x^2 - 2x + 1) - 4(y^2 + 2y + 1)$
$\quad = 12 + 1 - 4$
$(x - 1)^2 - 4(y + 1)^2 = 9$

$$\frac{(x - 1)^2}{9} - \frac{(y + 1)^2}{9/4} = 1$$

$C(1,-1)$, $a = 3$, $b = 3/2$

83. The parabola $y = -2x + bx - 5$ has only one x-intercept when the equation $-2x^2 + bx - 5 = 0$. This happens when the discriminant $b^2 - 4ac$ is equal to zero:

$b^2 - 4(-2)(-5) = 0$
$b^2 = 40$

$b = \pm\sqrt{40} = \pm 2\sqrt{10}$

85. $y = 4 - (x - 1)^2$, $y = x^2 - 2x - 3$
Set $x^2 - 2x - 3 = 4 - (x - 1)^2$
and solve it for x:

$x^2 - 2x - 3 = 4 - (x^2 - 2x + 1)$
$x^2 - 2x - 3 = 4 - x^2 + 2x - 1$
$2x^2 - 4x - 6 = 0$
$x^2 - 2x - 3 = 0$
$(x - 3)(x + 1) = 0$
$x = 3$, $x = -1$

Substituting into the first equation, obtain

$y = 4 - (3 - 1)^2 = 4 - 4 = 0$
and
$y = 4 - ((-1) - 1)^2 = 4 - 4 = 0$

The points of intersection are $(-1,0)$ and $(3,0)$

87. A(1,4), B(-5,2)

    The center of the circle is
C(-2,3) the midpoint of the
segment AB. The radius is

$$r = d(A,C)$$

$$= \sqrt{(1 - (-2))^2 + (4 - 3)^2}$$

$$= \sqrt{9 + 1} \; = \sqrt{10}$$

    Thus $(x + 2)^2 + (y - 3)^2 = 10$

89. The major axis of the orbit of
Mars has length $2a = 2.834 \times 10^8$,
thus $a = 1.417 \times 10^8$. If the
eccentricity $e = a/c = 9.3 \times 10^{-2}$,
then $c = ae$
  $= (1.417 \times 10^8)(9.3 \times 10^{-2})$
  $= 1.31781 \times 10^7$. At perihelion,
the distance from Mars to the Sun
is

$$a - c = 1.417 \times 10^8 - 1.31781 \times 10^7$$
$$= 1.285219 \times 10^8$$
$$\approx 1.285 \times 10^8 \text{ miles}$$

At aphelion, the distance from
Mars to the Sun is

$$a + c = 1.417 \times 10^8 + 1.31781 \times 10^7$$
$$= 1.548781 \times 10^8$$
$$\approx 1.549 \times 10^8 \text{ miles}$$

Chapter 3 Test

1. If $A(x,y)$, $B(-4,6)$, and $M(-3,2)$ is the midpoint of the segment AB, find the coordinates of A.

2. Find all values of a so that the distance between the points $(a,3)$ and $(-6,7)$ is $\sqrt{137}$.

3. The points $A(6,-2)$, $B(x,10)$, $C(3,-4)$ are the vertices of a right triangle with right angle at A. Find x.

4. Show that the points $A(-2,2)$, $B(4,-2)$, $C(9,1)$, $D(3,5)$ are the vertices of a parallelogram.

5. Find a and b such that $(-1/2,3)$ is the midpoint of the segment whose endpoints are $(a,2)$ and $(5,b)$.

6. Verify whether or not the points $A(-5,1)$, $B(-2,-3)$, $C(2,0)$ are the vertices of a right triangle.

7. Consider the points A, B, C of Problem 6 and let $D(-1,y)$. Find y so that A, B, C, D are the vertices of a square.

8. Find a so that the slope of the line through the points $(a,-8)$ and $(4,a)$ is -5.

9. Find the slope-intercept equation of the line through the points $(-2,5)$ and $(4,7)$.

10. Find an equation of a line through the point $(-5,4)$ and parallel to the line $2x - 5y + 1 = 0$.

11. Find the slopes of the lines $5x - 6y + 4 = 0$ and $4x + 8y - 3 = 0$. Are the lines parallel, perpendicular, or neither?

12. Find a so that the line $ax - 4y - 3 = 0$ is perpendicular to the line $5x - 3y - 2 = 0$

13. Find the slope-intercept equation of the line through $(-2,-5)$ and perpendicular to the line $2x - 3y - 1 = 0$.

14. Let $P(-4,3)$ and $Q(1,-5)$. Find a) the midpoint M of the segment PQ, b) the slope of the line through P and Q, c) an equation of the line through M and perpendicular to the line through P and Q.

15. The variables X and Y are linearly related. Find an equation relating these variables knowing that $Y = -5$ when $X = 2$ and $Y = 4$ when $X = -2$.

16. The point $A(3,-2)$ lies on the circle of center $C(-4,2)$. Find an equation of the circle.

17. If A(5,-3) and B(3,2) are endpoints of a diameter of a circle, find an equation of the circle.

18. Graph the equation $x^2 - 6x - y + 9 = 0$ and check for symmetry.

19. Graph the equation $x^2 + y^2 - 4x + 2y - 31 = 0$.

20. Find the points of intersection of the parabola $y = x^2 - 2x$ and the line $2x + y - 1 = 0$.

21. For what values of k does the parabola $y = -3x^2 + 2x + k$ have two x-intercepts?

22. Find all values of b for which the parabola $y = 3x^2 + bx + 3$ does not intercept the x-axis.

23. Determine the vertex, focus, principal axis, and directrix of the parabola $(y - 3)^2 = 2(x - 1)$.

24. Find an equation of the parabola whose vertex is (3,0) and directrix is $y = 2$.

25. Determine the center and the major and minor axes of the ellipse $4x^2 + 8x + 16y^2 - 12 = 0$.

26. An ellipse has center at (0,0), a focus at (-4,0), and the length of the minor axis is 3. Find its equation in standard form.

27. Find an equation of the ellipse with center (2,-1), a focus (5,-1) and eccentricity 3/5.

28. Find an equation of a hyperbola corresponding to the following data: center at (0,0), a focus at (3,0), b = 1.

29. If the foci of a hyperbola are (2,6) and (2,-4), and a = 3, write its equation in standard form.

30. Show that the equation $4x^2 - y^2 - 8x - 2y - 1 = 0$ represents a hyperbola. Find its center and vertices.

Chapter 4
Functions

Exercises 4.1

1. For any given number x, we can always compute $3x - 4$, so the domain of $f(x) = 3x - 4$ is the set of all real numbers. Since every real number y can be written as

$$y = 3(\frac{y + 4}{3}) - 4,$$

   it follows that the range of f is the set of all real numbers.

3. We can always compute

$$\frac{2x}{3} + \frac{1}{5}$$

   for any real number x, so the domain of

$$F(x) = \frac{2x}{3} + \frac{1}{5}$$

   is the set of all real numbers. On the other hand, every real number y can be written as

$$y = \frac{2}{3}[\frac{3}{2}(y - \frac{1}{5})] + \frac{1}{5},$$

   so the range of F is the set of all real numbers.

5. The domain of $H(x) = 4 - x^2$ is the set of all real numbers. Since the range of $f(x) = -x^2$ is the set of all nonpositive numbers, it follows that the range of $H(x) = 4 - x^2$ is the set of all numbers y such that $y \leq 4$.

7. Domain: all real numbers.

   Range: all numbers $\geq$ -1.

9. The domain of $u(x) = \sqrt{x} - 3$ is the set of all nonnegative numbers. Since $\sqrt{x}$ is always ≥ 0, it follows that $u(x) = \sqrt{x} - 3 \geq -3$. Now, if $y \geq -3$, then $y + 3 \geq 0$ and we can write

$$y = \sqrt{(y + 3)^2} - 3.$$

   Thus the range of $u(x)$ is the set of all numbers $\geq$ -3.

11. Domain: all numbers ≥ 0.
    Range: all numbers ≥ 1.

13. Domain: all numbers $\geq 1/2$.
    Range: all numbers ≥ 0.

15. $f(-2) = 3(-2) - 4 = -10$

    $f(0) = 3(0) - 4 = -4$

    $f(\frac{2}{3}) = 3 \cdot \frac{2}{3} - 4 = -2$

    $f(\sqrt{1.5}) = 3\sqrt{1.5} - 4 \approx -0.33$

17. $F(-3) = 4 - (-3)^2 = -5$

    $F(-2) = 4 - (-2)^2 = 0$

    $F(1) = 4 - 1^2 = 3$

    $F(\sqrt{3}) = 4 - (\sqrt{3})^2 = 1$

19. $h(2) = 3 - \sqrt{2} \approx 1.59$

    $h(16) = 3 - \sqrt{16} = -1$

    $h(3.5) = 3 - \sqrt{3.5} \approx 1.13$

    $h(-8)$ is not defined

    $h(1/4) = 3 - \sqrt{1/4}$

    $\qquad = 3 - 1/2 = 5/2$

21. $k(0) = \sqrt{0-4} = \sqrt{-4}$ not defined

    $k(5) = \sqrt{5 - 4} = \sqrt{1} = 1$

    $k(12) = \sqrt{12 - 4} = \sqrt{8} = 2\sqrt{2}$

    $k(7.25) = \sqrt{7.25 - 4}$

    $\qquad = \sqrt{3.25} \approx 1.80$

    $k(\frac{17}{4}) = \sqrt{\frac{17}{4} - 4} = \sqrt{\frac{17 - 16}{4}}$

    $\qquad = \sqrt{\frac{1}{4}} = 1/2$

Find the graphs for exercises 23 to 39 in the answer section of your text.

23. Domain: all real numbers

25. Domain: all real numbers

27. Domain: all real numbers

29. Domain: all real numbers

31. Domain: $[1, +\infty)$

33. Domain: $(-\infty, 4]$

35. Domain: $[2, +\infty)$

37. Domain: $[0, +\infty)$

39. Domain: $[0, +\infty)$

Exercises 4.2

1. Since $4 - 2x = 0$ when $x = 2$, it follows that the domain of

    $$f(x) = \frac{3}{4 - 2x}$$

    is the set of all real numbers except 2.

3. The square root function is only defined for nonnegative values of the radicand. So $4 - 3x \geq 0$, that is $x \leq 4/3$.

5. If $x = 1$, then $(x - 1)^2 = 0$. Thus $f(x) = x/(x - 1)^2$ is defined for all real numbers except the number 1.

7. $x^2 - 3x + 2 = (x - 1)(x - 2)$. So $x^2 - 3x + 2 = 0$ when $x = 1$ or $x = 2$. It follows that

    $$f(x) = \frac{2x + 1}{x^2 - 3x + 2}$$

    is defined for all real numbers except 1 and 2.

9. In order for $f(x) = \sqrt{1 - x^2}$ to be defined, we must have $1 - x^2 > 0$ and this holds for $-1 \leq x \leq 1$.

11. $f(1) = 3 \cdot 1 + 6 = 9$

 $f(0) = 3 \cdot 0 + 6 = 6$

 $f(\frac{2}{3}) = 3 \cdot \frac{2}{3} + 6 = 8$

 $f(\sqrt{1.5}) = 3 \cdot \sqrt{1.5} + 6 \approx 9.67$

13. $F(-1) = -(-1)^2 - (-1) - 1 = -1$

 $F(0) = -0^2 - 0 - 1 = -1$

 $F(1) = -1^2 - 1 - 1 = -3$

 $F(\sqrt{2}) = -(\sqrt{2})^2 - \sqrt{2} - 1 = -3 - \sqrt{2}$

 $\qquad = -(3 + \sqrt{2})$

15. $h(2) = \sqrt{2 \cdot 2 - 4} = \sqrt{4-4} = \sqrt{0} = 0$

 $h(10) = \sqrt{2 \cdot 10 - 4} = \sqrt{20 - 4}$

 $\qquad = \sqrt{16} = 4$

 $h(\frac{20}{9}) = \sqrt{2 \cdot \frac{20}{9} - 4}$

 $\qquad = \sqrt{\frac{40}{9} - 4} = \sqrt{\frac{4}{9}} = \frac{2}{3}$

 $h(\sqrt{3}) = \sqrt{2 \cdot \sqrt{3} - 4}$ not defined

17. $k(-4) = \dfrac{-4}{-4 + 3} = \dfrac{-4}{-1} = 4$

 $k(-2) = \dfrac{-2}{-2 + 3} = \dfrac{-2}{1} = -2$

 $k(0) = \dfrac{0}{0 + 3} = \dfrac{0}{3} = 0$

 $k(3.512) = \dfrac{3.512}{3.512 + 3} = \dfrac{3.512}{6.512}$

 $\qquad \approx 0.54$

19. a) $f(a) = 5 - 3a$

 b) $f(-a) = 5 - 3(-a) = 5 + 3a$

 c) $-f(a) = -(5 - 3a) = 3a - 5$

 d) $f(a) + f(b) = 5 - 3a + 5 - 3b$

 $\qquad = 10 - 3(a + b)$

 e) $f(a + b) = 5 - 3(a + b)$

21. a) $f(a) = a^2 - 4$

 b) $f(-a) = (-a)^2 - 4 = a^2 - 4$

 c) $-f(a) = -(a^2 - 4) = 4 - a^2$

 d) $f(a) + f(b) = a^2 - 4 + b^2 - 4$

 $\qquad = a^2 + b^2 - 8$

 e) $f(a + b) = (a + b)^2 - 4$

23. a) $f(a) = \dfrac{a - 1}{a + 1}$

 b) $f(-a) = \dfrac{-a - 1}{-a + 1} = \dfrac{-(a + 1)}{-(a - 1)}$

 $\qquad = \dfrac{a + 1}{a - 1}$

 c) $-f(a) = -\dfrac{a - 1}{a + 1} = \dfrac{-a + 1}{a + 1}$

 d) $f(a) + f(b) = \dfrac{a - 1}{a + 1} + \dfrac{b - 1}{b + 1}$

 $\qquad = \dfrac{2ab - 2}{(a + 1)(b + 1)}$

 $\qquad = \dfrac{2(ab - 1)}{(a + 1)(b + 1)}$

 e) $f(a + b) = \dfrac{a + b - 1}{a + b + 1}$

25. a) $h(\frac{1}{a}) = \dfrac{1}{1/a - 5}$

$\qquad = \dfrac{1}{(1 - 5a)/a}$

$\qquad = \dfrac{a}{1 - 5a}$

b) $\dfrac{1}{h(a)} = \dfrac{1}{\dfrac{1}{a - 5}} = a - 5$

c) $h(a) + h(1/a)$

$\qquad = \dfrac{1}{a - 5} + \dfrac{a}{1 - 5a}$

$\qquad = \dfrac{a^2 - 10a + 1}{(a - 5)(1 - 5a)}$

d) $h(a + \dfrac{1}{a}) = \dfrac{1}{a + 1/a - 5}$

$\qquad = \dfrac{1}{(a^2 - 5a + 1)/a}$

$\qquad = \dfrac{a}{a^2 - 5a + 1}$

27. a) $h(\frac{1}{a}) = \dfrac{1}{2(1/a) - 1}$

$\qquad = \dfrac{1}{(2 - a)/a} = \dfrac{a}{2 - a}$

b) $\dfrac{1}{h(a)} = \dfrac{1}{\dfrac{1}{2a - 1}} = 2a - 1$

c) $h(a) + h(\dfrac{1}{a})$

$\qquad = \dfrac{1}{2a - 1} + \dfrac{a}{2 - a}$

$\qquad = \dfrac{2a^2 - 2a + 2}{(2a - 1)(2 - a)}$

d) $h(a + \dfrac{1}{a}) = \dfrac{1}{2(a + \dfrac{1}{a}) - 1}$

$\qquad = \dfrac{1}{(2a^2 + 2 - a)/a}$

$\qquad = \dfrac{a}{2a^2 + 2 - a}$

29. a) $h(\frac{1}{a}) = 1 - (\frac{1}{a})^2 = 1 - \dfrac{1}{a^2}$

$\qquad = \dfrac{a^2 - 1}{a^2}$

b) $\dfrac{1}{h(a)} = \dfrac{1}{1 - a^2}$

c) $h(a) + h(\dfrac{1}{a})$

$\qquad = 1 - a^2 + \dfrac{a^2 - 1}{a^2}$

$\qquad = \dfrac{-a^4 + 2a^2 - 1}{a^2}$

$\qquad = \dfrac{-(a^2 - 1)^2}{a^2}$

d) $h(a + \dfrac{1}{a}) = 1 - (a + \dfrac{1}{a})^2$

$\qquad = 1 - a^2 - 2 - \dfrac{1}{a^2}$

$\qquad = \dfrac{-a^4 - a^2 - 1}{a^2}$

31. a) $(f + g)(x) = f(x) + g(x)$
$\qquad = 3x + 5 + x^2 - 3$
$\qquad = x^2 + 3x + 2,$
$\qquad$ all real numbers

b) $(f - g)(x) = f(x) - g(x)$
$\qquad = 3x + 5 - (x^2 - 3)$
$\qquad = -x^2 + 3x + 8,$
$\qquad$ all real numbers

c) $(f \cdot g)(x) = f(x) \cdot g(x)$
$\qquad = (3x + 5)(x^2 - 3)$
$\qquad = 3x^3 + 5x^2 - 9x - 15,$
$\qquad$ all real numbers

d) $(f/g)(x) = \dfrac{f(x)}{g(x)} = \dfrac{3x + 5}{x^2 - 3}$,

all real numbers $\neq \pm \sqrt{3}$

$= -\dfrac{6}{x}$,

domain: all $x \neq 0$

33. a) $(f + g)(x)$

$= \dfrac{3x}{2x - 1} + \dfrac{x + 1}{5x + 3}$

$= \dfrac{17x^2 + 10x - 1}{(2x - 1)(5x + 3)}$

b) $(f - g)(x)$

$= \dfrac{3x}{2x - 1} - \dfrac{x + 1}{5x + 3}$

$= \dfrac{13x^2 + 8x + 1}{(2x - 1)(5x + 3)}$

c) $(f \cdot g)(x)$

$= (\dfrac{3x}{2x - 1})(\dfrac{x + 1}{5x + 3})$

$= \dfrac{3x^2 + 3x}{(2x - 1)(5x + 3)}$

Domains: all real numbers
$\neq 1/2$ and $\neq -3/5$.

d) $(f/g)(x) = \dfrac{3x/(2x - 1)}{(x + 1)/(5x + 3)}$

$= \dfrac{3x(5x + 3)}{(2x - 1)(x + 1)}$

Domains: all real numbers
different from 1/2 and -1.

35. a) $(f + g)(x)$

$= 1 - \dfrac{3}{x} + 1 + \dfrac{3}{x} = 2$,

domain: all real numbers

b) $(f - g)(x)$

$= 1 - \dfrac{3}{x} - (1 + \dfrac{3}{x})$

c) $(f \cdot g)(x) = (1 - \dfrac{3}{x})(1 + \dfrac{3}{x})$

$= 1 - \dfrac{9}{x^2}$,

domain: all $x \neq 0$

d) $(f/g)(x) = \dfrac{1 - 3/x}{1 + 3/x} = \dfrac{x - 3}{x + 3}$,

domain: all $x \neq -3$

37. $\dfrac{f(a + h) - f(a)}{h} = \dfrac{8 - 8}{h}$

$= \dfrac{0}{h} = 0$

39. $\dfrac{f(a + h) - f(a)}{h}$

$= \dfrac{3 \cdot (a + h) + 5 - (3a + 5)}{h}$

$= \dfrac{3a + 3h + 5 - 3a - 5}{h}$

$= \dfrac{3h}{h} = 3$

41. $\dfrac{f(a + h) - f(a)}{h}$

$= \dfrac{(a + h)^2 + 4 - (a^2 + 4)}{h}$

$= \dfrac{\cancel{a^2} + 2ah + h^2 + \cancel{4} - \cancel{a^2} - \cancel{4}}{h}$

$= \dfrac{2ah + h^2}{h}$

$= 2a + h$

43. $$\frac{f(a + h) - f(a)}{h}$$

$$= \frac{3(a+h)^2 - (a+h) - (3a^2 - a)}{h}$$

$$= \frac{3a^2 + 6ah + 3h^2 - a - h - 3a^2 + a}{h}$$

$$= \frac{(6a + 3h - 1)h}{h} = 6a + 3h + 1$$

45. $$\frac{f(a + h) - f(a)}{h}$$

$$= \frac{2(a+h)^2 - 3(a+h) - 1 - (2a^2 - 3a - 1)}{h}$$

$$= \frac{2a^2 + 4ah + 2h^2 - 3a - 3h - 1 - 2a^2 + 3a + 1}{h}$$

$$= \frac{h(2h + 4a - 3)}{h} = 2h + 4a - 3$$

47. If x is the width of the rectangle, the its length is $2x + 4$. So the area is expressed by the formula $A(x) = x(2x + 4)$.

49. The area of a square of side ℓ is given by $A = \ell^2$. Since the perimeter is $p = 4\ell$, it follows that $A = p^2/16$.

51. If x is the width of the rectangular box, then the length is $4x$ and the height is $x/2$. Thus the volume is

$$V(x) = x(4x)(x/2) = 2x^3.$$

53. $r = h/4$

$$V = \pi r^2 h$$

$$= \pi(h/4)^2 h$$

$$= \pi h^3/16$$

55. $V = \pi r^2 h = 16\pi$, so that $r^2 = 16/h$ and $r = 4/\sqrt{h}$. Since $L = 2\pi rh$, it follows that

$$L(h) = 2\pi(4/\sqrt{h})h = 8\pi\sqrt{h}.$$

57. $$p(t) = 1,000 + (1,000) \cdot \frac{6}{100}t$$

that is $p(t) = 1,000 + 60t$.

59. Since $d_1 = 5d_2$ and

$$\frac{1}{f} = \frac{1}{d_1} + \frac{1}{d_2}$$

we have

$$\frac{1}{f} = \frac{1}{5d_2} + \frac{1}{d_2}.$$

Hence $f = \dfrac{5}{6}d_2$.

61. Let x be the number of rental increases. The revenue is the number of rented apartments times the monthly rent. Thus,

$$R = (250 + 10x)(100 - 2x).$$

63. If x denotes the number of nickels, then $x + 10$ is the number of dimes and $x + 15$ is the number of quarters. Thus,

$$A = 0.05x + 0.10(x + 10)$$
$$+ 0.25(x + 15)$$
$$= 0.4x + 4.75$$

65. Let w denote the width of the rectangular sheet of tin. The length of the box is $2w - 4$, the width is $w - 4$, and the height is 2. Thus,

$$V = 2(2w - 4)(w - 4).$$

67. a) Since $h = 3r$ and $L = 2\pi rh$, it follows that $L(h) = 2\pi h^2/3$

b) $L(0.25) = 2\pi(0.25)^2/3 \approx 0.13$
$L(1) = 2\pi/3 \approx 2.09$
$L(1.25) = 2\pi(1.25)^2/3 \approx 3.27$
$L(2.012) = 2\pi(2.012)^2/3 \approx 8.48$

69. $T(\ell) = 2\pi\sqrt{\ell/10}$

$T(1/2) = 2\pi\sqrt{(1/2)/10}$

$= 2\pi\sqrt{1/20} \approx 1.40$

$T(2) = 2\pi\sqrt{2/10} \approx 2.81$

$T(5) = 2\pi\sqrt{5/10} \approx 4.44$

Exercises 4.3

Find the graphs for exercises 1 through 43 in the answer section of your text.

45. $f(-x) = 3(-x)^4 + (-x)^2$
$= 3x^4 + x^2 = f(x),$
so $f(x)$ is an even function.

47. $F(-x) = 2(-x)^3 - 6(-x) = -2x^3 + 6x$
$= -(2x^3 - 6x) = -F(x),$
so $F(x)$ is an odd function.

49. $g(x)$ is even:
$g(-x) = (-x)^6 - 2(-x)^4 - 4$
$= x^6 - 2x^4 - 4 = g(x)$

51. $G(x)$ is even:
$G(-x) = |-x| + 2 = |x| + 2 = G(x)$

53. $h(-x) = |-x|^3 = |x|^3 = h(x),$
so h is even.

55. $H(-x) = 5/(-x) = -5/x = -H(x),$
so H is odd.

57. $k(-x) = 1/((-x)^2 - 1)$
$= 1/(x^2 - 1) = k(x),$
so k is even.

59. $f(x) = \dfrac{x^3 + 5x}{x^2 + 1}$

is an odd function:

$$f(-x) = \frac{(-x)^3 + 5(-x)}{(-x)^2 + 1}$$

$$= \frac{-x^3 - 5x}{x^2 + 1}$$

$$= - \frac{x^3 + 5x}{x^2 + 1}$$

$$= -f(x).$$

61. $f(x) = -5x + 3$ decreases on $(-\infty, +\infty)$.

63. $f(x) = x^2 - 4$ decreases on $(-\infty, 0)$ and increases on $(0, +\infty)$.

65. $f(x) = -|x - 3|$ increases on $(-\infty, 3)$ and decreases on $(3, +\infty)$.

67. $f(x) = x - |x|$ increases on $(-\infty, 0)$.

The graphs for exercises 71 and 73 are in the answer section of your text.

69. $V(x) = 16,000(1 + 0.08x)$

75. If f and g are even functions, then $f(-x) = f(x)$ and $g(-x) = g(x)$. Now,

$(f + g)(-x) = f(-x) + g(-x)$
$= f(x) + g(x)$
$= (f + g)(x),$

thus $f + g$ is an even function. A similar proof holds for the difference $f - g$.

77. If f and g are odd functions, then $f(-x) = -f(x)$ and $g(-x) = -g(x)$. Now,

$(f \cdot g)(-x) = f(-x) \cdot g(-x)$
$= [-f(x)] \cdot [-g(x)]$
$= f(x) \cdot g(x)$
$= (f \cdot g)(x),$

thus $f \cdot g$ is an even function. A similar argument applies for the quotient f/g.

79. $F(-x) = \frac{1}{2}[f(-x) + f(-(-x))]$

    $= \frac{1}{2}[f(-x) + f(x)] = F(x)$,

    so $F(x)$ is even.

    $G(-x) = \frac{1}{2}[f(-x) - f(-(-x))]$

    $= \frac{1}{2}[f(-x) - f(x)]$

    $= -\frac{1}{2}[f(x) - f(-x)] = -G(x)$,

    so G is an odd function.

Exercises 4.4

1. $u = kv$
   $18 = k(6)$
   $3 = k$
   $u = 3v$

3. $y = k/\sqrt{x}$

   $1/6 = k/\sqrt{4}$

   $\frac{1}{6} = \frac{k}{2}$

   $\frac{1}{3} = k$

   $y = 1/3\sqrt{x}$

5. $y = k/x^2$
   $27 = k/(1/9)^2$
   $27 = 81k$
   $1/3 = k$
   $y = 1/3x^2$

7. $w = kxy$
   $6 = k(3)(4)$
   $6 = 12k$
   $1/2 = k$

   $w = \frac{xy}{2}$

   When $x = 1$ and $y = 6$, obtain

   $w = \frac{(1)(6)}{2} = 3.$

9. $p = kq^2/r$

   $\frac{9}{4} = k\frac{9}{2}$

   $\frac{1}{2} = k$

   $p = \frac{q^2}{2r}$

   When $q = 2$ and $r = 3$, obtain

   $p = \frac{2^2}{2(3)} = \frac{4}{6} = \frac{2}{3}.$

11. $F = k\ell$

    When $F = 6$ pounds, $\ell = 2$ inches,
    so $6 = k(2)$
    $k = 3$
    and
    $F = 3\ell.$
    Now, when $F = 28$ pounds, we have
    $28 = 3\ell$
    so
    $\ell = 28/3$ inches.

13. If I denotes the intensity of
    illumination and d the distance
    from the point to the source, then
    $I = k/d^2$. When $d = 20$ ft, $I = 40$
    foot-candles, so $40 = k/(20)^2$ and
    $k = 16,000$. Hence $I = 16,000/d^2$.
    When $d = 8$ ft, we get
    $I = 16,000/8^2 = 250$ foot-candles.

15. $E = kAT^4$

17. $\ell = kv^2$.
$v = 40$ mi/h,
$\ell = 80$ ft $= 80/5280$ mi $= 1/66$ mi
$1/66 = k(40)^2$

If x denotes the length of the skid marks when $v = 60$ mi/h, then $x = k(60)^2$.

Dividing the last equation by the former, obtain

$$\frac{x}{1/66} = \frac{\cancel{k}(60)^2}{\cancel{k}(40)^2}$$

$$66x = \frac{3600}{1600} = \frac{9}{4}$$

$$x = \frac{9}{(4)(66)} = \frac{3}{88}\text{mi} = 180 \text{ ft}$$

19. a) $f = k\sqrt{T}/\ell$

b) $\dfrac{k\sqrt{4T}}{2\ell} = \dfrac{\cancel{2}k\sqrt{T}}{\cancel{2}\ell} = \dfrac{k\sqrt{T}}{\ell}$,

thus the frequency remains the same.

21. $V = kr^3$.

When $r = 3$ in, $V = 36\pi$ in^3, so
$36\pi = k(3)^3$
$27k = 36\pi$
$k = 36\pi/27 = 4\pi/3$.

Thus $V = \dfrac{4}{3}\pi r^3$. If $r = 2.25$ in,

then

$$V = \frac{4}{3}\pi(2.25)^3 \simeq 47.7 \text{ in}^3.$$

23. $w = k/d^2$, where w is the weight and d is the distance from the center of the earth. If $w = 175$ when $d = 3959$, then

$$175 = \frac{k}{(3959)^2}$$

$$k = (175)(3959)^2.$$

When the astronaut is 150 mi above the surface of the earth, $d = 3959 + 150 = 4109$, so

$$w = \frac{(175)(3959)^2}{(4109)^2}$$

$$= 162.46 \text{ lb} \simeq 162 \text{ lb}$$

25. $K = kmv^2$

a) When $m = 4$ and $v = 10$, $K = 200$, so
$200 = k(4)(10^2)$
$200 = k(400)$
$k = 1/2$.

Hence $K = \dfrac{1}{2}mv^2$

b) If $m = 6.80$ and $v = 18.5$, then

$$K = \frac{1}{2}(6.80)(18.5)^2$$

$$\simeq 1.16 \times 10^3$$

27. $T = k\sqrt{\ell/g}$

a) $2 = k\sqrt{\dfrac{g/\pi^2}{g}} = \dfrac{k}{\pi}$

$2\pi = k$

$T = 2\pi\sqrt{\ell/g}$

b) $\ell = 50$ cm $= 0.5$ m, $g = 9.8$ m/s^2

$$T = 2\pi\sqrt{\frac{0.5}{9.8}} \simeq 1.42s.$$

29. $T = kd^{3/2}$
$T = 365$ days, $d = 9.29 \times 10^7$ mi
$365 = k(9.29 \times 10^7)^{3/2}$

$$k = \frac{365}{(9.29 \times 10^7)^{3/2}}$$

$\approx 4.08 \times 10^{-10}$
$T = 4.08 \times 10^{-10}d^{3/2}$
If $d = 4.83 \times 10^8$ mi, obtain
$T = (4.08 \times 10^{-10})(4.83 \times 10^8)^{3/2}$
$\approx 4.33 \times 10^3$ days
≈ 11.9 years.

Review Exercises - Chapter 4

1. Since $5 - 4x = 0$ when $x = 5/4$, it follows that the domain of

$$f(x) = \frac{3x + 1}{5 - 4x}$$

is the set of all real numbers except 5/4.

3. In order for the square root to be defined, we must have $6 - 9x \geq 0$, that is $x \leq 2/3$.

5. $(x - 2)(3x + 5) \neq 0$ when $x \neq 2$ and $x \neq -5/3$. Thus the domain of the function is the set of all real numbers except 2 and -5/3.

7. $x^2 - x - 2 = (x - 2)(x + 1)$. Thus the function is defined for all real numbers $x \neq 2$ and $x \neq -1$.

9. $2x^2 - 18 = 2(x^2 - 9)$
$= 2(x - 3)(x + 3)$
Now $2x^2 - 18 \geq 0$ when $x \leq -3$ or $x \geq 3$.

11. The function is defined when

$$\frac{x - 1}{x + 2} \geq 0.$$

Solving this inequality, check that the solution set consists of all numbers x such that $x < -2$ and $x \geq 1$.

13. We must have

$$\frac{-x}{x^2 + 1} \geq 0.$$

Since $x^2 + 1 > 0$, it follows that the inequality holds for all $x \leq 0$. Thus the function is defined on the interval $(-\infty, 0]$.

15. $f(\dfrac{-7}{10}) = 5(\dfrac{-7}{10}) - 1$

$= \dfrac{-7}{2} - 1 = \dfrac{-9}{2}$

$f(\dfrac{2}{5}) = 5 \cdot \dfrac{2}{5} - 1 = 1$

$f(3) = 5 \cdot 3 - 1 = 14$
$f(3.12) = 5 \cdot (3.12) - 1 = 14.6$

17. $f(-5) = 3 - 2(-5)^2 = 3 - 50 = -47$
$f(0) = 3 - 2 \cdot 0^2 = 3$
$f(5) = 3 - 2 \cdot 5^2 = -47$

$f(\sqrt{10}) = 3 - 2(\sqrt{10})^2 = -17$

19. $f(-7) = \sqrt{4 - 3(-7)} = \sqrt{25} = 5$

$f(-15) = \sqrt{4 - 3(-15)} = \sqrt{49} = 7$

$f(\dfrac{11}{12}) = \sqrt{4 - 3 \cdot \dfrac{11}{12}}$

$= \sqrt{\dfrac{15}{12}} = \dfrac{\sqrt{5}}{2}$

$f(2.5) = \sqrt{4 - 3 \cdot (2.5)} = \sqrt{-3.5}$

not defined

21. $f(-3) = \dfrac{-2(-3)}{(-3)^2 + 1} = \dfrac{3}{5}$

    $f(0) = \dfrac{-2 \cdot 0}{0 + 1} = 0$

    $f(3) = \dfrac{-2 \cdot 3}{9 + 1} = \dfrac{-3}{5}$

    $f(4.25) = \dfrac{-2 \cdot (4.25)}{(4.25)^2 + 1}$

    $= -0.45$

23. $f(1) = \sqrt{\dfrac{1 - 1}{1 - 3}} = 0$

    $f(-10) = \sqrt{\dfrac{-10 - 1}{-10 - 3}} = \sqrt{\dfrac{11}{13}}$

    $f(7/2) = \sqrt{\dfrac{7/2 - 1}{7/2 - 3}} = \sqrt{\dfrac{5/2}{1/2}}$

    $= \sqrt{5}$

    $f(9.5) = \sqrt{\dfrac{9.5 - 1}{9.5 - 3}} = \sqrt{\dfrac{8.5}{6.5}}$

    $= 1.14$

25. $\dfrac{f(x + h) - f(x)}{h}$

    $= \dfrac{3 - 7(x + h) - (3 - 7x)}{h}$

    $= -7$

27. $\dfrac{f(x + h) - f(x)}{h}$

    $= \dfrac{5(x+h)^2 - (x+h) - (5x^2 - x)}{h}$

    $= 10x + 5h - 1$

29. $\dfrac{f(x + h) - f(x)}{h}$

    $= \dfrac{2/(x + h) - 2/x}{h}$

    $= \dfrac{1}{h} \cdot \dfrac{2x - 2(x + h)}{x(x + h)}$

    $= \dfrac{-2}{x(x + h)}$

31. $\dfrac{f(x + h) - f(x)}{h}$

    $= \dfrac{1}{h} \cdot \left(\dfrac{x + h - 1}{x + h + 1} - \dfrac{x - 1}{x + 1} \right)$

    $= \dfrac{2h}{h(x + 1)(x + h + 1)}$

    $= \dfrac{2}{(x + 1)(x + h + 1)}$

33. $f(a + b) = \dfrac{a + b + a}{a + b - a} = \dfrac{2a + b}{b}$

    $f(1/a) + f(1/b)$

    $= \dfrac{1/a + a}{1/a - a} + \dfrac{1/b + a}{1/b - a}$

    $= \dfrac{1 + a^2}{1 - a^2} + \dfrac{1 + ab}{1 - ab}$

    $= \dfrac{2(1 - a^3 b)}{(1 - a^2)(1 - ab)}$

    $f\left(\dfrac{1}{a} + \dfrac{1}{b}\right) = \dfrac{1/a + 1/b + a}{1/a + 1/b - a}$

    $= \dfrac{a + b + a^2 b}{a + b - a^2 b}$

35. $(f(a + b))^2 - (f(a - b))^2$
    $= (a + b)^2 - (a - b)^2$
    $= a^2 + 2ab + b^2 - a^2 + 2ab - b^2$
    $= 4ab = 4f(a)f(b)$

37. $A = bh/2$. If $b = h/3$, then
    $A(h) = h^2/6$.

39. The lateral surface area of a
    cylinder is given by the formula
    $L = 2\pi rh$. If $h = 5r$, then
    $L(r) = 2\pi r(5r)$
    $= 10\pi r^2$.

Find the art for exercises 41 through
53 in the answer section of your text.

55. $f(-x) = (-x)^3 - 6(-x) = -x^3 + 6x$
$$= -(x^3 - 6x) = -f(x),$$
so $f(x)$ is odd.

57. $f(-x) = (-x)((-x)^2 + 2)$
$$= -x(x^2 + 2) = -f(x),$$
thus $f(x)$ is odd.

59. $f(-x) = \dfrac{(-x)^2}{2} + |-x|^3$
$$= \dfrac{x^2}{2} + |x|^3 = f(x),$$
so $f(x)$ is even.

61. $f(-x) = \dfrac{|-x|}{(-x)^2 - 1} = \dfrac{x}{x^2 - 1}$
$$= f(x), \text{ so } f \text{ is even.}$$

63. Increasing for all x

65. Decreasing on the interval $(-\infty, 0)$ and increasing on $(0, +\infty)$

67. Increasing on $(-\infty, -3)$ and decreasing on $(-3, +\infty)$

69. Increasing on $(4, +\infty)$

71. If P denotes the pressure and d the depth, then $P = kd$.

73. If w denotes the weight, h the height, and r the radius of the cylinder, the $w = khr^2$.

75. We have $y = k\sqrt{x}$ and $3 = k\sqrt{4} = 2k$, thus $k = 3/2$. Hence $y = 3\sqrt{x}/2$. If $x = 12$, then $y = 3\sqrt{12}/2 = 3\sqrt{3}$.

77. $F = kwv^2$
If the velocity were doubled, then the new striking force would be
$F_1 = kw(2v)^2 = 4kwv^2 = 4F.$
Thus the striking force would increase by a factor of 4.

79. If n denotes the number of gallons and t the time, then $n = kt$. We know that the car uses 4 gallons in 1 hour and 20 minutes (= 80 minutes) of travelling, thus

$$4 = k(80).$$

If x is the number of gallons it uses in 4.5 hours (= 270 minutes) of travelling, then $x = k(270)$. Dividing this equation by the previous one, we get

$$\frac{x}{4} = \frac{k(270)}{k(80)}$$

$$x = \frac{(270)(4)}{80}$$

$$= 13.5 \text{ gallons}$$

Chapter 4 Test

Give the domain and range of each of the following

1. $f(x) = 3x^2 - 5$ 2. $g(x) = \sqrt{18 - 2x^2}$ 3. $F(x) = |x - 4|$

4. $f(x) = \dfrac{1}{\sqrt{3 - 2x}}$

5. If $f(x) = 5x - x^2$, find $f(-2)$, $f(1/2)$, $f(2)$ and $f(f(3))$.

6. If $F(x) = \dfrac{1}{\sqrt{x - 2}}$, calculate each of the following, if possible.

   a) $F(11)$ b) $F(0)$

   c) $F(9/4)$ d) $F(F(19/9))$

7. $f(x) = x + \dfrac{1}{x}$.

   Find a) $f(a) + f(\dfrac{1}{a})$ b) $f(a) - f(\dfrac{1}{a})$

   c) $f(a) \cdot f(\dfrac{1}{a})$ d) $f(a + \dfrac{1}{a})$

8. Let $f(x) = \dfrac{x}{x + 1}$.

   Find a) $f(a) - f(b)$ b) $f(a/b)$ c) $f(a)/f(b)$

   d) $f(\dfrac{a}{a + 1})$

9. If $F(x) = \dfrac{1}{x - 5}$ and $G(x) = 1 - \dfrac{5}{x}$, find a) $(F - G)(x)$,

   b) $(F \cdot G)(x)$, and state their domains of definition.

10. Let $f(x) = \dfrac{x + 4}{x}$ and $g(x) = x - \dfrac{4}{x}$. Find a) $(f + g)(x)$,

   b) $(f/g)(x)$, and state their domains of definition.

For each of the given functions find $\dfrac{f(x + h) - f(x)}{h}$ and simplify your

your answer as completely as possible.

11. $f(x) = 3 - 5x$ 12. $f(x) = \dfrac{1}{x}$

For each of the given functions find $\dfrac{g(2 + h) - g(2)}{h}$ and simplify your answer as completely as possible.

13. $g(x) = 4x^2 + 3$ 14. $g(x) = \dfrac{x - 1}{x}$

15. The height of a triangle is 6 units longer than twice its base. Express the area of the triangle as a function of its height.

16. The length of a rectangle is twice its width. Express its area as a function of the length of the diagonal.

17. The volume of a sphere of radius r is given by the formula $V = \dfrac{4}{3}\pi r^3$. Write V as a function of the diameter d of the sphere.

18. Paul can do a painting job in t hours, while Mary can do the same job in two hours less. If they work together, express the fraction of the painting job that they can do in one hour as a function of t.

Use the techniques of stretching, shrinking, horizontal and vertical translations to graph each of the following functions.

19. $f(x) = 2(x - 1)^2 + 3$ 20. $f(x) = 2 - |x - 3|$

21. $f(x) = \sqrt{4 - x} - 1$

22. $f(x) = 1 - \dfrac{|x|}{2}$

Determine whether each of the following functions is even, odd, or neither.

23. $f(x) = x^3 - 6x$ 24. $f(x) = |x|^3 - 6|x|$

25. $f(x) = \dfrac{x^2}{x^3 + x}$ 26. $f(x) = \dfrac{x}{x^3 - 5x}$

Find all intervals where the following functions are decreasing or increasing.

27. $f(x) = 3 - (x - 1)^2$ 28. $f(x) = |x + 2| - 1$

29. Express as an equation the following statement: "w is directly proportional to the square of x and inversely proportional to y and z." What happens to w if only x and y are doubled?

30. Write an equation for the following statement: "the cost of harvesting a field varies directly as the area of the field and the cost of labor, and inversely as the number of days of good weather."

Chapter 5

Polynomial and Rational Functions; Composite and Inverse Functions

Exercises 5.1

Find the corresponding graphs in the answer section of your text.

1. a) $f(x) = 3x^2$
   vertex: $(0, 0)$
   axis of symmetry: y-axis

   b) $f(x) = -\frac{1}{2}x^2$

   vertex: $(0, 0)$
   axis of symmetry: y-axis

   c) $f(x) = -4x^2$
   vertex: $(0, 0)$
   axis of symmetry: y-axis

   d) $f(x) = -2x^2$
   vertex: $(0, 0)$
   axis of symmetry: y-axis

3. a) $f(x) = (x - 2)^2$
   vertex: $(2, 0)$
   axis of symmetry: $x = 2$

   b) $f(x) = (x + 3)^2$
   vertex: $(-3, 0)$
   axis of symmetry: $x = -3$

   c) $f(x) = (x + 5)^2$
   vertex: $(-5, 0)$
   axis of symmetry: $x = -5$

   d) $f(x) = (x - 4)^2$
   vertex: $(4, 0)$
   axis of symmetry: $x = 4$

5. $\Delta = 0^2 - 4(3)(1) = -12 < 0$. Thus, the graph does not intercept the x-axis. Since $a = 3 > 0$, it follows the parabola opens upward.

7. $\Delta = (-8)^2 - 4(1)(20) = 64 - 80 = -16 < 0$. Thus, no x-intercept. Since $a = 1 > 0$ we conclude that the parabola opens upward.

9. $\Delta = 7^2 - 4(-2)(4) = 49 + 32 = 81 > 0$ and $a = -2 < 0$. thus, there are two x-intercepts and the parabola opens downward.

11. $\Delta = (-12)^2 - 4(9)(4) = 0$, $a = 9$. The parabola is tangent to the x-axis and opens upward.

13. $\Delta = (-2)^2 - 4(-1)(-3) = -8 < 0$, $a = -1$. No x-intercept, parabola opens downward.

15. $f(x) = x^2 - 9 = (x - 0)^2 - 9$. The parabola has vertex $V(0, -9)$ and the axis of symmetry is $x = 0$.

17. $f(x) = x^2 - 6x + 8$
$= (x^2 - 6x + 9) + 8 - 9$
$= (x - 3)^2 - 1$

The vertex is $(3, -1)$ and the axis of symmetry is the line $x = 3$.

19. $f(x) = x^2 - 6x + 9$
$= (x - 3)^2$

The parabola has vertex $(3, 0)$ and the line $x = 3$ is the axis of symmetry. Note that the curve is tangent to the x-axis at the point $(3, 0)$.

21. $f(x) = 2x^2 + 1$
$= 2(x - 0)^2 + 1$

So the graph of f is a parabola with vertex at $(0, 1)$ and axis of symmetry $x = 0$.

23. $f(x) = 2x^2 - x + 1$

$= 2(x^2 - \frac{1}{2}x + \frac{1}{16}) + 1 - \frac{1}{8}$

$= 2(x - \frac{1}{4})^2 + \frac{7}{8}$

Vertex $V(\frac{1}{4}, \frac{7}{8})$,

axis of symmetry $x = \frac{1}{4}$.

25. $f(x) = -x^2 + 2x$
$= -(x^2 - 2x)$
$= -(x^2 - 2x + 1) + 1$
$= -(x - 1)^2 + 1$

The graph of f is a parabola with vertex $(1, 1)$ and axis of symmetry $x = 1$. Note that $a = -1$, so the parabola opens downward.

27. $f(x) = -2x^2 + 8x - 11$
$= -2(x^2 - 4x) - 11$
$= -2(x^2 - 4x + 4) - 11 + 8$
$= -2(x - 2)^2 - 3$

The vertex has coordinates $(2, -3)$, the axis of symmetry is $x = 2$, and the parabola opens downward $(a = -2)$.

29. $f(x) = x^2 + 8x + c$

The graph is tangent to the x-axis if and only if the equation $x^2 + 8x + c = 0$ has only one solution. This occurs when the discriminant is zero:

$\Delta = 64 - 4c = 0$,

That is when $c = 16$.

31. $f(x) = 2x^2 + bx + 2$

The graph does not cross the x-axis when the equation $2x^2 + bx + 2 = 0$ does not have real solutions. This happens when $\Delta \leq 0$, that is

$\Delta = b^2 - 4(2)(2) = b^2 - 16 \leq 0$.

Solving the last inequality, we obtain $-4 \leq b \leq 4$.

33. Let x and y be two positive numbers such that $x + y = 36$ and let $P = xy$ be their product. By solving the first equation for y and substituting into the expression of P, we obtain

$$
\begin{aligned}
P &= x(36 - x) \\
&= -x^2 + 36x \\
&= -(x^2 - 36x) \\
&= -(x^2 - 36x + 324) + 324 \\
&= -(x - 18)^2 + 324
\end{aligned}
$$

Thus P represents a parabola opening downward. It follows that the maximum of P is attained at the vertex $(18,324)$. So the numbers are 18 and 18 and the maximum product is 324.

35. If x and y are the dimensions of the field, then $2x + 2y = 1260$ or $x + y = 630$. We want the area $A = xy = x(630 - x)$ to be a maximum. By completing the square, we can rewrite this quadratic function as follows

$$
\begin{aligned}
A &= -x^2 + 630x \\
&= -(x^2 - 630x + (315)^2) + (315)^2 \\
&= -(x - 315)^2 + 99,225
\end{aligned}
$$

Thus the maximum occurs when $x = 315$ (and $y = 630 - 315 = 315$) and the area of the field is $99,225 \ m^2$.

37. Let ℓ be the length and w be the width of the rectangular plot.

$$
\begin{aligned}
A &= \ell w \\
P &= 3w + \ell = 90 \\
\ell &= 90 - 3w \\
A &= (90 - 3w)w \\
&= 90w - 3w^2 \\
&= -3(w^2 - 30w) \\
&= -3(w^2 - 30w + 225) + 675 \\
&= -3(w - 15)^2 + 675
\end{aligned}
$$

The maximum occur when $w = 15$ and is equal to 675. The dimensions of the plot are 15 ft. and 45 ft. The maximum area is 675 ft^2.

39. $$
\begin{aligned}
P(x) &= 0.8x - 0.002x^2 \\
&= -0.002(x^2 - 400x) \\
&= -0.002(x^2 - 400x + (200)^2) \\
&\quad + (0.002)(200)^2 \\
&= -0.002(x - 200)^2 + 80
\end{aligned}
$$

The function $P(x)$ attains its maximum when $x = 200$. Thus to maximize his profits Peter must sell 200 hot dogs, and the maximum profit is $P(200) = \$80$.

41. $$
\begin{aligned}
p &= 80 - x \\
R(x) &= (80 - x)x \\
&= 80x - x^2 \\
&= -(x^2 - 80x + 1600) + 1600 \\
&= -(x - 40)^2 + 1600
\end{aligned}
$$

The maximum is attained when $x = 40$ (thousand). The price is 40¢ and the maximum revenue is

$$40,000 \cdot \$.40 = \$16,000$$

43. $$
\begin{aligned}
s &= -16t^2 + 88t \\
&= -16(t^2 - 5.5t) \\
&= -16(t^2 - 5.5t + (2.75)^2) + 121 \\
&= -16(t - 2.75)^2 + 121
\end{aligned}
$$

The maximum distance, attained after 2.75 seconds, is $s(2.75) = 121$ ft. When the object hits the ground, $s = 0$, thus we have to solve the equation.

$$
\begin{aligned}
-16t^2 + 88t &= 0 \\
-16t(t - 5.5) &= 0 \\
t = 0, \quad t &= 5.5
\end{aligned}
$$

Thus the object will hit the ground after 5.5 seconds.

45. $A(t) = 800 - 160t + 16t^2$
$$= 16(t^2 - 10t) + 800$$
$$= 16(t^2 - 10t + 25)$$
$$+ 800 - 400$$
$$= 16(t - 5)^2 + 400$$

V(5, 400)
5 days, 400.

47. Let x denote the number of increases by 50 posters. If $100 + 50x$ posters are printed, the price of printing each poster is $20 - 2x$ cents. Thus the revenue is given by

$$R(x) = (100 + 50x)(20 - 2x)$$
$$= 2000 + 800x - 100x^2$$
$$= -100(x^2 - 8x) + 2000$$
$$= -100(x^2 - 8x + 16) + 2000$$
$$+ 1600$$
$$= -100(x - 4)^2 + 3600$$

It follows that maximum revenue occurs when $x = 4$, that is when $100 + 50(4) = 300$ posters are printed.

49. $f(x) = ax^2 + bx + c \quad (a \neq 0)$

$$= a(x^2 + \frac{b}{a} + \frac{c}{a})$$

$$= a[x^2 + \frac{b}{a} + (\frac{b}{2a})^2]$$

$$+ c - a(\frac{b}{2a})^2$$

$$= a(x + \frac{b}{2a})^2 + c - \frac{b^2}{4a}$$

$$= a(x + \frac{b}{2a})^2 + \frac{4ac - b^2}{4a}$$

If we set $h = -b/2a$ and

$k = \dfrac{4ac - b^2}{4a} = \dfrac{-\Delta}{4a}$, we obtain

$f(x) = a(x - h)^2 + k$

Exercises 5.2

Find the corresponding graphs in the answer section of the text.

1. $f(x) = (x - 2)^3$

Shift the graph of $y = x^3$ two units to the right.

3. $f(x) = (x + 3)^4$

Shift the graph of $y = x^4$ three units to the left.

5. $f(x) = x(x + 3)^2$
   $f(0) = 0, \quad f(-3) = 0$

Note that if $x > 0$, then $f(x) > 0$, so the graph lies above the x-axis. If $x < 0$, then $f(x) < 0$ and the graph lies below the x-axis.

7. $f(x) = x^2(x - 1)$

Since x^2 is always nonnegative, it follows that $f(x) > 0$ (graph above the x-axis) when $x > 1$ and $f(x) < 0$ (graph below the x-axis) when $x < 1$. The x-intercepts are 0 and 1, and the y-intercept is 0.

9. $f(x) = x^3 - 4x = x(x^2 - 4)$
   $$= x(x - 2)(x + 2)$$

The factored polynomial shows that -2, 0, 2 are the x-intercepts. According to the sign table $f(x) < 0$ in $(-\infty, -2)$ and $(0, 2)$, and $f(x) > 0$ in $(-2, 0)$ and $(2, +\infty)$.

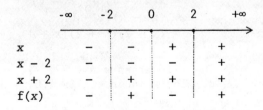

	$-\infty$	-2	0	2	$+\infty$
x	$-$	$-$	$+$	$+$	
$x - 2$	$-$	$-$	$-$	$+$	
$x + 2$	$-$	$+$	$+$	$+$	
$f(x)$	$-$	$+$	$-$	$+$	

11. $f(x) = x(x - 2)(x - 4)$

The x-intercepts are 0, 2, and 4.
The y-intercept is $f(0) = 0$.

	$-\infty$	0	2	4	$+\infty$
x		$-$	$+$	$+$	$+$
$x - 2$		$-$	$-$	$+$	$+$
$x - 4$		$-$	$-$	$-$	$+$
$f(x)$		$-$	$+$	$-$	$+$

13. $f(x) = x^2(x - 1)(x + 2)$

The x-intercepts are 0, 1, and -2.
The y-intercept is -2 (because
$f(0) = -2$). Make a sign table to
find the intervals where $f(x) > 0$
or $f(x) < 0$.

15. $f(x) = x^4 - 2x^2 - 8$
    $= (x^2 + 2)(x^2 - 4)$
    $= (x^2 + 2)(x - 2)(x + 2)$

The x-intercepts are -2 and 2.
The y-intercept is $f(0) = -8$.

	$-\infty$	-2	2	$+\infty$
$x^2 + 2$		$+$	$+$	$+$
$x - 2$		$-$	$-$	$+$
$x + 2$		$-$	$+$	$+$
$f(x)$		$+$	$-$	$+$

Since f is an even function, it
follows that the graph is
symmetric with respect to the
y-axis.

17. $f(x) = x^3 + x^2 - 4x - 4$
    $= (x^3 - 4x) + (x^2 - 4)$
    $= x(x^2 - 4) + (x^2 - 4)$
    $= (x^2 - 4)(x + 1)$
    $= (x - 2)(x + 2)(x + 1)$

The x-intercepts are -2, -1, and
2. The y-intercept is $f(0) = -4$.

	$-\infty$	-2	-1	2	
$x + 2$		$-$	$+$	$+$	$+$
$x - 2$		$-$	$-$	$-$	$+$
$x + 1$		$-$	$-$	$+$	$+$
$f(x)$		$-$	$+$	$-$	$+$

19. $f(x) = 2x^3 + 3x^2 - 2x - 3$
    $= (2x^3 - 2x) + (3x^2 - 3)$
    $= 2x(x^2 - 1) + 3(x^2 - 1)$
    $= (x^2 - 1)(2x + 3)$
    $= (x + 1)(x - 1)(2x + 3)$

The x-intercepts are -3/2, -1, and
1. The y-intercept is $f(0) = -3$.

	$-\infty$	-3/2	-1	1	$+\infty$
$x + 1$		$-$	$-$	$+$	$+$
$x - 1$		$-$	$-$	$-$	$+$
$2x + 3$		$-$	$+$	$+$	$+$
$f(x)$		$-$	$+$	$-$	$+$

21. $f(x) = 2x^3 + 4$

Shift the graph of $y = 2x^3$ four
units up.

23. $f(x) = 3x^4$

Note that $f(-x) = 3(-x)^4 = 3x^4 = f(x)$, that is f is an even
function, so the graph is
symmetric with respect to the y-
axis.

25. $f(x) = 2x^3 - \dfrac{1}{3}$

Shift the graph of $y = 2x^3$ one
third of a unit down. Since the
graph of $y = 2x^3$ is symmetric
about the origin it follows that
the graph of f is symmetric about
the point (0, -1/3).

27. $f(x) = x^3 - 4$

Shift the graph of $y = x^3$ four
units down. We have symmetry
about the point (0, -4).

29. $f(x) = \frac{1}{2}(x + 1)^3$

Shift the graph of $y = 1/2x^3$ one unit to the left. The curve is symmetric about the point $(-1, 0)$.

31. $f(x) = (x - 2)^4 + 1$

Shift the graph of $y = x^4$ two units to the right and then one unit up. The graph is symmetric with respect to the line $x = 2$.

33. $f(x) = (2x - 1)^3 + 4$

Rewrite f as follows

$$f(x) = [2(x - \frac{1}{2})]^3 + 4$$

$$= 8(x - \frac{1}{2})^3 + 4$$

To obtain the graph of f shift the graph of $y = 8x^3$ one half of a unit to the right and four units up. The curve is symmetric about the point $(1/2, 4)$.

35. $f(x) = x^3 + x$

The following table gives several points on the graph.

x	-2	-1	0	1	2
$f(x)$	-10	-2	0	2	10

37. $f(x) = x^3 + x - 1$

x	-2	-1	0	1	2
$f(x)$	-11	-3	-1	1	9

39. $f(x) = 1 - x - x^3$

x	-2	-1	0	1	2
$f(x)$	11	3	1	-1	-9

Exercises 5.3

Find the corresponding graphs in the answer section of your text.

1. $f(x) = \frac{1}{2x + 1}$

The given function is defined for all real numbers x different from -1/2. Since the numerator is always different from zero and the denominator is zero when x is equal to -1/2, it follows that the line $x = -1/2$ is a vertical asymptote. On the other hand, the degree of the numerator is smaller than the degree of the denominator, so the line $y = 0$ (x-axis) is a horizontal asymptote.

3. $f(x) = \frac{2x - 4}{3x + 2}$

The function is defined for all $x \neq -2/3$. When $x = -2/3$, the denominator is zero but the numerator is different from zero, so the line $x = -2/3$ is a vertical asymptote. The degrees of the numerator and denominator are the same, therefore the line $y = 2/3$ is a horizontal asymptote.

5. $f(x) = \frac{2x^2 + 3}{x^2 - 4x + 1}$

The reader can check that $2 - \sqrt{3}$ and $2 + \sqrt{3}$ are the zeros of the denominator. The lines $x = 2 - \sqrt{3}$ and $x = 2 + \sqrt{3}$ are vertical asymptotes. The line $y = 2$ is the horizontal asymptote of the graph.

7. $f(x) = \dfrac{2x + 3}{x^2 - 3x + 5}$

The denominator is always different from zero, because the discriminant $\Delta = (-3)^2 - 4(5)(1) = -11 < 0$. Thus $f(x)$ is defined for all real numbers x and its graph has no vertical asymptote. The degree of the numerator is smaller than the degree of the denominator, so $y = 0$ is a horizontal asymptote.

9. $f(x) = \dfrac{3x^2 - 4x + 1}{5x - 3}$

The function is defined for all real numbers x different from 3/5. The line $x = 3/5$ is a vertical asymptote. The graph has no horizontal asymptote because the degree of the numerator is greater than the degree of the denominator.

11. $f(x) = \dfrac{5}{x}$

The graph of f is obtained by stretching the graph of $y = 1/x$ vertically by a factor of 5.

13. $f(x) = \dfrac{-3}{x + 5}$

To obtain the graph of f, first stretch the graph of $y = 1/x$ vertically by a factor of 3, next reflect the graph across the x-axis to get the graph of $y = -3/x$, finally shift horizontally this graph 5 units to the left.

15. $f(x) = 2 - \dfrac{3}{x}$

After obtaining the graph of $y = -3/x$ as indicated in Exercise 13, shift it two units up.

17. $f(x) = \dfrac{1}{x^2} + 3$

Just shift the graph of $y = 1/x^2$ three units up.

19. $f(x) = \dfrac{2}{(x - 1)^2} + 3$

Stretch the graph of $y = 1/x^2$ by a factor of 2, shift the graph of $y = 2/x^2$ one unit to the right, and shift the graph of $y = 2/(x - 1)^2$ three units up.

21. $f(x) = \dfrac{-3}{x - 5}$

The domain of f is the set of all real numbers $x \neq 5$. The line $x = 5$ is a vertical asymptote and the line $y = 0$ a horizontal asymptote. The y-intercept is $f(0) = 3/5$. Since $f(x) \neq 0$ for all x in the domain of definition, it follows that the graph has no x-intercepts. Moreover, when $x < 5$, $f(x) > 0$, thus the graph lies above the x-axis; when $x > 5$, $f(x) < 0$, and the graph lies below the x-axis.

x	0	2	4	6	8
f(x)	3/5	1	3	-3	-1

23. $f(x) = \dfrac{3}{1 - 2x}$

The domain is the set of all real numbers $x \neq 1/2$. The line $x = 1/2$ is a vertical asymptote and the line $y = 0$ is a horizontal asymptote. The graph has no x-intercept because $f(x) \neq 0$ for all x in the domain of definition. The y-intercept is $f(0) = 3$. When $x < 1/2$, $f(x) > 0$ (graph above the x-axis); when $x > 1/2$, $f(x) < 0$ (graph below the x-axis).

x	-1	0	1	2
f(x)	1	3	-3	-1

25. $f(x) = \dfrac{x - 4}{2x + 1}$

Domain: all real numbers $x \neq -1/2$. The line $x = -1/2$ is a vertical asymptote and the line $y = 1/2$ is a horizontal asymptote. If $x = 0$, then $f(0) = -4$, and -4 is the y-intercept. $f(x) = 0$ when $x = 4$, so 4 is the x-intercept.

x	-5	-2	-1	4	1	0
$f(x)$	1	2	5	0	-1	-4

27. $f(x) = \dfrac{x^2}{x^2 - 4} = \dfrac{x^2}{(x - 2)(x + 2)}$

Domain: all real numbers $x \neq -2$ and $x \neq 2$. The lines $x = -2$ and $x = 2$ are vertical asymptotes. $y = 1$ is the horizontal asymptote. 0 is the x- and y-intercepts. The given function is even $f(-x) = f(x)$, so its graph is symmetric with respect to the y-axis.

x	-3	-1	0	1	3
$f(x)$	1.8	-1/3	0	-1/3	1.8

29. $f(x) = \dfrac{x^2}{x^2 + 1}$

The domain of f is the set of all real numbers. Since $x^2 + 1$ will never equal 0, there is no vertical asymptote. The horizontal asymptote is $y = 1$. 0 is the x- and y-intercepts. The function is even and the graph is symmetric with respect to the y-axis. $f(x) > 0$ for all $x \neq 0$, so the graph lies above the x-axis.

x	-2	-1	0	1	2
$f(x)$	4/5	1/2	0	1/2	4/5

31. $f(x) = \dfrac{3x}{x^2 - 4} = \dfrac{3x}{(x - 2)(x + 2)}$

The domain is the set of all real numbers $x \neq \pm 2$. The lines $x = -2$ and $x = 2$ are vertical asymptotes. $y = 0$ is a horizontal asymptote. 0 is the x- and y-intercepts. The given function is odd $(f(-x) = -f(x))$, so its graph is symmetric with respect to the origin. The following sign table shows the intervals where $f(x) > 0$ and where $f(x) < 0$.

	$-\infty$	-2	0	2	$+\infty$
$3x$	$-$	$-$	$+$	$+$	
$x^2 - 4$	$+$	$-$	$-$	$+$	
$f(x)$	$-$	$+$	$-$	$+$	

x	-4	-3	-1	0	1	3	4
$f(x)$	-1	-1.8	1	0	-1	1.8	1

33. $f(x) = \dfrac{3x - 1}{x^2 - 4} = \dfrac{3x - 1}{(x - 2)(x + 2)}$

The domain is the set of all real numbers $x \neq \pm 2$. The lines $x = -2$ and $x = 2$ are vertical asymptotes. $y = 0$ is a horizontal asymptote. When $x = 1/3$, $f(1/3) = 0$, so $1/3$ is the x-intercept. When $x = 0$, $f(0) = 1/4$, so $1/4$ is the y-intercept.

	$-\infty$	-2	$1/3$	2	$+\infty$
$3x - 1$	$-$	$-$	$+$	$+$	
$x^2 - 4$	$+$	$-$	$-$	$+$	
$f(x)$	$-$	$+$	$-$	$+$	

x	-3	-1	0	1/3	1	3
$f(x)$	-2	4/3	1/4	0	-2/3	1.6

35. $f(x) = \dfrac{2x + 1}{(x - 1)(x + 3)}$

Defined for all real numbers $x \neq 1$ and $x \neq -3$. Vertical asymptotes: $x = 1$ and $x = -3$. Horizontal asymptote: $y = 0$. x-intercept: $x = -1/2$; y-intercept: $-1/3$.

	$-\infty$	-3	$-1/2$	1	∞
$2x + 1$	$-$	$-$		$+$	$+$
$x - 1$	$-$	$-$		$-$	$+$
$x + 3$	$-$	$+$		$+$	$+$
$f(x)$	$-$	$+$		$-$	$+$

x	-2	$-1/2$	0	2
$f(x)$	1	0	$-1/3$	1

37. $f(x) = \dfrac{2x^2}{x^2 - x - 6}$

$= \dfrac{2x^2}{(x - 3)(x + 2)}$

Defined for all real numbers $x \neq -2$ and $x \neq 3$. Vertical asymptotes: $x = -2$ and $x = 3$. Horizontal asymptote: $y = 2$. 0 is the x- and y-intercepts.

	$-\infty$	-2	0	3	∞
$2x^2$	$+$	$+$	$+$	$+$
$x + 2$	$-$	$+$	$+$	$+$
$x - 3$	$-$	$-$	$-$	$+$
$f(x)$	$+$	$-$	$-$	$+$

x	-3	-1	0	1	2	6
$f(x)$	3	$-1/2$	0	$-1/3$	-2	3

39. $f(x) = \dfrac{x^2}{x^2 - 2x + 1} = \dfrac{x^2}{(x - 1)^2}$

Defined for all $x \neq 1$. $x = 1$ is a vertical asymptote. $y = 1$ is a horizontal asymptote. 0 is the x- and y-intercepts. Note that $f(x) \geq 0$ for all x in the domain of definition.

x	-2	-1	0	$1/2$
$f(x)$	$4/9$	$1/4$	0	1

41. $F(x) = 1/x^3$

x	-2	-1	1	2
$F(x)$	$-1/8$	-1	1	$1/8$

43. $f(x) = \dfrac{3x^2 - 1}{5x^2 + 3x + 2}$

Since the degree of the numerator is equal to the degree of the denominator, it follows that the line $y = 3/5$ is a horizontal asymptote.

45. $f(x) = \dfrac{x^2 + 1}{3x^4 - 5}$

The degree of the numerator (2) is smaller than the degree of the denominator (4). So the line $y = 0$ is a horizontal asymptote.

47. $f(x) = \dfrac{3x^{99} - 1}{x^{100} + x^{50}}$

The degree of the numerator (99) is smaller than the degree of the denominator (100). Thus the line $y = 0$ is a horizontal asymptote.

49. $f(x) = \dfrac{x^n}{3x^6 - 5x + 1}$,

n a positive integer.

a) $n = 6$. In this case, both the numerator and denominator have the same degree, so $y = 1/3$ is a horizontal asymptote.

b) $n < 6$. In this case, the degree of the numerator is smaller than the degree of the denominator, so $y = 0$ is a horizontal asymptote.

Exercises 5.4

1. $f(x) = 2x + 3$, $g(x) = 3x - 4$
    $(f \circ g)(x) = f(g(x)) = f(3x - 4)$
    $\qquad = 2(3x - 4) + 3 = 6x - 5$
    $(g \circ f)(x) = g(f(x)) = g(2x + 3)$
    $\qquad = 3(2x + 3) - 4 = 6x + 5$
    $(f \circ f)(x) = f(f(x)) = f(2x + 3)$
    $\qquad = 2(2x + 3) + 3 = 4x + 9$

3. $f(x) = 3x + 1$, $g(x) = \sqrt{x + 5}$

    $(f \circ g)(x) = f(g(x)) = f(\sqrt{x + 5})$

    $\qquad = 3\sqrt{x + 5} + 1$
    $(g \circ f)(x) = g(f(x)) = g(3x + 1)$

    $\qquad = \sqrt{(3x + 1) + 5} = \sqrt{3x + 6}$
    $(f \circ f)(x) = f(3x + 1)$
    $\qquad = 3(3x + 1) + 1$
    $\qquad = 9x + 4$

5. $f(x) = 2x - 5$, $g(x) = \dfrac{1}{x^2 + 2}$

    $(f \circ g)(x) = f(g(x))$

    $\qquad = f(\dfrac{1}{x^2 + 2}) = \dfrac{2}{x^2 + 2} - 5$

    $\qquad = \dfrac{2 - 5(x^2 + 2)}{x^2 + 2} = -\dfrac{5x^2 + 8}{x^2 + 2}$

    $(g \circ f)(x) = g(f(x))$

    $\qquad = g(2x - 5) = \dfrac{1}{(2x - 5)^2 + 2}$

    $\qquad = \dfrac{1}{4x^2 - 20x + 27}$

    $(f \circ f)(x) = f(f(x))$
    $\qquad = f(2x - 5) = 2(2x - 5) - 5$
    $\qquad = 4x - 15$

7. $f(x) = x^3 + 1$, $g(x) = \sqrt[3]{x - 1}$
    $(f \circ g)(x) = f(g(x))$

    $\qquad = f(\sqrt[3]{x - 1}) = (\sqrt[3]{x - 1})^3 + 1$
    $\qquad = x - 1 + 1 = x$
    $(g \circ f)(x) = g(f(x))$

$\qquad = g(x^3 + 1) = \sqrt[3]{x^3 + 1 - 1}$
$\qquad = \sqrt[3]{x^3} = x$
$(f \circ f)(x) = f(g(x))$
$\qquad = f(x^3 + 1) = (x^3 + 1)^3 + 1$
$\qquad = x^9 + 3x^6 + 3x^3 + 1 + 1$
$\qquad = x^9 + 3x^6 + 3x^3 + 2$

9. $(f \circ g)(\sqrt{6}) = f(g(\sqrt{6}))$
    $\qquad = f((\sqrt{6})^2 - 4)$
    $\qquad = f(6 - 4) = f(2)$

    $\qquad = \sqrt{2 + 2} = 2$

11. $(f \circ g)(1.8) = f(g(1.8))$
    $\qquad = f((1.8)^2 - 4) = f(-0.76)$

    $\qquad = \sqrt{-0.76 + 2} = \sqrt{1.24} \approx 1.11$

13. $(g \circ f)(\sqrt{3.25}) = g(f(\sqrt{3.25}))$

    $\qquad = g(\sqrt{\sqrt{3.25} + 2})$

    $\qquad \approx g(\sqrt{3.8028}) = (\sqrt{3.8028})^2 - 4$
    $\qquad = 3.8028 - 4 = -0.1972$
    $\qquad \approx -0.197$

15. $(g \circ f)(34) = g(f(34))$

    $\qquad = g(\sqrt{34 + 2}) = g(\sqrt{36})$
    $\qquad = g(6) = 6^2 - 4 = 32$

17. $(f \circ f)(23) = f(f(23))$

    $\qquad = f(\sqrt{23 + 2})$

    $\qquad = f(\sqrt{25}) = f(5)$

    $\qquad = \sqrt{5 + 2}$
    $\qquad = \sqrt{7} \approx 2.6$

19. $(g \circ g)(-\sqrt{2}) = g(g(-\sqrt{2}))$
    $\qquad = g((-\sqrt{2})^2 - 4) = g(-2)$
    $\qquad = (-2)^2 - 4 = 0$

21. $F(x) = \sqrt{5x - 2}$

    Let $f(x) = \sqrt{x}$ and $h(x) = 5x - 2$.
    Then $(f \circ h)(x) = f(h(x))$
    $= f(5x - 2) = \sqrt{5x - 2} = F(x)$. So
    $F = f \circ h$.

23. $H(x) = 25x^2 - 20x + 5$

Let $g(x) = x^2 + 1$ and $h(x)$
$= 5x - 2$. Then

$(g \circ h)(x) = g(h(x)) = g(5x - 2)$
$= (5x - 2)^2 + 1$
$= 25x^2 - 20x + 4 + 1$
$= 25x^2 - 20x + 5 = H(x)$

So $H = g \circ h$.

25. $U(x) = x^4 + 2x^2 + 2$

Let $g(x) = x^2 + 1$. Then

$(g \circ g)(x) = g(g(x)) = g(x^2 + 1)$
$= (x^2 + 1)^2 + 1 = x^4 + 2x^2 + 1 + 1$
$= x^4 + 2x^2 + 2 = U(x)$

So $U = g \circ g$.

27. $f(x) = \dfrac{1}{2 - x}$, $x \neq 2$.

Set $y = \dfrac{1}{2 - x}$ and solve for x:

$2 - x = \dfrac{1}{y}$

$x = 2 - \dfrac{1}{y} = \dfrac{2y - 1}{y}$

Interchanging x and y, obtain

$y = \dfrac{2x - 1}{x}$

Therefore the inverse of f is given by

$f^{-1}(x) = \dfrac{2x - 1}{x}$, $x \neq 0$.

29. $f(x) = x^2 - 4$, $x \geq 0$.

Set $y = x^2 - 4$ and note that $y \geq -4$ because $x \geq 0$. Solving for x, get

$x^2 = y + 4$
$x = \sqrt{y + 4}$

Interchange x and y and obtain the inverse function of f:

$f^{-1}(x) = \sqrt{x + 4}$, $x \geq -4$.

31. $f(x) = \sqrt{4x - 1}$, $x \geq 1/4$.

Set $y = \sqrt{4x - 1}$
and note that $y \geq 0$ for all
$x \geq 1/4$
Solve for x and get

$y^2 = 4x - 1$
$4x = y^2 + 1$

$x = \dfrac{y^2 + 1}{4}$

Interchange x and y to obtain the inverse of $f(x)$:

$f^{-1}(x) = \dfrac{x^2 + 1}{4}$, $x \geq 0$

33. $f(x) = \sqrt[3]{2x - 4}$, $x \in \Re$

Set $y = \sqrt[3]{2x - 4}$ and solve for x

$y^3 = 2x - 4$

$x = \dfrac{y^3 + 4}{2}$

The inverse function is

$f^{-1}(x) = \dfrac{x^3 + 4}{2}$, $x \in \Re$

35. $f(x) = \dfrac{x}{x - 1}$, $x \neq 1$

We have

$(f \circ f)(x) = f(f(x)) = f(\dfrac{x}{x - 1})$

$= \dfrac{\dfrac{x}{x - 1}}{\dfrac{x}{x - 1} - 1} = \dfrac{\dfrac{x}{x - 1}}{\dfrac{x - x + 1}{x - 1}} = x$

So f is its own inverse and the graphs of f and f^{-1} are identical.

37. $F(x) = \dfrac{x + 1}{2x - 1}$

$$(F \circ F)(x) = \frac{\left(\dfrac{x + 1}{2x - 1}\right) + 1}{2\left(\dfrac{x + 1}{2x - 1}\right) - 1}$$

[Multiply by $\dfrac{2x - 1}{2x - 1}$]

$$= \frac{x + 1 + 2x - 1}{2(x + 1) - (2x - 1)}$$

$$= \frac{3x}{2x + 2 - 2x + 1}$$

$$= \frac{3x}{3} = x$$

Therefore F is its own inverse. The graphs of F and F^{-1} are identical. Moreover, the graph of F is symmetric with respect to the line $y = x$.

Find the graphs for exercises 39-43 in the answer section of your text.

39. $f(x) = 4x + 5$, $g(x) = \dfrac{x - 5}{4}$,

$x \in \Re$. We have

$$(f \circ g)(x) = f(g(x)) = f\left(\frac{x - 5}{4}\right)$$

$$= 4\left(\frac{x - 5}{4}\right) + 5 = x - 5 + 5 = x$$

and
$(g \circ f)(x) = g(f(x)) = g(4x + 5)$

$$= \frac{4x + 5 - 5}{4} = \frac{4x}{4} = x$$

Since $(f \circ g)(x) = (g \circ f)(x) = x$ for all x, it follows that f and g are inverses of each other.

41. $f(x) = \dfrac{1}{x - 1}$, $x \neq 1$,

$g(x) = \dfrac{x + 1}{x}$, $x \neq 0$.

We have

$$(f \circ g)(x) = f(g(x)) = f\left(\frac{x + 1}{x}\right)$$

$$= \frac{1}{\dfrac{x + 1}{x} - 1} = \frac{1}{\dfrac{x + 1 - 1}{x}} = x,$$

for all $x \neq 0$, and

$$(g \circ f)(x) = g(f(x)) = g\left(\frac{1}{x - 1}\right)$$

$$= \frac{\dfrac{1}{x - 1} + 1}{\dfrac{1}{x - 1}} = \frac{\dfrac{1 + x - 1}{x - 1}}{\dfrac{1}{x - 1}}$$

$$= \frac{\dfrac{x}{x - 1}}{\dfrac{1}{x - 1}} = x,$$

for all $x \neq 1$. Therefore f and g are inverses of each other.

43. $f(x) = \sqrt{3x - 1}$, $x \geq 1/3$,

$g(x) = \dfrac{x^2 + 1}{3}$, $x \geq 0$.

We have $(f \circ g)(x) = f(g(x))$

$$= f\left(\frac{x^2 + 1}{3}\right) = \sqrt{3\left(\frac{x^2 + 1}{3}\right) - 1}$$

$$= \sqrt{x^2 + 1 - 1} = \sqrt{x^2} = x,$$

for all $x \geq 0$, and
$(g \circ f)(x) = g(f(x))$

$$= g(\sqrt{3x - 1}) = \frac{(\sqrt{3x - 1})^2 + 1}{3}$$

$$= \frac{3x - 1 + 1}{3} = \frac{3x}{x}, \text{ for all}$$

$x \geq 1/3$.

Thus f and g are inverses of each other.

45. $D(p) = 300 - 0.002p^2$ and
$p(c) = 3c - 20$
$(D \circ p)(c) = 300 - 0.002(3c - 20)^2$
$= 300 - 0.002(9c^2 - 120c + 400)$
$= 300 - 0.018c^2 + 0.24c - 0.8$
$= -0.018c^2 + 0.24c + 299.2$

47. By the definition of the inverse
f^{-1} of the function f, we have
$f(a) = b$ if and only if $f^{-1}(b) = a$.
The first relation means that the
point (a, b) belongs to the graph
of f, while the second relation
means that (b, a) belongs to the
graph of f^{-1}.

49. $f(x) = mx + b$, $g(x) = nx + d$
We have

$(f \circ g)(x) = f(g(x))$
$= f(nx + d) = m(nx + d) + b$
$= mnx + nb + d$
and
$(g \circ f)(x) = g(f(x))$
$= g(mx + b) = n(mx + b) + d$
$= mnx + nb + d$

If $f \circ g = g \circ f$, then we must
have $mnx + md + b = mnx + nb + d$
that is $md + b = nb + d$.

Review Exercises - Chapter 5

Find the corresponding graphs in
the answer section of your text.

1. $f(x) = 8 - 2x^2 = -2(x - 0)^2 + 8$

3. $f(x) = 3x^2 + x - 4$

$= 3(x^2 + \frac{1}{3}x) - 4$

$= 3(x^2 + 2(\frac{1}{6})x + \frac{1}{36}) - 4 - \frac{1}{12}$

$= 3(x + \frac{1}{6})^2 - \frac{49}{12}$

5. $f(x) = -x^2 + 2x - 5$
$= -(x^2 - 2x) - 5$
$= -(x^2 - 2x + 1) - 5 + 1$
$= -(x - 1)^2 - 4$

7. $f(x) = 3x^2(x + 2)$

The x-intercepts are -2 and 0.
When $x < -2$, $f(x) < 0$ and the
graph is below the x-axis. When
$x > -2$, $f(x) \geq 0$, so the graph is
above the x-axis. The y-intercept
is 0.

x	-3	-2	-1	0	1
f(x)	-27	0	3	0	9

9. $f(x) = (x - 2)(x - 3)(x + 2)$

The x-intercepts are -2, 2, and 3.
The y-intercept is 12. The sign
table

	$-\infty$	-2	2	3	∞
x - 2		−	−	+	+
x - 3		−	−	−	+
x + 2		−	+	+	+
f(x)		−	+	−	+

shows that the graph of f(x) lies
below the x-axis for x in the
intervals $(-\infty, -2)$ and $(2, 3)$, and
above the x-axis for x in the
intervals $(-2, 2)$ and $(3, \infty)$.

11. $f(x) = 2x^3 - x^2 - 8x + 4$
     $= (2x^3 - 8x) - (x^2 - 4)$
     $= 2x(x^2 - 4) - (x^2 - 4)$
     $= (x^2 - 4)(2x - 1)$
     $= (x - 2)(x + 2)(2x - 1)$

The x-intercepts are -2, 1/2, 2.
The y-intercept is 4.

	-∞	-2	1/2	2	∞
$x - 2$	−	−	−	+	
$x + 2$	−	+	+	+	
$2x - 1$	−	−	+	+	
$f(x)$	−	+	−	+	

So when $x < -2$ or $1/2 < x < 2$, the graph is below the x-axis. When $-2 < x < 1/2$ or $x > 2$, the graph is above the x-axis.

x	-2	-1	0	1/2	1	2	3
$f(x)$	0	9	4	0	-3	0	25

13. $f(x) = x(x + 1)^3$

The x-intercepts are 0 and -1. The y-intercept is 0. When $x < -1$ or $x > 0$, $f(x) > 0$ and the graph is above the x-axis. When $-1 < x < 0$, $f(x) < 0$ and the graph is below the x-axis.

x	-3	-2	-1	0	1
$f(x)$	24	2	0	0	8

15. $f(x) = 16x^2 - x^4$
     $= x^2(16 - x^2)$
     $= x^2(4 - x)(4 + x)$

The x-intercepts are -4, 0, and 4. The y-intercept is 0. When $x < -4$ or $x > 4$, $f(x) < 0$ and the graph is below the x-axis. When $-4 < x < 0$ or $0 < x < 4$, $f(x) > 0$ and the graph is above the x-axis. The given function is even, so the graph is symmetric with respect to the y-axis.

x	0	±1	±2	±3	±4
$f(x)$	0	15	48	63	0

17. $f(x) = x^3 + 2x + 1$

The y-intercept is 1.

x	-3	-2	-1	0	1	2
$f(x)$	-32	-11	-2	1	4	13

19. $f(x) = \dfrac{5}{x + 4}$

The function is defined for all real numbers $x \neq -4$. The line $x = -4$ is a vertical asymptote and the line $y = 0$ is a horizontal asymptote. When $x < -4$, $f(x) < 0$ and the graph is below the x-axis. When $x > -4$, $f(x) > 0$ and the graph is above the x-axis.

x	-5	-3	0	1
$f(x)$	-5	5	5/4	1

21. $f(x) = \dfrac{x + 1}{x - 4}$

The domain of f is the set of all real numbers $x \neq 4$. The line $x = 4$ is a vertical asymptote and the line $y = 1$ is a horizontal asymptote. The x-intercept is -1 and the y-intercept is -1/4. When $x < -1$ or $x > 4$, $f(x) > 0$ and the graph is above the x-axis. When $-1 < x < 4$, $f(x) < 0$ and the graph is below the x-axis.

x	-2	-1	0	3	5
$f(x)$	1/6	0	-1/4	-4	6

23. $f(x) = \dfrac{2x + 3}{x - 2}$

Domain: all $x \neq 2$. Vertical asymptote: $x = 2$. Horizontal asymptote: $y = 2$. x-intercept: -3/2, y-intercept: -3/2. When $x < -3/2$ or $x > 2$, $f(x) > 0$, when $-3/2 < x < 2$, $f(x) < 0$.

x	-5	-3/2	0	1	3
$f(x)$	1	0	-3/2	-5	9

25. $f(x) = \dfrac{x^2}{3x^2 + 1}$

The domain of f is the set of all real numbers. There are no vertical asymptotes. The line $y = 1/3$ is a horizontal asymptote. The origin is the x- and y-intercepts. When $x < 0$ or $x > 0$, $f(x) > 0$ and the graph is above the x-axis. The given function is even and the graph is symmetric with respect to the x-axis.

x	0	±1	±2
$f(x)$	0	1/4	4/13

27. $f(x) = \dfrac{2x}{x^2 - 9} = \dfrac{2x}{(x - 3)(x + 3)}$

The function is defined for all $x \neq \pm 3$. The lines $x = -3$ and $x = 3$ are vertical asymptotes, and $y = 0$ is a horizontal asymptote. The origin is the x- and y-intercept.

	$-\infty$	-3	0	3	∞
$2x$	$-$	$-$	$+$	$+$	
$x - 3$	$-$	$-$	$-$	$+$	
$x + 3$	$-$	$+$	$+$	$+$	
$f(x)$	$-$	$+$	$-$	$+$	

This table shows that $f(x) < 0$, when $x < -3$ or $0 < x < 3$, and $f(x) > 0$, when $-3 < x < 0$ or $x > 3$. The function is odd, so the graph is symmetric with respect to the origin.

x	-4	-1	0	1	4
$f(x)$	-8/7	1/4	0	-1/4	8/7

29. $f(x) = \dfrac{x - 1}{x^2 - 16} = \dfrac{x - 1}{(x - 4)(x + 4)}$

The function is defined for all $x \neq \pm 4$. Vertical asymptotes: $x = -4$ and $x = 4$. Horizontal asymptote: $y = 0$. x-intercept: 1, y-intercept: 1/16. When $x < -4$ or $1 < x < 4$, the graph is below the x-axis. When $-4 < x < 1$ or $x > 4$, the graph is above the x-axis.

x	-5	-1	0	1	2	5
$f(x)$	-2/3	2/15	1/16	0	-1/12	4/9

31. $f(x) = \dfrac{x^2}{x^2 + 5x + 4}$

$= \dfrac{x^2}{(x + 4)(x + 1)}$

The domain is the set of all $x \neq -4$ and $x \neq -1$. Vertical asymptotes: $x = -4$, $x = -1$. Horizontal asymptote: $y = 1$. 0 is the x- and y-intercepts. The sign table below indicates the location of the graph with respect to the x-axis.

	$-\infty$	-4	-1	0	∞
x^2	$+$	$+$	$+$	$+$	
$x + 4$	$-$	$+$	$+$	$+$	
$x + 1$	$-$	$-$	$+$	$+$	
$f(x)$	$+$	$-$	$+$	$+$	

x	-5	-2	-4/5	0	1
$f(x)$	25/4	-2	1	0	1/10

33. $f(x) = \dfrac{3x - 4}{2x + 5}$

Since the numerator and denominator have the same degree (equal to 1), $y = 3/2$ is the horizontal asymptote.

35. $f(x) = \dfrac{4x^2 - 5}{3x^2 - 6x - 1}$

Both numerator and denominator have the same degree, so the line $y = 4/3$ is the horizontal asymptote.

37. $f(x) = \dfrac{3x + 7}{x^3 - 6x + 1}$

The degree of the numerator is smaller than the degree of the denominator, so the line $y = 0$ is the horizontal asymptote.

39. $f(x) = \dfrac{1}{3x + 1}$ $g(x) = \sqrt{2x - 3}$

$(f \circ g)(x) = f(g(x))$

$= f(\sqrt{2x - 3}) = \dfrac{1}{3\sqrt{2x - 3} + 1}$

$(g \circ f)(x) = g(f(x))$

$= g(\dfrac{1}{3x + 1}) = \sqrt{2(\dfrac{1}{3x + 1}) - 3}$

$= \sqrt{\dfrac{2 - 3(3x + 1)}{3x + 1}} = \sqrt{\dfrac{-(9x + 1)}{3x + 1}}$

$(f \circ f)(x) = f(f(x))$

$= f(\dfrac{1}{3x + 1}) = \dfrac{1}{3(\dfrac{1}{3x + 1}) + 1}$

$= \dfrac{1}{\dfrac{3 + (3x + 1)}{3x + 1}} = \dfrac{3x + 1}{3x + 4}$

41. $f(x) = \dfrac{x + 1}{x - 3}$ $g(x) = \dfrac{3x + 1}{x - 1}$

$(f \circ g)(x) = f(g(x))$

$= f(\dfrac{3x + 1}{x - 1}) = \dfrac{(\dfrac{3x + 1}{x - 1}) + 1}{(\dfrac{3x + 1}{x - 1}) - 3}$

$= \dfrac{\dfrac{(3x + 1) + (x - 1)}{x - 1}}{\dfrac{(3x + 1) - 3(x - 1)}{x - 1}} = \dfrac{4x}{4} = x$

$(g \circ f)(x) = g(f(x))$

$= g(\dfrac{x + 1}{x - 3}) = \dfrac{3(\dfrac{x + 1}{x - 3}) + 1}{(\dfrac{x + 1}{x - 3}) - 1}$

$= \dfrac{\dfrac{3(x + 1) + (x - 3)}{x - 3}}{\dfrac{(x + 1) - (x - 3)}{x - 3}} = \dfrac{4x}{4} = x$

$(f \circ f)(x) = f(f(x))$

$= f(\dfrac{x + 1}{x - 3}) = \dfrac{\dfrac{x + 1}{x - 3} + 1}{\dfrac{x + 1}{x - 3} - 3}$

$= \dfrac{\dfrac{(x + 1) + (x - 3)}{x - 3}}{\dfrac{(x + 1) - 3(x - 3)}{x - 3}} = \dfrac{2x - 2}{-2x + 10}$

$= \dfrac{2x - 2}{10 - 2x} = \dfrac{x - 1}{5 - x}$

43. $f(x) = \dfrac{2x + 1}{x}$, $x \neq 0$,

$g(x) = \dfrac{1}{x - 2}$, $x \neq 2$.

We have

$(f \circ g)(x) = f\left(\dfrac{1}{x - 2}\right)$

$= \dfrac{2\left(\dfrac{1}{x - 2}\right) + 1}{\dfrac{1}{x - 2}}$

$= \dfrac{\dfrac{2 + (x - 2)}{x - 2}}{\dfrac{1}{x - 2}} = x$

and

$(g \circ f)(x) = g\left(\dfrac{2x + 1}{x}\right)$

$= \dfrac{1}{\left(\dfrac{2x + 1}{x}\right) - 2}$

$= \dfrac{1}{\dfrac{(2x + 1) - 2x}{x}}$

$= \dfrac{x}{(2x + 1) - 2x} = \dfrac{x}{1} = x$

Since $(f \circ g)(x) = (g \circ f)(x) = x$, it follows that g and f are inverses of each other.

45. $F(x) = \sqrt{3 - x}$, $x \leq 3$,
$G(x) = 3 - x^2$, $x \geq 0$.
We have
$(F \circ G)(x) = F(G(x))$

$= F(3 - x^2) = \sqrt{3 - (3 - x^2)}$

$= \sqrt{3 - 3 + x^2} = \sqrt{x^2} = x$,
for all $x \geq 0$, and
$G \circ F(x) = G(F(x))$

$= G(\sqrt{3 - x}) = 3 - (\sqrt{3 - x})^2$
$= 3 - (3 - x) = x$,
for all $x \leq 3$.

Therefore F and G are inverse of each other.

47. $f(x) = \dfrac{2}{3x - 5}$, $x \neq 5/3$

Set $y = \dfrac{2}{3x - 5}$ and solve for x:

$3x - 5 = \dfrac{2}{y}$

$3x = \dfrac{2}{y} + 5 = \dfrac{2 + 5y}{y}$

$x = \dfrac{2 + 5y}{3y}$

Next, interchange x and y, and get

$y = \dfrac{2 + 5x}{3x}$

Thus, the inverse is

$f^{-1}(x) = \dfrac{2 + 5x}{3x}$, $x \neq 0$.

49. $f(x) = \sqrt{3x - 4}$, $x \geq 4/3$. Set $y = \sqrt{3x - 4}$ and note that $y \geq 0$ when $x \geq 4/3$. Now, solve for x

$y^2 = 3x - 4$
$3x = y^2 + 4$

$x = \dfrac{y^2 + 4}{3}$, $y \geq 0$.

Interchange x an y and obtain the inverse:

$f^{-1}(x) = \dfrac{x^2 + 4}{3}$, $x \geq 0$.

51. $f(x) = 9 - x^2$, $-3 \le x \le 0$. Set $y = 9 - x^2$ and observe that $0 \le y \le 9$ when $-3 \le x \le 0$. Solve for x and obtain

$$x^2 = 9 - y$$

$x = -\sqrt{9 - y}$, because $x < 0$. Interchange x an y to obtain the inverse

$$f^{-1}(x) = -\sqrt{9 - x}, \ 0 \le x \le 9.$$

53. Let f and g be odd functions and consider f ∘ g. We have

$$
\begin{aligned}
(f \circ g)(-x) &= f(g(-x)) \\
&\quad [\text{because g is odd}] \\
&= f(-g(x)) \\
&\quad [\text{because f is odd}] \\
&= -f(g(x)) \\
&= -(f \circ g)(x)
\end{aligned}
$$

Thus f ∘ g is an odd function.

55. $F(x) = \dfrac{x^n + 1}{1 - x^n}$, n a positive

integer. Since both numerator and denominator have the same degree, it follows that $y = 1/-1 = -1$ is a horizontal asymptote.

57. $P = 100 - 5D^2$, $D(t) = 3 + 2\sqrt{t}$

$$
\begin{aligned}
P(t) = (P \circ D)(t) &= P(D(t)) \\
&= P(3 + 2\sqrt{t}) \\
&= 100 - 5(3 + 2\sqrt{t})^2 \\
&= 100 - 5(9 + 12\sqrt{t} + 4t) \\
&= 55 - 60\sqrt{t} - 20t
\end{aligned}
$$

59. Let $f(x) = mx + b$ $(m \ne 0)$ and $g(x) = nx + d$ $(d \ne 0)$. If f and g are inverses of each other, then $(f \circ g)(x) = (g \circ f)(x) = x$, for all x. But

$$
\begin{aligned}
(f \circ g)(x) &= f(nx + d) \\
&= m(nx + d) + b = mnx + (md + b)
\end{aligned}
$$
and
$$
\begin{aligned}
(g \circ f)(x) &= g(mx + b) \\
&= n(mx + b) + d = mnx + (nb + d)
\end{aligned}
$$

So we must have
$$mnx + (md + b) = x$$
and
$$mnx + (nb + d) = x$$
for all x. If we set $x = 0$, we obtain

$$md + b = nb + d = 0$$

Thus $mnx = x$, for all x. Now, if we set $x = 1$, we obtain $mn = 1$. Therefore, f and g are inverses of each other if $mn = 1$ and $md + b = nb + d = 0$.

Chapter 5 Test

Let $f(x) = a(x - h)^2 + k$. For each of the following values of the constants a, h, and k, find the vertex, axis of symmetry, x- and y-intercept of the parabola represented by $f(x)$. Does the parabola open upward or downward?

1. $a = 2$, $h = 1$, $k = -32$
2. $a = -1$, $h = -5$, $k = 9$
3. $a = -2$, $h = 7$, $k = 0$
4. $a = 3$, $h = 0$, $k = 5$

By completing the square write each of the following quadratic functions in the form $f(x) = a(x - h)^2 + k$. Determine the vertex and principal axis.

5. $f(x) = x^2 - 10x + 16$
6. $f(x) = 2x^2 + 12x + 15$
7. Let $f(x) = kx^2 + kx - 1$, $k \neq 0$. Find all values of k for which the graph of f has exactly two x-intercepts.

8. For what values of b is the graph of $f(x) = -3x^2 + bx - 4$ tangent to the x-axis?

9. Let $f(x) = ax^2 - 6x + 1$. For what values of a does the graph of f open upward and cross the x-axis?

10. The cost function for manufacturing clock radios is given by $C(x) = x^2 - 50x + 1000$, where x is the number (in hundreds) of units manufactured per week, and C is the cost in hundreds of dollars. What weekly output corresponds to minimum cost? What is the minimum cost?

11. The difference of two numbers is 64 and their product is a minimum. What are the two numbers?

12. Let $c(x) = 1 - \dfrac{x^2}{2}$

   a) Is $c(x)$ an even function, odd function, or neither?
   b) Find the x- and y-intercepts of its graph.
   c) Determine the intervals on which $c(x) > 0$ and the intervals on which $c(x) < 0$.
   d) Sketch the graph of $c(x)$.

For each of the following functions, name the line or point of symmetry.

13. $f(x) = (x + 2)^3 + 4$
14. $f(x) = (x - 5)^4 - 2$

15. A homeowner wants to make a rectangular flower bed surrounded on three sides by an ornamental fence. (The fourth side will have no fence at all.) He has $420 to spend on his project and knows that the fencing costs $15 per foot. What dimensions of the flower bed will give him the maximum area?

For each of the following functions, find the domain, vertical and horizontal asymptotes, x- and y-intercepts.

16. $f(x) = \dfrac{5}{2x - 3}$
17. $f(x) = \dfrac{2x + 5}{x + 6}$

18. $f(x) = \dfrac{4x - 2}{2x^2 - 5x + 3}$ 19. $f(x) = \dfrac{x^2 - 1}{4x^2 - 4x + 1}$

20. Graph the function $f(x) = \dfrac{2x + 5}{x + 6}$

21. Graph the function $f(x) = \dfrac{x^2}{x^2 - 9}$

22. A manufacturer can produce at most 200 pounds of margarine per date at a cost, in dollars, given by

$$C(x) = \dfrac{204}{60 + x}$$

where x is the number of pounds produced per day. Find a) the domain of $C(x)$, b) $C(20)$, $C(42)$, $C(100)$, and $C(180)$.

23. If $f(x) = 2x + 1$ and $g(x) = \sqrt{x - 4}$, find $(f \circ g)(x)$, $(g \circ f)(x)$, and $(g \circ g)(x)$.

24. Using the results of Problem 23, find $(f \circ g)(5)$, $(g \circ f)(38/25)$, and $(g \circ g)(29)$.

25. Let $f(x) = 3 - 4x^2$ be defined on the interval $(-\infty, 0]$. Find $f^{-1}(x)$.

26. Let $f(x) = \dfrac{2x}{3x - 2}$, $x \neq 2/3$. Show that $f^{-1}(x) = f(x)$.

27. Show that

$$f(x) = \dfrac{4x}{2x - 1}, \ x \neq 1/2, \text{ and } g(x) = \dfrac{x}{2x - 4}, \ x \neq 2,$$

are inverses of each other.

28. Suppose that the cost, $C(x)$, in ten-thousand of dollars, to clean a polluted lake is given by

$$C(x) = \dfrac{8x}{100 - x}$$

where x is the percentage of pollutant removed from the lake. Find the domain of definition of $C(x)$ and sketch its graph.

29. In Problem 28, find the cost to remove 50%, 75%, and 90% of pollutant from the lake. Would it be possible to remove all pollutant?

30. The total number, q, of coffee makers produced daily by a manufacturer is given by

$$q = 50n - \frac{n^2}{2}$$

where n is the number of employees. The revenue, R, as a function of q is given by R = 12q. Find the revenue as a function of the number of employees.

Chapter 6
Exponential and Logarithmic Functions

Exercises 6.1

Find the corresponding graphs in the answer section of your text.

1. $f(x) = 3^x$

x	-2	-1	0	1	2
3^x	1/9	1/3	1	3	9

3. $f(x) = (3/2)^x$

x	-2	-1	0	1	2
$(3/2)^x$	4/9	2/3	1	3/2	9/4

5. $f(x) = 2^x + 3$

    The graph is obtained by shifting the graph of $y = 2^x$ three units up.

7. $f(x) = 3^x - 1$

    The graph is obtained by shifting the graph of $y = 3^x$ one unit down.

9. $f(x) = 2^{x-1}$

    The graph is obtained by shifting the graph of $y = 2^x$ one unit to the right.

11. $f(x) = 3^{x+2}$

    Translate the graph of $y = 3^x$ two units to the left.

13. $f(x) = 3^{2-x}$

    Translate the graph of $y = 3^{-x}$ two units to the right.

15. $f(x) = 1 - 2^x$

    Reflect the graph of $y = 2^x$ along the x-axis to obtain the graph of $g(x) = -2^x$. Shift the graph of g one unit up to obtain the graph of f.

17. $f(x) = 4^x$
    $f(2) = 4^2 = 16$
    $f(-2) = 4^{-2} = 1/16$
    $f(1/2) = 4^{1/2} = 2$

19. $f(x) = 5(9^x)$
    $f(1/2) = 5(9^{1/2}) = 5(3) = 15$
    $f(-1/2) = 5(9^{-1/2}) = 5(1/3) = 5/3$
    $f(1) = 5(9^1) = 45$

21. $f(x) = 3 + 2(\frac{1}{8})^{x+1}$

$f(-2) = 3 + 2(\frac{1}{8})^{-2+1}$

$= 3 + 2(\frac{1}{8})^{-1}$

$= 3 + 2(8) = 19$

$f(0) = 3 + 2(\frac{1}{8})^{0+1}$

$= 3 + 2(\frac{1}{8}) = 3 + \frac{1}{4} = 13/4$

$f(-2/3) = 3 + 2(\frac{1}{8})^{-2/3+1}$

$= 3 + 2(\frac{1}{8})^{1/3}$

$= 3 + 2(\frac{1}{2}) = 4$

23. $e^2 \simeq 7.389$

25. $e^{1/2} = e^{0.5} \simeq 1.649$

27. $e^{-1.2} \simeq 0.301$

29. $e^{1/20} = e^{0.05} \simeq 1.051$

31. $f(x) = e^x$

x	-3	-2	-1	0	1
e^x	0.05	0.135	0.368	1	2.718

33. $f(x) = e^{x^2}$

x	-2	-1	0	1	2
e^{x^2}	54.6	2.72	1	2.72	54.6

35. $f(x) = \dfrac{e^x + e^{-x}}{2}$

x	0	1	2	3
$f(x)$	1	1.54	3.76	10.1

37. $3^{\sqrt{2}} \simeq 4.729$

39. $4^{-\sqrt{5}} \simeq 0.045$

41. $(\sqrt{2})^{\sqrt{2}} \simeq 1.633$

43. $\pi^{\pi} \simeq 36.462$

45. $7(1/2)^{-0.15} \simeq 7.767$

47. $f(x) = x(2^x)$
$f(-0.51) = (-0.51)(2^{-0.51}) \simeq -0.358$
$f(0.01) = (0.01)(2^{0.01}) \simeq 0.01007$
$f(0.51) = (0.51)(2^{0.51}) \simeq 0.726$

49. $f(x) = [1 + (1/x)]^x$
$f(4) = [1 + (1/4)]^4$
$= (5/4)^4 \simeq 2.44141$
$f(10) = [1 + (1/10)]^{10}$
$= (11/10)^{10} \simeq 2.59374$
$f(10^3) = [1 + (1/10^3)]^{10^3}$
$= (1001/1000)^{1000}$
$= 2.71692$
$f(10^5) = [1 + (1/10^5)]^{10^5}$
$= (1.00001)^{10^5}$
$\simeq 2.71827$

Exercises 6.2

1. Let t be the number of hours and let $N(t)$ be the number of bacteria

t	$N(t)$
0	1800
12	3600
24	7200
36	14,400

3. Let t be the number of years after 1910 and let $P(t)$ be the population after t years.

Year	t	$P(t)$
1910	0	15 million
1935	25	30 million
1960	50	60 million
1985	75	120 million

5. Let t be the number of minutes and let $Q(t)$ be the amount of polonium-128.

t	$Q(t)$
0	60g
3	30g
6	15g
9	7.5g

7. If interest is compounded quarterly, then

$$p(t) = p_o(1 + \frac{r}{N})^{Nt}$$

where $p_o = 10,000$, $r = 0.075$, $N = 4$ and t is time in years.

a) $t = 1$, so

$$p(1) = 10,000(1 + \frac{0.075}{4})^{4(1)}$$

$$= 10,000(1.01875)^4$$
$$\approx 10,771.36$$

b) $t = 2$, so

$$p(2) = 10,000(1 + \frac{0.075}{4})^{4(2)}$$

$$= 10,000(1.01875)^8$$
$$\approx 11,602.22$$

If interest is compounded continuously, then $p(t) = p_o e^{rt}$, where $p_o = 10,000$ and $r = 0.075$. So

$$p(1) = 10,000 \; e^{0.075(1)}$$
$$\approx 10,778.84$$

and

$$p(2) = 10,000 \; e^{0.075(2)}$$
$$= 10,000 \; e^{0.15}$$
$$\approx 11,618.34$$

9. We have $p(t) = p_o(1 + r)t$, where $r = 0.06$, $t = 4$ and $p(4) = 8,000$. Thus

$$8,000 = p_o(1 + 0.06)^4$$
$$8,000 = p_o(1.06)^4$$

$$p_o = \frac{8,000}{(1.06)^4} \approx 6,336.75$$

11. We have $p(t) = p_o e^{rt}$, where $p_o = 4 \times 10^6$, $r = 0.0224$, and $t = 170$ years ($= 1960 - 1790$) Thus

$$p(170) = (4 \times 10^6)e^{0.0224(170)}$$
$$= 1.8024091 \times 10^8$$
$$\approx 1.80 \times 10^8$$
$$= 180 \text{ million}$$

13. $P(t) = 20,000(1.02)^t$ where t is the number of years after 1980. So

$$P(3) = 20,000(1.02)^3$$
$$= 20,000(1.061208)$$
$$\approx 21,224$$

15. If the number of bacteria doubles every hour and the initial number is n_o, then the formula

$$n(t) = n_o 2^t$$

gives the number of bacteria after t hours. For $n_o = 15,000$ and $t = 3$, we have

$$n(3) = 15,000 \cdot 2^3$$
$$= 15,000(8)$$
$$= 120,000$$

17. $p(x) = 760\, e^{-0.11445x}$

    a) $p(2) = 760\, e^{-0.11445(2)}$
            $= 760\, e^{-0.2289}$
            $= 604.51013$
            ≈ 605 millimeters of mercury

    b) $p(5) = 760\, e^{-0.11445(5)}$
            ≈ 429 millimeters of mercury

19. $Q(t) = Q_o(1/2)^{t/T}$, where $T = 1620$ years and $Q_o = 100$mg, so $Q(t) = 100(1/2)^{t/1620}$.

    a) $Q(630) = 100(1/2)^{630/1620}$
             $= 100(0.7637175)$
             ≈ 76 mg

    b) $Q(3240) = 100(1/2)^{3240/1620}$
               $= 100(1/2)^2$
               $= 100/4 = 25$ mg

21. $I(d) = 180\, e^{-1.4d}$

    $I(4) = 180\, e^{(-1.4)4}$
          $= 180\, e^{-5.6}$
          $= 0.6656154$
          ≈ 0.67 lumens

23. $Q(t) = Q_o(1/2)^{t/T}$, where $T = 5750$ is the half-life of ^{14}C. If $Q(t) = 0.0625\, Q_o$, then

$$0.0625\, Q_o = Q_o(1/2)^{t/5750}$$

$$\left(\frac{1}{2}\right)^{t/5750} = \frac{625}{10000} = \frac{1}{16}$$

$$\left(\frac{1}{2}\right)^{t/5750} = \left(\frac{1}{2}\right)^4$$

so

$$\frac{t}{5750} = 4$$

$$t = 5750(4)$$
$$= 23{,}000 \text{ years}$$

25. $A(t) = C(1/2)^{t/15}$

$$10 = 40(1/2)^{t/15}$$

$$\left(\frac{1}{2}\right)^{t/15} = \frac{10}{40} = \frac{1}{4} = \left(\frac{1}{2}\right)^2$$

$$\frac{t}{15} = 2$$
$$t = 30s.$$

$$5 = 40(1/2)^{t/15}$$

$$\left(\frac{1}{2}\right)^{t/15} = \frac{1}{8} = \left(\frac{1}{2}\right)^3$$

$$t = 45s$$

27. $L = L_o e^{-kt}$, $k = 0.2$
If $t = 3.5$, then

$$L = L_o e^{-0.2(3.5)}$$
$$= 0.4965853 L_o$$
$$\approx 0.50 L_o.$$

Therefore, 50% of the initial amount decays in 3.5 days.

29. $C = C_e(1 - e^{-kt})$

$$\frac{C_e}{2} = C_e(1 - e^{-k(10)})$$

$$\frac{1}{2} = 1 - e^{-10k}$$

$$e^{-10k} = \frac{1}{2}$$

$$C = C_e(1 - e^{-k(20)})$$
$$= C_e(1 - e^{(-10k)2})$$

$$= C_e(1 - (\frac{1}{2})^2)$$

$$= C_e(1 - \frac{1}{4})$$

$$= \frac{3C_e}{4}$$

Exercises 6.3

1. $y = \log_2 16$ if and only if $2^y = 16$, thus $y = 4$.

3. $y = \log_2(\frac{1}{8})$ if and only if $2^y = \frac{1}{8}$ or $2^y = 2^{-3}$, so $y = -3$.

5. $y = \log 10^4$ exactly when $10^y = 10^4$, thus $y = 4$.

7. $y = \log(0.1)^2$
$10^y = (0.1)^2$
$10^y = (10^{-1})^2$
$10^y = 10^{-2}$
$y = -2$

9. $y = \ln(\frac{1}{e})^3$

$e^y = (\frac{1}{e})^3$

$e^y = e^{-3}$
$y = -3$

11. $2^5 = 32$ if and only if $\log_2 32 = 5$.

13. $4^{-2} = \frac{1}{16}$ exactly when

$\log_4(1/16) = -2$.

15. $(\frac{1}{3})^4 = \frac{1}{81}$ means that

$\log_{1/3}(1/81) = 4$

17. $a^b = c$ is the same as $\log_a c = b$. (Here we are assuming that $a > 0$ and $a \neq 1$.)

19. $\log_2 128 = 7$ means that $128 = 2^7$.

21. $\log_2(\frac{1}{1024}) = -10$ is equivalent

to $2^{-10} = 1/1024$.

23. $\log 100 = 2$ or $100 = 10^2$

25. $\log q = r$ is the same as $q = 10r$

27. $\log 10 = \log(2 \times 5)$
$= \log 2 + \log 5$
$= 0.30 + 0.70$
$= 1$

29. $\log 18 = \log(2 \times 3^2)$
$= \log 2 + \log 3^2$
$= \log 2 + 2 \log 3$
$\approx 0.30 + 2(0.48)$
$= 0.30 + 0.96$
$= 1.26$

31. $\log 20 = \log(2^2 \times 5)$
$= \log(2^2) + \log 5$
$= 2 \log 2 + \log 5$
$= 2(0.30) + 0.70$
$= 0.60 + 0.70$
$= 1.30$

33. $\log \left(\dfrac{3}{5}\right) = \log 3 - \log 5$

$\simeq 0.48 - 0.70$

$= -0.22$

35. $\log \left(\dfrac{10}{3}\right) = \log 10 - \log 3$

$\simeq 1 - 0.48$

$= 0.52$

37. $\log 8 = \log(2^3)$

$= 3 \log 2$

$\simeq 3(0.30)$

$= 0.90$

39. $\log \sqrt{10} = \log(10^{1/2})$

$= \dfrac{1}{2} \log 10$

$= 1/2$

41. $\log_6 x = 3$ means, by definition, that $x = 6^3 = 216$.

43. $\log_{3/2} x = 4$

$x = (3/2)^4 = 81/16$

45. $\log_x 4 = 5$

$x^5 = 4$

$x = \sqrt[5]{4}$

47. $\log_{2x} 8 = 3$

$(2x)^3 = 8$

$8x^3 = 8$

$x^3 = 1$

$x = 1$

49. $\log_7 49 = x$

$7^x = 49 = 7^2$

$x = 2$

51. $\log_3(2x + 1) = 2$

$2x + 1 = 3^2$

$2x + 1 = 9$

$2x = 8$

$x = 4$

53. $\log_2(1 - 2x) = -3$

$1 - 2x = 2^{-3}$

$1 - 2x = \dfrac{1}{8}$

$-2x = \dfrac{1}{8} - 1 = -\dfrac{7}{8}$

$x = 7/16$

55. $\log 8 + \log 6 - \log 12$

$= (\log 8 + \log 6) - \log 12$

$= \log (8 \times 6) - \log 12$

$= \log 48 - \log 12$

$= \log \left(\dfrac{48}{12}\right)$

$= \log 4$

57. $3 \log_5 2 + 2 \log_5 3 - \left(\dfrac{1}{2}\right)\log_5 4$

$= \log_5 2^3 + \log_5 3^2 - \log_5 \sqrt{4}$

$= \log_5 8 + \log_5 9 - \log_5 2$

$= \log_5 \dfrac{8 \times 9}{2}$

$= \log_5 36$

59. $2 \ln x - \dfrac{1}{4} \ln y + \ln z$

$= \ln x^2 - \ln y^{1/4} + \ln z^4$

$= \ln \left(\dfrac{x^2 z^4}{y^{1/4}}\right)$

61. $2 \log x + \frac{1}{3} \log y - 4 \log x$

   $= \log x^2 + \log y^{1/3} - \log x^4$

   $= \log \dfrac{x^2 y^{1/3}}{x^4}$

   $= \log (\sqrt[3]{y}/x^2)$

63. $\log(x + 1) - \log (x - 2) + 2 \log x$
   $= \log(x + 1) - \log(x - 2)$
   $\quad + \log x^2$

   $= \log \dfrac{(x + 1)x^2}{x - 2}$

65. $3 \log(x + 2) + 2 \log x$
   $\quad - 4 \log(x + 1)$
   $= \log(x + 2)^3 + \log x^2$
   $\quad - \log(x + 1)^4$

   $= \log \dfrac{(x + 2)^3 x^2}{(x + 1)^4}$

67. $\log \dfrac{x^2}{(x - 2)^3}$

   $= \log x^2 - \log(x - 2)^3$
   $= 2 \log x - 3 \log(x - 2)$

69. $\log [x^2(x - 2)^4)]$
   $= \log x^2 + \log(x - 2)^4$
   $= 2 \log x + 4 \log(x - 2)$

71. $\log \sqrt{(x - 2)^3 x^4}$
   $= \log((x - 2)^3 x^4)^{1/2}$

   $= \dfrac{1}{2} \log((x - 2)^3 x^4)$

   $= \dfrac{1}{2}(\log(x - 2)^3 + \log x^4)$

   $= \dfrac{1}{2}(3 \log(x - 2) + 4 \log x)$

   $= \dfrac{3}{2} \log(x - 2) + 2 \log x$

73. $\log(\dfrac{x^2(x + 1)^3}{(x - 2)^5})$

   $= \log[x^2(x + 1)^3] - \log(x - 2)^5$
   $= \log x^2 + \log(x + 1)^3$
   $\quad - \log (x - 2)^5$
   $= 2 \log x + 3 \log(x + 1)$
   $\quad - 5 \log(x - 2)$

75. $\log(\dfrac{1}{x}) \sqrt[3]{\dfrac{x + 1}{(x - 2)^2}}$

   $= \log(\dfrac{1}{x}) + \log \sqrt[3]{\dfrac{x + 1}{(x - 2)^2}}$

   $= \log(\dfrac{1}{x}) + \dfrac{1}{3} \log \dfrac{x + 1}{(x - 2)^2}$

   $= \log 1 - \log x + \dfrac{1}{3} \log(x + 1)$

   $\quad - \dfrac{1}{3} \log (x - 2)^2$

   $= -\log x + \dfrac{1}{3} \log (x + 1)$

   $\quad - \dfrac{2}{3} \log (x - 2)$

77. $\log \dfrac{x + \sqrt{x^2 - 1}}{x - \sqrt{x^2 - 1}}$

   $= \log \dfrac{(x + \sqrt{x^2 - 1})(x + \sqrt{x^2 - 1})}{(x - \sqrt{x^2 - 1})(x + \sqrt{x^2 - 1})}$

   $= \log \dfrac{(x + \sqrt{x^2 - 1})^2}{x^2 - (x^2 - 1)}$

   $= \log \dfrac{(x + \sqrt{x^2 - 1})^2}{1}$

   $= 2 \log (x + \sqrt{x^2 - 1})$

79. $\dfrac{\log(x + h) - \log x}{h}$

$$= \dfrac{\log\left(\dfrac{x + h}{x}\right)}{h}$$

$$= \frac{1}{h} \log\left(\frac{x + h}{x}\right)$$

$$= \frac{1}{h} \log\left(1 + \frac{h}{x}\right)$$

$$= \log\left(1 + \frac{h}{x}\right)^{1/h}$$

Exercises 6.4

1. $\log 3.72 \simeq 0.5705$

3. $\log 3250 \simeq 3.5119$

5. $\log 71300 = \log(7.13 \times 10^4)$
    $= \log 7.13 + \log 10^4$
    $\simeq 0.8531 + 4 = 4.8531$

7. $\log 0.0225$
    $= \log(2.25 \times 10^{-2})$
    $= \log(2.25) + \log 10^{-2}$
    $\simeq 0.3522 + (-2)$
    $= -1.6478$

9. $\log 51.2 = \log(5.12 \times 10)$
    $= \log 5.12 + \log 10$
    $\simeq 0.7093 + 1 = 1.7093$

11. $\log x = 3.5391 = 0.5391 + 3$
    $\simeq \log(3.46) + \log 10^3$
    $= \log(3.46 \times 10^3)$
    $= \log 3460.$

    So
    $x \simeq 3460$

13. $\log x = 4.9253 = 0.9253 + 4$
    $\simeq \log 8.4198 + \log 10^4$
    $= \log(8.4198 \times 10^4)$
    $= \log 84,198.$

    So
    $x \simeq 84,198.$

15. $\log x = -2.2636$
    $= -3 + 0.7364$
    $\simeq \log 10^{-3} + \log 5.45$
    $= \log(5.45 \times 10^{-3})$

    So
    $x \simeq 5.45 \times 10^{-3}$

17. $\log x = 1.3181 = 1 + 0.3181$
    $\simeq \log 10 + \log 2.08$
    $= \log(2.08 \times 10).$

    Thus
    $x \simeq 20.8.$

19. $\log x = 5.3243 - 8$
    $= 0.3243 + 5 - 8$
    $= 0.3243 - 3$
    $\simeq \log 2.11 + \log 10^{-3}$
    $= \log(2.11 \times 10^{-3})$

    so
    $x \simeq 2.11 \times 10^{-3}$

21. a) We have
    $pH = -\log(1.1 \times 10^{-5})$
    $= -\log 1.1 - \log 10^{-5}$
    $= -\log 1.1 - (-5)$
    $= -\log 1.1 + 5$
    $\simeq -0.0414 + 5$
    $= 4.9586 \simeq 5.0$

    b) $pH = -\log(3.1 \times 10^{-7})$
    $= -\log 3.1 - \log 10^{-7}$
    $= -\log 3.1 + 7$
    $\simeq -0.4914 + 7$
    $= 6.5086 = 6.5$

23. <u>Sea water</u>: pH = 7.8. Thus
$$7.8 = -\log[H^+] \text{ or}$$

$$\log[H^+] = -7.8 = 0.2 - 8$$
$$\simeq \log 1.5849 + \log 10^{-8}$$
$$= \log(1.5849 \times 10^{-8})$$

thus
$$[H^+] \simeq 1.5849 \times 10^{-8}$$
$$\simeq 1.6 \times 10^{-8} \text{ moles per liter.}$$

<u>Soil</u>: pH = 5. Thus
$$5 = -\log[H^+] \text{ or}$$

$$\log[H^+] = -5$$
$$H^+ = 10^{-5} \text{ moles per liter.}$$

25. $pH = -\log[H^+]$
$6.0 < pH < 8.0$
$6.0 < -\log[H^+] < 8.0$
$-6 > \log [H^+] > -8$

Write in exponential form

$$10^{-6} > [H^+] > 10^{-8}$$

27. $L = 10 \log(I/I_o)$, where
$I_o = 10^{-16}$ watt/cm^2.
If $I = 10^{-4}$ watt/cm^2, then

$$L = 10 \log(10^{-4}/10^{-16})$$
$$= 10 \log(10^{12})$$
$$= 10(12) = 120 \text{ decibels}$$

Since $120 > 90$, the decibel level is dangerous.

29. $R = \log(I/I_o)$.
If $I = 10^{6.5}I_o$, then

$$R = \log(10^{6.5}I_o/I_o)$$
$$= \log(10^{6.5})$$
$$= 6.5 \log 10$$
$$= 6.5$$

31. $R = \log(\dfrac{I}{I_o})$, $R = 5.6$

$$\log(\dfrac{I}{I_o}) = 5.6$$

$$\dfrac{I}{I_o} = 10^{5.6}$$

$$I = 10^{5.6}I_o.$$

33. Let I_1 be the earthquake with intensity 7.5 and let I_2 be the earthquake with intensity 6.5. Proceeding as in Exercise 31, obtain the following relations

$$\dfrac{I_1}{I_o} = 10^{7.5}$$

and

$$\dfrac{I_2}{I_o} = 10^{6.5}.$$

Dividing the first one by the second, obtain

$$\dfrac{I_1}{I_2} = 10^{7.5 - 6.5} = 10,$$

so
$$I_1 = 10I_2.$$

35. $M = -2.5 \log(kI)$.

Let I_1 denote the intensity of the light from Sirius. We have

$$-1.42 = -2.5 \log(kI_1)$$

$$\log(kI_1) = \frac{1.42}{2.5} = 0.568$$

$$kI_1 = 10^{0.568}$$

If I_2 denotes the intensity of the light from Canopus, then

$$-0.72 = -2.5 \log(kI_2)$$

$$\log(kI_2) = \frac{0.72}{2.5} = 0.288$$

$$kI_2 = 10^{0.288}$$

The ratio of the intensities of light from Sirius and Canopus is

$$\frac{I_1}{I_2} = \frac{10^{0.568}}{10^{0.288}}$$

$$= 10^{0.568 - 0.288}$$
$$= 10^{0.28}$$
$$= 1.9054607 \simeq 1.91.$$

37. From Exercise 36, we know that Canopus has magnitude -0.72, so

$$-0.72 = -2.5 \log(kI).$$

If M is the magnitude of a star whose light intensity is 1/50 that of Canopus, we have

$$M = -2.5 \log(\frac{kI}{50})$$

Subtracting from this equation the first one, we get

$$M + 0.72 = -2.5[\log(\frac{kI}{50})$$

$$- \log(kI)]$$

$$=-2.5 \log(\frac{kI}{50} \times \frac{1}{kI})$$

$$=-2.5 \log(\frac{1}{50})$$

$$\simeq 4.25$$

So

$$M \simeq 4.25 - 0.72 = 3.53.$$

39. $F(t) = 65 - 15 \log(t + 1)$

a) $t = 0$
$$F(0) = 65 - 15 \log(0 + 1)$$
$$= 65 - 15 \log(1)$$
$$= 65$$

b) $t = 2, \ t = 6$
$$F(2) = 65 - 15 \log(2 + 1)$$
$$= 65 - 15 \log(3)$$
$$\simeq 58$$

$$F(6) = 65 - 15 \log(6 + 1)$$
$$= 65 - 15 \log(7)$$
$$\simeq 52.$$

Exercises 6.5

1. $2^{x-1} = 16$
$2^{x-1} = 2^4$
$x - 1 = 4$
$x = 5$

3. $3^{2x-6} = 27^{3-x}$
$3^{2x-6} = (3^3)^{3-x}$
$3^{2x-6} = 3^{9-3x}$
$2x - 6 = 9 - 3x$
$5x = 15$
$x = 3$

5. $\left(\dfrac{1}{4}\right)^{x+4} = 8^{2x+4}$

$(1/2^2)^{x+4} = (2^3)^{2x+4}$
$(2^{-2})^{x+4} = (2^3)^{2x+4}$
$2^{-2x-8} = 2^{6x+12}$
$-2x - 8 = 6x + 12$
$-8x = 20$

$$x = -\frac{20}{8} = -\frac{5}{2}$$

7. $(\sqrt{2})^{6x+1} = 16^{x+2}$
$(2^{1/2})^{6x+1} = (2^4)^{x+2}$
$2^{3x+1/2} = 2^{4x+8}$

$3x + \dfrac{1}{2} = 4x + 8$

$-x = 8 - \dfrac{1}{2} = \dfrac{15}{2}$

$x = -15/2$

9. $7^x = 2$
$\log(7^x) = \log 2$
$x \log 7 = \log 2$

$x = \dfrac{\log 2}{\log 7} \simeq 0.356$

11. $3^{-x} = 8$
$\log(3^{-x}) = \log 8$
$-x \log 3 = \log 8$

$x = -\dfrac{\log 8}{\log 3} \simeq -1.893$

13. $10^{5x+1} = 21$
$\log(10^{5x+1}) = \log 21$
$(5x+1)\log 10 = \log 21$
$5x + 1 = \log 21$

$x = \dfrac{\log 21 - 1}{5} \simeq 0.064$

15. $2^{3x-1} = 3^{1-2x}$
$\log(2^{3x-1}) = \log(3^{1-2x})$
$(3x-1)\log 2 = (1-2x)\log 3$
$3x \log 2 - \log 2$
$\quad = \log 3 - 2x \log 3$
$3x \log 2 + 2x \log 3$
$\quad = \log 3 + \log 2$
$x(3 \log 2 + 2 \log 3)$
$\quad = \log 3 + \log 2$

$x = \dfrac{\log 3 + \log 2}{3 \log 2 + 2 \log 3}$

$\simeq 0.4190$

17. $\log x - \log 2 = 1$

$\log\left(\dfrac{x}{2}\right) = 1$

$\dfrac{x}{2} = 10$

$x = 20$

19. The equation

$$\log(2x + 4) = 1 + \log(x - 2)$$

makes sense only when $2x + 4 > 0$ and $x - 2 > 0$, that is, when $x > 2$. Now

$\log(2x + 4) = 1 + \log(x - 2)$
$\log(2x + 4) - \log(x - 2) = 1$

$\log\left(\dfrac{2x + 4}{x - 2}\right) = 1$

$\dfrac{2x + 4}{x - 2} = 10$

$2x + 4 = 10(x - 2)$
$2x - 10x = -20 - 4$
$-8x = -24$
$x = 3$

21. $\log(x + 1) = 1 - \log(x + 2)$

We must have $x + 1 > 0$ and $x + 2 > 0$, that is $x > -1$. Solving the given equation, we get

$$\log(x + 1) + \log(x + 2) = 1$$
$$\log[(x + 1)(x + 2)] = 1$$
$$(x + 1)(x + 2) = 10$$
$$x^2 + 3x + 2 = 10$$
$$x^2 + 3x - 8 = 10$$

Using the quadratic formula, we obtain the solutions

$$x = \frac{-3 \pm \sqrt{41}}{2}.$$

Since $\dfrac{-3 - \sqrt{41}}{2} < -1$

and $\dfrac{-3 + \sqrt{41}}{2} > -1$, it follows

that $x = \dfrac{-3 + \sqrt{41}}{2}$ is the only

solution of the given equation.

23. $(\log x)^2 + \log x^2 = 0$

Here x must be a positive number. We have

$$(\log x)^2 + 2 \log x = 0$$
$$\log x(\log x + 2) = 0$$

$\log x = 0$	$\log x + 2 = 0$
$x = 10^0$	$\log x = -2$
$x = 1$	$x = 10^{-2}$

Thus the given equation has two solutions: 1 and 10^{-2}.

25. $\log(\log x) = 1$

From the definition of logarithms it follows that $\log x = 10^1 = 10$. Applying once again, the definition, obtain $x = 10^{10}$.

27. $\dfrac{e^x + e^{-x}}{2} = 1$

$$e^x + e^{-x} = 2$$
   [Multiply both sides by e^x]

$$e^{2x} + 1 = 2e^x$$
$$e^{2x} - 2e^x + 1 = 0$$

   [Set $u = e^x$]

$$u^2 - 2u + 1 = 0$$
$$(u - 1)^2 = 0.$$

So $u = 1$, that is, $e^x = 1$, hence $x = 0$.

29. $\dfrac{e^x - e^{-x}}{e^x + e^{-x}} = \dfrac{1}{2}$

$$2e^x - 2e^{-x} = e^x + e^{-x}$$
$$e^x - 3e^{-x} = 0$$

   [Multiply both sides by e^x]

$$e^{2x} - 3 = 0$$
$$e^{2x} = 3$$
$$2x \ln e = \ln 3$$
$$2x = \ln 3$$

$$x = \frac{\ln 3}{2} \simeq 0.5493$$

31. Applying the change of base formula, get

$$\log_5 4 = \frac{\ln 4}{\ln 5} = \frac{\ln 2^2}{\ln 5}$$

$$= \frac{2 \ln 2}{\ln 5} \simeq \frac{2(0.7)}{1.6}$$

$$= \frac{1.4}{1.6} = 0.875$$

33. $\log 18 = \dfrac{\ln 18}{\ln 10} = \dfrac{\ln(2 \times 3^2)}{\ln(2 \times 5)}$

$\qquad = \dfrac{\ln 2 + \ln 3^2}{\ln 2 + \ln 5}$

$\qquad = \dfrac{\ln 2 + 2 \ln 3}{\ln 2 + \ln 5}$

$\qquad \simeq \dfrac{0.7 + 2(1.1)}{0.7 + 1.6} = \dfrac{2.9}{2.3}$

$\qquad \simeq 1.26$

35. $\log_6 30 = \dfrac{\ln 30}{\ln 6} = \dfrac{\ln(2 \times 3 \times 5)}{\ln(2 \times 3)}$

$\qquad = \dfrac{\ln 2 + \ln 3 + \ln 5}{\ln 2 + \ln 3}$

$\qquad \simeq \dfrac{0.7 + 1.1 + 1.6}{0.7 + 1.1} = \dfrac{3.4}{1.8}$

$\qquad \simeq 1.889$

37. $\log_{15} 24 = \dfrac{\ln 24}{\ln 15} = \dfrac{\ln(2^3 \times 3)}{\ln(3 \times 5)}$

$\qquad = \dfrac{3 \ln 2 + \ln 3}{\ln 3 + \ln 5}$

$\qquad \simeq \dfrac{3(0.7) + 1.1}{1.1 + 1.6}$

$\qquad = \dfrac{3.2}{2.7} \simeq 1.185$

39. Applying the change of base formula, get

$\qquad \log_2 9 = \dfrac{\log 9}{\log 2} = \dfrac{\log 3^2}{\log 2}$

$\qquad\qquad = \dfrac{2 \log 3}{\log 2} \simeq \dfrac{2(0.5)}{0.3}$

$\qquad\qquad = \dfrac{1}{0.3} \simeq 3.33$

41. $\log_3 12 = \dfrac{\log 12}{\log 3}$

$\qquad\qquad = \dfrac{\log(2^2 \times 3)}{\log 3}$

$\qquad = \dfrac{\log 2^2 + \log 3}{\log 3}$

$\qquad = \dfrac{2 \log 2 + \log 3}{\log 3}$

$\qquad = \dfrac{2(0.3) + 0.5}{0.5} = \dfrac{1.1}{0.5}$

$\qquad = 2.2$

43. $\log_6 24 = \dfrac{\log 24}{\log 6}$

$\qquad = \dfrac{\log(2^3 \times 3)}{\log(2 \times 3)}$

$\qquad = \dfrac{3 \log 2 + \log 3}{\log 2 + \log 3}$

$\qquad \simeq \dfrac{3(0.3) + 0.5}{0.3 + 0.5} = \dfrac{1.4}{0.8}$

$\qquad \simeq 1.75$

45. $\log_4 (1/6) = \log_4 1 - \log_4 6$

$\qquad\qquad = 0 - \log_4 6$

$\qquad = -\dfrac{\log 6}{\log 4}$

$\qquad = -\dfrac{\log(2 \times 3)}{\log 2^2}$

$\qquad = -\dfrac{\log 2 + \log 3}{2 \log 2}$

$\qquad \simeq -\dfrac{0.3 + 0.5}{2(0.3)} = -\dfrac{0.8}{0.6}$

$\qquad \simeq -1.33$

47. $\log_7 8 = \dfrac{\log 8}{\log 7}$

$\qquad \simeq 1.0686$

49. $\log_2 5.4 = \dfrac{\log 5.4}{\log 2}$

$\qquad = 2.4329594$

$\qquad \simeq 2.4330$

51. $\log_3 1.23 = \dfrac{\log 1.23}{\log 3} \approx \dfrac{0.0899}{0.4771}$

≈ 0.1884

53. $\log_4(2.3 \times 10^{-2})$

$= \dfrac{\log(2.3 \times 10^{-2})}{\log 4}$

$= \dfrac{\log(2.3) - 2}{\log 4}$

$\approx \dfrac{0.3617 - 2}{0.6021}$

$= \dfrac{-1.6383}{0.6021} \approx -2.7210$

55. Applying the change of base formula, get

$\log_2 11 = \dfrac{\log_4 11}{\log_4 2} \approx \dfrac{1.7}{0.5} = 3.4$

57. $n(t) = n_o 2^{2t}$, $n_o = 10,000$

a) if $n(t) = 320,000$, then
$320,000 = 10,000 \cdot 2^{2t}$
$2^{2t} = 32$
$2^{2t} = 2^5$

so

$2t = 5$
$t = 5/2 = 2.5$

b) if $n(t) = 2,560,000$, then
$2,560,000 = 10,000 \cdot 2^{2t}$
$2^{2t} = 256$
$2^{2t} = 2^8$

so

$2t = 8$
$t = 4$

59. $Q(t) = Q_o \left(\dfrac{1}{2}\right)^{t/25}$

$\dfrac{\cancel{Q_o}}{10} = \cancel{Q_o}\left(\dfrac{1}{2}\right)^{t/25}$

$\dfrac{1}{10} = \left(\dfrac{1}{2}\right)^{t/25}$

$\left(\dfrac{1}{2}\right)^{t/25} = \dfrac{1}{10}$

$\dfrac{t}{25} \log\left(\dfrac{1}{2}\right) = \log\left(\dfrac{1}{10}\right)$

$\dfrac{t}{25} \log 2 = \log 10$

$\dfrac{t}{25} = \dfrac{1}{\log 2}$

$t = \dfrac{25}{\log 2} \approx 83$ minutes

61. The rate of decay is given by

$r = -(\ln 2)/T$

where T is the half-life. So

$r = -\dfrac{\ln 2}{6 \times 10^{11}}$

$\approx -1.1552 \times 10^{-12}$

63. $p(h) = 14.7 \, e^{-0.12h}$

At sea level the atmospheric pressure is p(0) = 14.7.

a) if p(h) = 14.7/2, then

$$\frac{14.7}{2} = 14.7 \, e^{-0.12h}$$

$$e^{-0.12h} = \frac{1}{2}$$

$$-0.12h = \ln(\frac{1}{2}) = -\ln 2$$

$$h = \frac{\ln 2}{0.12} \simeq 5.78 \text{ miles}$$

b) if p(h) = 14.7/3, then

$$\frac{14.7}{3} = 14.7 \, e^{-0.12h}$$

$$e^{-0.12h} = \frac{1}{3}$$

$$-0.12h = \ln(\frac{1}{3}) = -\ln 3$$

$$h = \frac{\ln 3}{0.12} \simeq 9.16 \text{ miles}$$

65. $PV = Ae^{-rt}$
a) $r = 0.08$, A = \$5,000, and $t = 5$
$$PV = 5,000 \, e^{-0.08(5)}$$
$$= 5,000 \, e^{-0.4}$$
$$= 5,000(0.67032)$$
$$\simeq 3,351.60$$

b) PV = 4,282, $t = 10$, $A = 8,000$
$$4,282 = 8,000 \, e^{-10r}$$
$$e^{-10r} = 0.53525$$
$$-10r = \ln(0.53525)$$
$$-10r = -0.6250213$$
$$r \simeq 0.0625 \text{ or } 6.25\%$$

67. $T = A + Ce^{-kt}$
$A = -5^0$

If $t = 0$, then T = 100
$$100 = -5 + C$$
so
$$C = 105$$
$$T = -5 + 105e^{-kt}$$

After 4 minutes, T = 16, so
$$16 = -5 + 105e^{-4k}$$
$$105e^{-4k} = 21$$

$$e^{-4k} = \frac{21}{105} = \frac{1}{5} = 0.2$$

Taking natural logarithms, get

$$-4k = \ln 0.2 = -1.609$$
$$k = 0.402$$
$$T = -5 + 105e^{-0.402t}$$

Find T when $t = 6$ min:

$$T = -5 + 105e^{-0.402(6)}$$
$$= -5 + 105e^{-2.412}$$
$$= 4.4° \text{ C}$$

69. $P(t) = P_o(1 + \dfrac{r}{N})^{Nt}$

$P_o = 5,000$, $r = 0.12$, $N = 12$

$$10,000 = 5,000(1 + \frac{0.12}{12})^{12t}$$

$$2 = (1.01)^{12t}$$
$$\log 2 = \log[(1.01)^{12t}]$$
$$\log 2 = 12t \, \log(1.01)$$

$$\frac{\log 2}{12 \, \log(1.01)} = t$$

$$t \simeq 5.8 \text{ years.}$$

Review Exercises - Chapter 6

1. a) If $y = 5^{\log_5 8}$ then, by definition of logarithm, get $\log_5 y = \log_5 8$ and so $y = 8$.

   b) If $y = 3^{\log_3 7}$, then $\log_3 y = \log_3 7$ and so $y = 7$.

   c) $y = e^{\ln 25}$ means that $\ln y = \ln 25$ and so $y = 25$.

   d) $y = 10^{\log 2}$ is equivalent to $\log y = \log 2$ and so $y = 2$.

3. a) $f(x) = (\frac{1}{4})^x$

   $f(3) = (\frac{1}{4})^3 = \frac{1}{4^3} = \frac{1}{64}$

   $f(2) = (\frac{1}{4})^2 = \frac{1}{4^2} = \frac{1}{16}$

   $f(-2) = (\frac{1}{4})^{-2} = \frac{1}{4^{-2}}$

   $= 4^2 = 16$

   b) $g(x) = x^2(2^{-x})$
   $g(-0.5) = (-0.5)^2 2^{-(-0.5)}$

   $= 0.25(2)^{0.5} = 0.25\sqrt{2}$
   ≈ 0.3536

   $g(0.01) = (0.01)^2 2^{-0.01}$
   $= (10^{-2})^2 2^{-0.01}$
   $\approx 0.9931 \times 10^{-4}$
   $= 9.9931 \times 10^{-5}$

   $g(0.5) = (0.5)^2 2^{-0.5}$

   $= 0.25(\frac{1}{\sqrt{2}}) \approx 0.1768$

5. $q(t) = ab^t$

   If $q(0) = 10$, then $10 = ab^0$, that is, $a = 10$. Replacing into the given equation, obtain

   $q(t) = 10b^t$.

   If $q(1) = 20$, then $20 = 10b^1$, that is, $b = 2$. Substituting 2 for b into the last equation, get

   $q(t) = 10(2^t)$.

   Next

   $q(3) = 10(2^3)$
   $= 80$.

7. $S(t) = S_0(1 + \frac{r}{2})^t$

   a) If $S(0) = 16$, then

   $16 = S_0(1 + \frac{r}{2})^0$ and $S_0 = 16$.

   Rewrite the original equation as follows

   $S(t) = 16(1 + \frac{r}{2})^t$.

   If $S(3) = 250$, then

   $250 = 16(1 + \frac{r}{2})^3$

   $(1 + \frac{r}{2})^3 = \frac{250}{16} = \frac{125}{8}$

   $= (\frac{5}{2})^3$

   $1 + \frac{r}{2} = \frac{5}{2}$

   $r = 3$.

   Substituting $\frac{5}{2}$ for $1 + \frac{r}{2}$ into the last expression, get

   $S(t) = 16(\frac{5}{2})^t$

   b) $S(2) = 16(\frac{5}{2})^2 = 16(\frac{25}{4}) = 100$

9. a) $f(x) = 2^x - 2^{-x}$

    Since $f(-x) = 2^{-x} - 2^x$
    $= -(2^x - 2^{-x}) = -f(x)$, it
    follows that $f(x)$ is an odd
    function, so its graph is
    symmetric with respect to the
    origin. The following is a
    table of function values
    for f:

x	0	1	2	3
$f(x)$	0	1.5	3.75	7.875

    b) $f(x) = \dfrac{2}{e^x - e^{-x}}$

    The domain of f is the set of
    all real numbers $x \neq 0$. You
    may check that $f(-x) = -f(x)$,
    so the function is odd and the
    graph is symmetric with
    respect to the origin. The
    following table gives us
    approximate values for f:

x	1/2	1	2
$f(x)$	1.92	0.85	0.28

11. $\log_4(x + 2)$ is equivalent to
    $x + 2 = 4^2$
    $x + 2 = 16$
    $x = 14$

13. $\log_8(2x + 1) = 2/3$
    $2x + 1 = 8^{2/3} = (2^3)^{2/3}$
    $2x + 1 = 4$
    $x = 3/2$

15. $\log(2x + 4) = \log 10 - \log 6$
    $\log(2x + 4) = \log(10/6)$
    $2x + 4 = 10/6$
    $2x + 4 = 5/3$
    $6x + 12 = 5$
    $x = -7/6$

17. $2 \log_3 x = 3 \log_3 4$
    $\log_3 x^2 = \log_3 4^3$

$x^2 = 4^3$
$x^2 = 2^6$
$x = 8$

(Note that -8 is an extraneous
solution.)

19. $\log x - \log(x - 1) = 2 \log 4$
    First observe that x must be > 1,
    otherwise $\log(x - 1)$ is not
    defined. Next, we have

    $$\log x - \log(x - 1) = 2 \log 4$$

    $$\log \frac{x}{x - 1} = \log 16$$

    $$\frac{x}{x - 1} = 16$$

    $$x = 16x - 16$$
    $$15x = 16$$

    $$x = \frac{16}{15} > 1.$$

Thus $\dfrac{16}{15}$ is the solution.

21. $4^{2x - 1} = 16$
    $4^{2x - 1} = 4^2$
    $2x - 1 = 2$
    $2x = 3$

    $x = \dfrac{3}{2}$

23. $4^{2x + 1} = 8^{x - 3}$
    $(2^2)^{2x + 1} = (2^3)^{x - 3}$
    $2^{4x + 2} = 2^{3x - 9}$
    $4x + 2 = 3x - 9$
    $x = -11$

25. $(\dfrac{1}{3})^{1 - 2x} = 9^{3x - 2}$

    $(3^{-1})^{1 - 2x} = (3^2)^{3x - 2}$
    $3^{-1 + 2x} = 3^{6x - 4}$
    $-1 + 2x = 6x - 4$
    $-4x = -3$
    $x = 3/4$

27. $(\sqrt[3]{2})^{x+3} = 4^{x-2}$
 $(2^{1/3})^{x+3} = (2^2)^{x-2}$

 $\dfrac{x}{3} + 1 = 2x - 4$

 $x + 3 = 6x - 12$
 $15 = 5x$
 $x = 3$

29. $3^{2x+1} = 4^{3x-2}$
 $\log 3^{2x+1} = \log 4^{3x-2}$
 $(2x+1)\log 3 = (3x-2)\log 4$
 $2x \log 3 - 3x \log 4$
 $\quad = -2 \log 4 - \log 3$
 $x(2 \log 3 - 3 \log 4)$
 $\quad = -2 \log 4 - \log 3$
 $x(\log 9 - \log 64)$
 $\quad = -\log 16 - \log 3$
 $x \log(9/64) = -\log 48$

 $x = -\dfrac{\log 48}{\log(9/64)} \simeq 1.9734$

31. $\dfrac{3^x + 3^{-x}}{2} = 1$

 $3^x + 3^{-x} = 2$

 Multiply both sides by 3^x, to get

 $3^{2x} + 1 = 2(3^x).$

 Set $u = 3^x$ and obtain

 $u^2 - 2u + 1 = 0$
 $(u - 1)^2 = 0$
 $u = 1$

 So $3^x = 1$ and $x = 0$.

33. $\ln 7 + \ln 12 - \ln 5$
 $= \ln(7 \times 12) - \ln 5$

 $= \ln(\dfrac{7 \times 12}{5}) = \ln(\dfrac{84}{5})$

35. $3 \ln 4 - 2 \ln 5 + 3 \ln 8$
 $= \ln 4^3 - \ln 5^2 + \ln 8^3$

 $= \ln(\dfrac{4^3}{5^2}) + \ln 8^3$

 $= \ln(\dfrac{4^3 8^3}{5^2}) = \ln(\dfrac{32768}{25})$

37. $7 \log a - \dfrac{1}{2} \log b + \dfrac{3}{4} \log c$

 $= (\log a^7 - \log b^{1/2}) + \log c^{3/4}$

 $= \log(\dfrac{a^7}{b^{1/2}}) + \log c^{3/4}$

 $= \log(\dfrac{a^7 c^{3/4}}{b^{1/2}})$

 $= \log(\dfrac{a^7 \sqrt[4]{c^3}}{\sqrt{b}})$

39. $2 \log(x+1) - 3 \log(x+2)$
 $\quad + 4 \log x$

 $= (\log(x+1)^2 - \log(x+2)^3)$
 $\quad + \log x^4$

 $= \log(\dfrac{(x+1)^2}{(x+2)^3}) + \log x^4$

 $= \log[\dfrac{x^4 (x+1)^2}{(x+2)^3}]$

41. $\ln(\dfrac{x(x-1)^2}{(2x+1)^3})$

 $= \ln(x(x-1)^2) - \ln(2x+1)^3$
 $= \ln x + \ln(x-1)^2$
 $\quad - \ln(2x+1)^3$
 $= \ln x + 2 \ln(x-1)$
 $\quad - 3 \ln(2x+1)$

43. $\ln(\dfrac{\sqrt{x(x-1)^3}}{(2x+1)^4})$

$= \ln(\sqrt{x(x-1)^3} - \ln(2x+1)^4$
$= \ln[(x(x-1)^3)^{1/2}]$
$\quad - \ln(2x+1)^4$
$= \ln(x^{1/2}(x-1)^{3/2})$
$\quad - \ln(2x+1)^4$
$= \ln x^{1/2} + \ln(x-1)^{3/2}$
$\quad - \ln(2x+1)^4$

$= \dfrac{1}{2} \ln x + \dfrac{3}{2} \ln(x-1)$

$\quad - 4 \ln(2x+1)$

45. a) $\log(32.14 \times 10^{-3})$
$\quad = \log(32.14) + \log 10^{-3}$
$\quad = \log(32.14) + (-3)$
$\quad \simeq 1.5070 - 3$
$\quad \simeq -1.4930$

b) $\log(257.1 \times 10^{-5})$
$\quad = \log(257.1) + \log 10^{-5}$
$\quad = \log(257.1) - 5$
$\quad \simeq 2.4101 - 5$
$\quad \simeq -2.5899$

c) $\log 0.006254 \simeq -2.2038$

d) $\log 4516000 \simeq 6.6548$

47. a) Lemon juice: $[H^+] = 5.1 \times 10^{-3}$
$\quad pH = -\log(5.1 \times 10^{-3})$
$\quad\quad = 2.2924298$
$\quad\quad \simeq 2.3$

b) Beer: $[H^+] = 5 \times 10^{-5}$
$\quad pH = -\log(5 \times 10^{-5})$
$\quad\quad = 4.30103$
$\quad\quad \simeq 4.3$

49. $L = 10 \log(I/I_o)$, $I_o = 10^{-16}$,
$\quad I \simeq 10^{-8}$, so
$\quad L = 10 \log(10^{-8}/10^{-16})$
$\quad\quad = 10 \log(10^8)$
$\quad\quad = 80$ db.

51. $R = \log(I/I_o)$

Let I_1 be the intensity of the earthquake of magnitude 5.5 and I_2 the intensity of the earthquake of magnitude 7.

$5.5 = \log(I_1/I_o)$

$\dfrac{I_1}{I_o} = 10^{5.5}$

$7 = \log(I_2/I_o)$

$\dfrac{I_2}{I_o} = 10^7$

So

$\dfrac{I_1}{I_2} = \dfrac{10^{5.5}}{10^7} = 10^{-1.5}$

$I_1 = 10^{-1.5}I_2$
$\quad \simeq 3.16 \times 10^{-2}I_2$

53. $M = -2.5 \log(kI)$

If I_1 denotes the light intensity of Vega, a star whose magnitude is 0.04, then we have

$(*)\quad 0.04 = -2.5 \log(kI_1)$

If M_2 is the magnitude of a star whose light intensity is $I_2 = I_1/50$, then

$M_2 = -2.5 \log(kI_2)$
$\quad = -2.5 \log(kI_1/50)$
$\quad = -2.5[\log(kI_1) - \log 50]$
$\quad = -2.5 \log(kI_1) + 2.5 \log 50$

[Taking into account (*)]

$\quad = 0.04 + 2.5 \log 50$
$\quad = 0.04 + 2.5(1.69897)$
$\quad \simeq 4.29$

55. $e^r = 2^{1/T}$, $r = 0.021$

$e^{0.021} = 2^{1/T}$

$\ln e^{0.021} = \ln 2^{1/T}$

$0.021 = \frac{1}{T} \ln 2$

$T = \frac{\ln 2}{0.021}$

$T \simeq 33$ years

57. $Q(t) = Q_o (\frac{1}{2})^{t/T}$ where $T = 1.3 \times 10^9$

years and $Q_o = 1.25 \times 10^{-3}$ gm.
Thus

$Q(t) = 1.25 \times 10^{-3} (\frac{1}{2})^{t/1.3 \times 10^9}$

$Q(5 \times 10^9)$

$= 1.25 \times 10^{-3} (\frac{1}{2})^{5 \times 10^9 / 1.3 \times 10^9}$

$= 1.25 \times 10^{-3} (0.5)^{3.8461538}$

$= 8.69 \times 10^{-5}$ gm

59. $p(t) = p_o e^{rt}$, $r = 0.075$,

$p(8) = \$12,500$

$12,500 = p_o e^{0.075(8)}$

$12,500 = p_o e^{0.6}$

$p_o = 12,500 \cdot e^{-0.6}$

$= 12,500(0.5488116)$

$\simeq \$6,860.15$

61. $V(t) = C(1 - \frac{2}{N})^t$, $C = \$1,200$,

$N = 6$, so

$V(t) = 1200(1 - \frac{2}{6})^t$

$= 1200(\frac{2}{3})^t$

$V(4) = 1200(\frac{2}{3})^4$

$= 1200(\frac{16}{81})$

$\simeq \$237.04$

$V(6) = 1200(\frac{2}{3})^6$

$= 1200(\frac{64}{729})$

$\simeq \$105.35$

63. $T(t) = A + Ce^{-kt}$, where $A = 70°$, so

$T(t) = 70 + Ce^{-kt}$

When $t = 0$, $T(0) = 325°$, so

$325 = 70 + C$

$C = 255.$

Thus

$T(t) = 70 + 255e^{-kt}$

Since $T(1) = 250$, it follows that

$250 = 70 + 255e^{-k}$

$255e^{-k} = 180$

$e^{-k} = \frac{180}{255} = \frac{36}{51}$

Therefore,

$T(t) = 70 + 255(\frac{36}{51})^t$

Find T when $t = 5$:

$T(5) = 70 + 255(\frac{36}{51})^5$

$= 114.68912$

$\simeq 114.7°$ F

65. $\log_a(m/n) = \log_a m - \log_a n$.
To prove it set $u = \log_a m$ and $v = \log_a n$. We have

$$m = a^u \text{ and } n = a^v$$

hence

$$\frac{m}{n} = \frac{a^u}{a^v} = a^{u-v}$$

that is

$$\log_a(m/n) = u - v$$
$$= \log_a m - \log_a n.$$

67. $\log_{ab} x = \dfrac{\log_a x}{1 + \log_a b}$

Rewrite the right-hand side as follows

$$\frac{\log_a x}{1 + \log_a b} = \frac{\log_a x}{\log_a a + \log_a b}$$

$$= \frac{\log_a x}{\log_a(ab)}$$

$$= \log_{ab} x$$

by the change of base formula.

69. $\log_a x = 2 \log_{a^2} x$

Set

$$\log_a x = y$$

and

$$2 \log_{a^2} x = \log_{a^2} x^2 = z$$

By the definition of logarithms, we get

$$x = a^y \text{ and } x^2 = (a^2)^z = a^{2z}$$

Squaring the first relation and using the second one, we obtain

$$a^{2y} = x^2 = a^{2z},$$
hence

$$2y = 2z$$
$$y = z.$$

Chapter 6 Test

1. Let $f(x) = 6^x$. Without using a table or calculator, find $f(-2)$, $f(2)$, $f(-3/2)$, $f(4/3)$. Give exact answers. Do not approximate.

2. Given $f(x) = 2 + \dfrac{4^{x+2}}{8}$, find $f(-2)$, $f(-1/2)$, $f(0)$, and $f(2)$.

3. Use Table 1 to approximate the following numbers

   a) $e^{-0.70}$ b) $e^{1/4}$
   c) $e^{-3/4}$ d) $e^{-2.40}$

4. Use your calculator to complete the following table

x	1.7	1.73	1.732
2^x			

   Round off your answers to four decimal places. Also find $2^{\sqrt{3}}$ and compare your results.

5. If $g(x) = 4^{-x} + 3$, find $g(-0.5)$, $g(3/2)$, and $g(-5/2)$.

6. Let $f(x) = AB^x$ where A and B are positive constants. If $f(0) = -5$ and $f(2) = -20$, find A and B. Next find $f(-3)$.

7. Assume that a colony of bacteria growing exponentially doubles every 4 hours. If the initial number of bacteria is 1,600 write a formula giving the number of bacteria after t hours. Find a) the number of bacteria after 8 hours, b) the number of bacteria after 10 hours.

8. The amount $A(t)$ of Carbon 14 remaining after t years is given by

$$A(t) = A_o \left(\frac{1}{2}\right)^{t/5600}$$

   a) What does A_o represent?
   b) What is the half-life of Carbon 14?
   c) If 160 mg of Carbon 14 are present initially (when $t = 0$), find the amount remaining after 22,400 years.

9. According to Newton's law of cooling, the temperature $T(t)$ of a body at time t after being introduced in an environment having constant temperature A is given by

$$T(t) = A + Ce^{-kt}.$$

   Suppose that a bowl containing boiling water, at 212° F, is placed in a freezer at 32° F. If after 25 minutes the temperature of the water is 122° F, find its temperature after 50 minutes.

10. The atmospheric pressure, $p(h)$, in pounds per square inch, is given by

$$p(h) = 13.7 \, e^{-0.21h},$$

where h is the height (in miles) above the sea level. Find the atmospheric pressure at 2 miles. At what height is the atmospheric pressure half of the pressure at sea level?

11. The population of a city is declining according to the relationship

$$P(t) = P_o e^{-0.02t}$$

where t is time measured in years. If at time $t = 0$ the population is 600,000, find the population after 10 years. How long will it take for the population to be reduced to 300,000?

12. During 1985, Claudia is planning to make a contribution of \$2,000 towards her Individual Retirement Account. If the bank pays interest at a fixed rate of 10% per year compounded quarterly, when will the initial \$2,000 be worth \$4,000?

13. Rewrite each of the following expressions in logarithmic form

   a) $6^4 = 1296$ b) $4^{-5} = 1/1024$
   c) $a = 3^8$ d) $x^{-3} = 15$

14. Rewrite each of the following expressions in exponential form

   a) $\log_7 343 = 3$ b) $\log_{100} 10 = 1/2$
   c) $\log_4 x = -2$ d) $\log_5 18 = y$

15. Write $\log_2 \dfrac{4\sqrt{10}}{50}$ as a sum or difference of logarithms. Simplify your

   result as much as possible.

16. Use the properties of logarithms to write

$$3 \log_4 2 + \frac{1}{2} \log_4 25 - 2 \log_4 10 - 1$$

as a single logarithm with a coefficient of 1.

17. Write the expression

$$\frac{3}{4} \log(3x + 1) + \frac{1}{2} \log(2x + 7) - \log(x + 4)$$

as a single logarithm with a coefficient of 1. Assume that all variables represent nonnegative real numbers.

18. Write $\ln(\dfrac{8p^2q^5}{r^4})$ as a sum or difference of logarithms.

19. Use Table 2 or a calculator to approximate the following numbers to four decimal places.

    a) $\log 7.36$ b) $\log 0.0217$

20. Use Table 2 or a calculator to approximate x

    a) $\log x = 3.5843$ b) $\log x = -1.1273$

21. The relationship

$$L = 10 \log(I/I_o)$$

   gives the sound level, in decibels, of a sound intensity I, where I_o is the minimum intensity detectable by our ears. Find the sound level, in decibels, of the clatter of a chain saw if $I = 3.2 \times 10^{10} I_o$.

22. Compare the intensities I_1 and I_2 of two earthquakes whose magnitudes on the Richter scale are 8.2 and 7.2 respectively.

23. Let $\log_2 3 \approx 1.58$ and $\log_2 5 \approx 2.32$. Approximate the numbers

    a) $\log_2 10$ b) $\log_2 36$

24. Given $\ln 2 \approx 0.7$, $\ln 3 \approx 1.1$, and $\ln 5 \approx 1.6$, approximate the number $\log_5 6$ without the use of tables or calculators.

25. Find the exact solution of the equation $2^{5x-1} = 4^{x+2}$.

26. Solve the logarithmic equation $\log_3(4x + 7) - 2 = \log_3(x - 2)$.

27. Solve the exponential equation $2^{4x-3} = 3^{2x+1}$. Give your answer in exact simplified form. Do not approximate. Do not use a calculator.

28. Solve the equation $\log x + \log(x - 2) = 2 \log(x + 1)$.

29. Find all solutions of the equation $\log(x + 1)^3 = [\log(x + 1)]^2$.

30. Compute the product $(\log_2 3)(\log_3 4)(\log_4 5)$ without the use of tables or calculators. [Hint. Express each factor in terms of logarithms to the base 2.]

Chapter 7
Right Triangle Trigonometry

Exercises 7.1

1. $510°$, $-210°$

3. $300°$, $-420°$

5. $\pi/3$, $-5\pi/3$

7. Second quadrant

9. First quadrant

11. Second quadrant

13. $150° = \dfrac{150\pi}{180} = \dfrac{5\pi}{6}$ rad

15. $-240° = \dfrac{-240\pi}{180}$

    $= -\dfrac{4\pi}{3}$ rad

17. $-120° = \dfrac{-120\pi}{180}$

    $= -\dfrac{2\pi}{3}$ rad

19. $\dfrac{5\pi}{6} = \dfrac{5\pi}{6} \cdot \dfrac{180}{\pi} = 150°$

21. $-\dfrac{\pi}{6} = \dfrac{\pi}{6} \cdot \dfrac{180}{\pi} = -30°$

23. $-\dfrac{9\pi}{4} = -\dfrac{9\pi}{4} \cdot \dfrac{180}{\pi} = -405°$

25. $\theta = \dfrac{s}{r} = \dfrac{20}{4} = 5$ rad

27. $\theta = \dfrac{600}{24} = 25$ rad

29. $\theta = \dfrac{2\pi}{6} = \dfrac{\pi}{3}$ rad $= 60°$

31. $\theta = \dfrac{2\pi}{12} = \dfrac{\pi}{6}$ rad $= 30°$

33. $s = \theta r = 5(12) = 60$ cm

35. $\theta = 30° = \dfrac{\pi}{6}$ rad

    $r = \dfrac{s}{\theta} = \dfrac{12}{\pi/6} = \dfrac{72}{\pi}$ m

37. $\theta = \dfrac{s}{r}$

    $\theta = 4.05$ rad, $s = 9.72$ cm

    $4.05 = \dfrac{9.72}{r}$

    $r = \dfrac{9.72}{4.05} = 2.4$ cm

39. $30°44' = 30 + \dfrac{44}{60} = 30.73°$

41. $85°30'45'' = 85 + \dfrac{30}{60} + \dfrac{45}{3600}$

    $= 85.5125°$
    $\simeq 85.51°$

43. 4 rad $= 4(\dfrac{180}{\pi}) \simeq 229.183°$

45. 1.5 rad $= 1.5(\dfrac{180}{\pi})$

    $\simeq 85.9444°$

47. 5 rad $= 5(\dfrac{180}{\pi})$

    $= 286.4789°$
    $\simeq 286°28'44''$

49. 1.2 rad $= 68.754935$
    $\simeq 68°45'18''$

51. 0.15 rad $= 8.594367$
    $\simeq 8°35'40''$

53. $45° = 45(\dfrac{\pi}{180})$

    $= 0.785398$
    $\simeq 0.7854$ rad

55. $38° = 38(\dfrac{\pi}{180})$

    $\simeq 0.6632$ rad

57. $126° \simeq 2.1991$ rad

59. $38°18' = 38.3°$
    $\simeq 0.6685$ rad

61. $67°40' = 67.\overline{6}°$

    $= 67.\overline{6}(\dfrac{\pi}{180})$

    $\simeq 1.1810$ rad

63. $12°30'15'' = 12.5041\overline{6}°$

    $= 12.50416° \ (\dfrac{\pi}{180})$

    $\simeq 0.2182$ rad

65. In one hour the hour hand turns 1/12 of a revolution, thus

    $\dfrac{1}{12} (2\pi) = \dfrac{\pi}{6}$ rad.

    In half hour it turns 1/24 of a revolution, thus

    $\dfrac{1}{24} (2\pi) = \dfrac{\pi}{12}$ rad.

    In 15 hours it turns 1 1/4 revolutions, thus

    $\dfrac{5}{4} (2\pi) = \dfrac{5\pi}{2}$ rad.

67. a) The rate of 32 rpm corresponds to 32/60 = 8/15 revolutions per second. Since the angular speed of one revolution per second is 2π rad/s, it follows that the angular speed of the stone is

$$\omega = \frac{8}{15}(2\pi) = \frac{16\pi}{15} \text{ rad/s}.$$

The radius of the circle is 1 m, so the linear speed is

$$v = \frac{16\pi}{15} \text{ m/s}$$

$$\simeq 3.35 \text{ m/s}.$$

b) In 5 seconds the length of the arc described by the stone is

$$s = 5(\frac{16\pi}{15}) = \frac{16\pi}{3}\text{m}.$$

69. a) 78 rpm = 78/60 revolutions per second.

$$w = \frac{78}{60}(2\pi) = \frac{13\pi}{5} \text{ rad/s}$$

b) $v = rw$, $r = 6$ in

$$v = 6(\frac{13\pi}{5}) = \frac{78\pi}{5} \text{ in/s}$$

71. In 365 days the angle swept by the line is 2π radians.

a) In 10 days the angle is

$$\frac{10}{365}(2\pi) = \frac{4\pi}{73} \text{ radians.}$$

b) In one month the angle is

$$\frac{30}{365}(2\pi) = \frac{12\pi}{73} \text{ radians.}$$

73. $\theta = \dfrac{s}{r}$

$$s = 120 \text{ cm}, \quad r = \frac{30}{2} = 15 \text{ cm}$$

$$\theta = \frac{120}{15} = 8 \text{ rad}$$

75. a) $s = \dfrac{\pi}{6}(6.371 \times 10^3)$

$$= 3.336 \times 10^3 \text{ km}$$

b) $s = \dfrac{\pi}{3}(6.371 \times 10^3)$

$$= 6.672 \times 10^3 \text{ km}$$

c) $s = 1(6.371 \times 10^3)$
$$= 6.371 \times 10^3 \text{ km}$$

77. $s = \theta r$
$r = 3.959 \times 10^3$ mi

$$1' = (\frac{1}{60})^\circ = 2.9 \times 10^{-4} \text{ rad}$$

$$s = (3.959 \times 10^3)(2.9 \times 10^{-4})$$
$$\simeq 1.15 \text{ mi}$$

79. $s = \theta r$
$r = 3.95 \times 10^3$ m
$12° + 41° = 53° = 0.9250$ rad
$s = (3959)(0.9250) \simeq 3662$ mi

Exercises 7.2

1. $\sin \alpha = \dfrac{4}{5}$ $\quad \cos \alpha = \dfrac{3}{5}$

$\tan \alpha = \dfrac{4}{3}$ $\quad \cot \alpha = \dfrac{3}{4}$

$\sec \alpha = \dfrac{5}{3}$ $\quad \csc \alpha = \dfrac{5}{4}$

3. By the Pythagorean Theorem the hypotenuse is

$$c = \sqrt{15^2 + 8^2} = 17. \quad \text{So}$$

$$\sin \alpha = \frac{8}{17} \qquad \cos \alpha = \frac{15}{17}$$

$$\tan \alpha = \frac{8}{15} \qquad \cot \alpha = \frac{15}{8}$$

$$\sec \alpha = \frac{17}{15} \qquad \csc \alpha = \frac{17}{8}$$

5. The hypotenuse is

$$c = \sqrt{2^2 + 3^2} = \sqrt{13}. \quad \text{So}$$

$$\sin \alpha = \frac{3}{\sqrt{13}} = \frac{3\sqrt{13}}{13},$$

$$\cos \alpha = \frac{2}{\sqrt{13}} = \frac{2\sqrt{13}}{13},$$

$$\tan \alpha = \frac{3}{2}, \quad \cot \alpha = \frac{2}{3},$$

$$\sec \alpha = \frac{\sqrt{13}}{2}, \quad \csc \alpha = \frac{\sqrt{13}}{3}$$

7. $\sin \alpha = 0.5573$
   $\cos \alpha = 0.8303$
   $\tan \alpha = 0.6712$
   $\cot \alpha = 1.4900$
   $\sec \alpha = 1.2043$
   $\csc \alpha = 1.7944$

9. The hypotenuse is 3.616.
   So $\sin \alpha = 0.556$,
   $\cos \alpha = 0.831$,
   $\tan \alpha = 0.670$,
   $\cot \alpha = 1.493$,
   $\sec \alpha = 1.204$,
   $\csc \alpha = 1.797$

11. $\dfrac{5}{3} = \dfrac{b}{4}$

$$b = 20/3$$

13. $\dfrac{2.03}{6.35} = \dfrac{a}{2.14}$

$$a = 2.14(\frac{2.03}{6.35}) = 0.68$$

15. $\sin \alpha = 5/8$. Sketch the corresponding right angle and find that the length of the other leg is $\sqrt{39}$. Thus

$$\cos \alpha = \frac{\sqrt{39}}{8}, \quad \tan \alpha = \frac{5\sqrt{39}}{39},$$

$$\cot \alpha = \frac{\sqrt{39}}{5}, \quad \sec \alpha = \frac{8\sqrt{39}}{39}$$

$$\csc \alpha = \frac{8}{5}$$

17. $\cos \alpha = \dfrac{3}{4}$. Thus $\sin \alpha = \dfrac{\sqrt{7}}{4}$,

$$\tan \alpha = \frac{\sqrt{7}}{3}, \quad \cot \alpha = \frac{3\sqrt{7}}{7},$$

$$\sec \alpha = \frac{4}{3}, \quad \csc \alpha = \frac{4\sqrt{7}}{7}.$$

19. $\tan \alpha = \dfrac{3}{5}$. $\sin \alpha = \dfrac{3\sqrt{34}}{34}$,

$$\cos \alpha = \frac{5\sqrt{34}}{34}, \quad \cot \alpha = \frac{5}{3},$$

$$\sec \alpha = \frac{\sqrt{34}}{5}, \quad \csc \alpha = \frac{\sqrt{34}}{3}$$

21. $\cot \alpha = \sqrt{2}$. $\sin \alpha = \dfrac{\sqrt{3}}{3}$,

$$\cos \alpha = \frac{\sqrt{6}}{3}, \quad \tan \alpha = \frac{\sqrt{2}}{2},$$

$$\sec \alpha = \frac{\sqrt{6}}{2}, \quad \csc \alpha = \sqrt{3}$$

23. $\csc \alpha = \dfrac{3}{2}$. Thus

$$\sin \alpha = \frac{2}{3}, \quad \cos \alpha = \frac{\sqrt{5}}{3},$$

$$\tan \alpha = \frac{2\sqrt{5}}{5}, \quad \cot \alpha = \frac{\sqrt{5}}{2},$$

$$\sec \alpha = \frac{3\sqrt{5}}{5}$$

25. $\sin 48° = \cos 42°$

27. $\tan \dfrac{5\pi}{12} = \cot \dfrac{\pi}{12}$

29. $\sec 78° = \csc 12°$

31. $\sin \alpha = \cos (90° - \alpha) = 0.5299$

33. $\sin (\dfrac{\pi}{2} - \alpha) = \dfrac{4}{5} = \cos \alpha$.

Sketch the corresponding right triangle and find that $\sin \alpha = 3/5$.

35. $\sin (\dfrac{\pi}{2} - \alpha) = \dfrac{3}{4} = \cos \alpha$.

Sketch a right triangle and find that $\tan \alpha = \sqrt{7}/3$.

37. $\tan (90° - \alpha) = 7/3 = \cot \alpha$.

Sketch a right triangle, find the hypotenuse, and obtain

$$\sin \alpha = 3\sqrt{58}/58.$$

39. $\tan 15°30' = 0.2773$

41. $\cos 56°40' = 0.5495$

43. $\sin 0.1047 = 0.1045$

45. $\tan 0.3927 = 0.4142$

47. $\sin 22.6° = 0.3843$

49. $\tan 12.05° = 0.2135$

51. $\sin 23°15' = \sin 23.25° = 0.3947$

53. $\tan 48°12' = \tan 48.2° = 1.1184$

55. $\tan \alpha = \dfrac{a}{b} = \dfrac{a/c}{b/c} = \dfrac{\sin \alpha}{\cos \alpha}$

$$\sec \alpha = \frac{c}{b} = \frac{1}{b/c} = \frac{1}{\cos \alpha}$$

57. $\sin \alpha = \dfrac{a}{c}$, $\cos \alpha = \dfrac{b}{c}$, and $c^2 = a^2 + b^2$.

$$(\sin \alpha)^2 + (\cos \alpha)^2 = \frac{a^2}{c^2} + \frac{b^2}{c^2}$$

$$= \frac{a^2 + b^2}{c^2}$$

$$= \frac{c^2}{c^2} = 1.$$

59. $\tan \alpha = \dfrac{\sin \alpha}{\cos \alpha}$

$$(\tan \alpha)^2 + 1 = \frac{(\sin \alpha)^2}{(\cos \alpha)^2} + 1$$

$$= \frac{(\sin \alpha)^2 + (\cos \alpha)^2}{(\cos \alpha)^2}$$

$$= \frac{1}{(\cos \alpha)^2} = (\sec \alpha)^2$$

Exercises 7.3

1. $\alpha = 60°$, $a = 10$

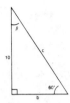

$\beta = 30°$

$$\frac{10}{b} = \tan 60° \approx 1.732$$

$b \approx 5.77 \approx 6$

$$\frac{10}{c} = \sin 60° \approx 0.8660$$

$c = 11.55 \approx 12$

3. Since $a = b = 3$, it follows that $\tan \alpha = 1$, thus $\alpha = 45°$ and $\beta = 45°$. Next,

$$c = \sqrt{a^2 + b^2} = \sqrt{3^2 + 3^2} = \sqrt{18}$$

$$= 3\sqrt{2}.$$

5. $\beta = 25°$, $a = 7$
    $\alpha = 90° - 25° = 65°$

$$\frac{b}{7} = \tan 25° = 0.4663$$

$b = 7(0.4663)$
$\quad = 3.2642$
$\quad \approx 3$

$$\frac{c}{7} = \sec 25° = 1.103$$

$c = 7.7236$
$c \approx 8$

7. $\alpha = 32°10'$, $a = 15$
    $\beta = 90° - 32°10' = 57°50'$

$$\frac{b}{15} = \tan 57°50' = 1.590$$

$b = 23.850$
$\quad = 24$

$$\frac{c}{15} = \sec 57°50' = 1.878$$

$c = 28.175$
$c \approx 28$

9. $\alpha = 75°20'$, $b = 12$
    $\beta = 90° - 75°20' = 14°40'$

$$\frac{a}{12} = \tan 75°20' = 3.821$$

$a = 45.852$
$\quad \approx 46$

$$\frac{c}{12} = \sec 75°20' = 3.950$$

$c = 47.4$
$\quad \approx 47$

11. $a = 5$, $b = 12$

$$c = \sqrt{5^2 + 12^2} = 13$$

$$\sin \alpha = \frac{a}{c} = \frac{5}{13}$$

$\alpha = 22.619865°$
$\quad \approx 22°37'11''$

$$\sin \beta = \frac{b}{c} = \frac{12}{13}$$

$\beta = 67.380135°$
$\quad \approx 67°22'49''$

13. $\alpha = 30°20'$, $b = 10.5$
    $\beta = 59°40'$

$$\frac{a}{b} = \frac{a}{10.5} = \tan 30°20'$$

$$a = 10.5(0.5851335)$$
$$= 6.1439019$$
$$\simeq 6.1$$

$$\frac{b}{c} = \frac{10.5}{c} = \cos 30°20'$$

$$c = 10.5 \sec 30°20'$$
$$= 10.5(1.1586118)$$
$$\simeq 12.2$$

15. $\beta = 72°31'$, $c = 18.5$
    $\alpha = 17°29'$

$$\frac{a}{18.5} = \cos 72°31'$$

$$a = 18.5(0.3004283)$$
$$= 5.5579247$$
$$\simeq 5.6$$

$$\frac{b}{18.5} = \sin 72°31'$$

$$b = 18.5(0.9538043)$$
$$\simeq 17.6$$

17. $\alpha = 32°26'$, $a = 125.6$
    $\beta = 57°34'$

$$\frac{a}{b} = \frac{125.6}{b} = \tan 32°26'$$

$$b = 125.6 \cot 32°26'$$
$$= (125.6)(1.5737234)$$
$$= 197.65966$$
$$\simeq 197.7$$

$$\frac{a}{c} = \frac{125.6}{c} = \sin 32°26'$$

$$c = 125.6 \csc 32°26'$$
$$= (125.6)(1.8645657)$$
$$\simeq 234.2$$

19. $\alpha = 36°15'$, $b = 218.3$
    $\beta = 53°45'$

$$\frac{a}{218.3} = \tan 36°15'$$

$$a = (218.3)(0.7332303)$$
$$\simeq 160.1$$

$$\frac{218.3}{c} = \cos 36°15'$$

$$c = 218.3 \sec 36°15'$$
$$\simeq 270.7$$

21. We have

$$\frac{h}{2 \times 5280} = \sin 5°$$

$$h = 10560(0.0871557)$$
$$= 920.36464$$
$$\simeq 920 \text{ ft.}$$

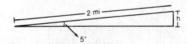

23.

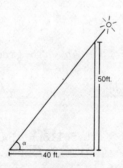

$$\tan \alpha = \frac{50}{40} = \frac{5}{4}$$

$$\alpha \simeq 51°20'$$

25.

$$\cos \alpha = \frac{45}{60} = 0.75$$

$$\alpha = 41.409622°$$
$$\simeq 41°25'$$

27. We have that

$$\tan 75° = \frac{w}{30}.$$

Thus $w = 111.96$
$$\simeq 112 \text{ ft.}$$

29.

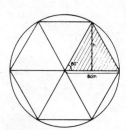

The area of the hexagon is A = $6 \cdot A_T$ where A_T is the area of the shaded triangle.

Now, $A_T = \frac{1}{2} 8h = 4h$. To find h

notice that $\frac{h}{8} = \sin 60° = \frac{\sqrt{3}}{2}$,

so $h = 4\sqrt{3}$ cm. It follows that
$A = 6 \cdot A_T = 6(4 \cdot 4\sqrt{3}) = 96\sqrt{3}$ cm^2.

31. We have (see figure on page 322 of textbook),

$$1 = \tan 45° = \frac{h}{1500 + x},$$

So $1500 + x = h$. On the other hand,

$$\cot 60° = \frac{\sqrt{3}}{3} = \frac{x}{h}$$

so that $x = \frac{\sqrt{3}}{3}h$.

Replacing above, we obtain

$$h = 1500 + \frac{\sqrt{3}}{3}h$$

$$h - \frac{\sqrt{3}}{3} h = 1500$$

$$h(\frac{3 - \sqrt{3}}{3}) = 1500$$

$$h = \frac{4500}{3 - \sqrt{3}} = 750(3 + \sqrt{3})$$

$$\simeq 3549\text{m}$$

33.

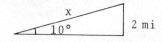

Let x be the distance that the airliner flies to reach the altitude of 4 mi. We have

$$\csc 10° = \frac{x}{2}, \text{ so}$$

$$x = 11.5175 \text{ mi.}$$

Since the speed is constant and equal to 400 mi/h, we have

$$t = \frac{11.5175}{400}$$

$$= 0.0288 \ h = 1.727 \text{ min.}$$
$$t \simeq 2 \text{ min}$$

35. Let x be the height of the steeple and let h be the distance from the top of the steeple to the ground (see figure on page 323 of the textbook). Then

$$\tan 22° = \frac{h}{300}$$

so $h = 300 \times \tan 22°$
$= 300(0.4040)$
$\simeq 121.20$ ft.

Also

$$\tan 15° = \frac{h - x}{300}$$

$h - x = 300 \times \tan 15°$
$= 300(0.2679)$
$\simeq 80.38$ ft.

It follows that $x = 121.20 - 80.38$
$= 40.82 \simeq 41$ ft and $h \simeq 121$ ft.

37. We have $\tan 40° = \dfrac{h}{50}$, thus

$h = 50 \times \tan 40°$
$= 50(0.8391)$
$= 41.9550$
$\simeq 42$ m

39.

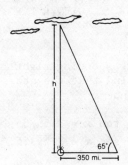

$$\tan 65° = \frac{h}{350}$$

$h \simeq 750.57$
$\simeq 751$ m

Review Exercises - Chapter 7

1. $72° = 72° \cdot \dfrac{\pi}{180°}$ rad

$= \dfrac{2\pi}{5}$ rad

$15° = 15° \cdot \dfrac{\pi}{180°}$ rad

$= \dfrac{\pi}{12}$ rad

$-50° = -50° \cdot \dfrac{\pi}{180}$ rad

$= \dfrac{-5\pi}{18}$ rad

$135° = 135° \cdot \dfrac{\pi}{180°}$ rad

$= \dfrac{3\pi}{4}$ rad

$-85° = -85° \cdot \dfrac{\pi}{180°}$ rad

$= \dfrac{-17\pi}{36}$ rad

3. $\dfrac{2\pi}{9} = \dfrac{2(180°)}{9} = 40°$

$\dfrac{7\pi}{5} = \dfrac{7(180°)}{5} = 252°$

$\dfrac{3\pi}{15} = \dfrac{\pi}{5} = \dfrac{180°}{5} = 36°$

$\dfrac{7\pi}{18} = \dfrac{7(180°)}{18} = 70°$

$\dfrac{\pi}{12} = \dfrac{180°}{12} = 15°$

5. $25°15' = 0.4407$ rad
$120°36' = 2.1049$ rad
$75°10' = 1.3119$ rad
$68°10'25" = 1.1899$ rad
$210°5'12" = 3.6667$ rad

7. $4 \text{ rad} = \dfrac{4(180°)}{\pi} = 229.18°$

$10 \text{ rad} = \dfrac{10(180°)}{\pi} = 572.96°$

$1.5 \text{ rad} = \dfrac{1.5(180°)}{\pi} = 85.94°$

$4.8 \text{ rad} = \dfrac{4.8(180°)}{\pi} = 275.02°$

$2.25 \text{ rad} = \dfrac{2.25(180)°}{\pi} = 128.92°$

9.   Since $s = r\theta$ we have that

$\theta = \dfrac{s}{r} = \dfrac{60}{8} = \dfrac{15}{2}$

   $= 7.5$ radians

11.   Since $s = r\theta$ it follows that

$r = \dfrac{s}{\theta} = \dfrac{12.16}{6.4}.$

Thus $r = 1.9$ ft.

13. $s = r\theta$
$r = 12 \quad \theta = 7.5° = 0.1309 \text{ rad}$
$s = 12(0.1309) = 1.57$ cm

15. $r = 2.5$, 2400 rpm

a) $\omega = 2400(2\pi)$
     $= 4800\pi$ rad/min
     $= 80\pi$ rad/sec

b) $d = \omega r$
     $= 90\pi(2.5) = 628.318$
     ≈ 628.3 m

17.   Radius of Earth: 4000 mi
Radius of space shuttle's orbit:
4175 mi
2π radians in 90 minutes

$\omega = \dfrac{2\pi}{90} = \dfrac{\pi}{45}$ rad/min

$v = \dfrac{\pi}{45} (4175)$

   $= 291.5$ mi/min
   ≈ 17488 mi/h

19.   Diameter: 22 in.
Rate: 16 rpm
$\omega = 16(2\pi) = 32\pi$ rad/min
$v = 32\pi(11) = 352\pi$ in/min
Thus in one minute the bike
travels 352π in ≈ 1106 in.

21. $a = 4$, $b = 7$, so $c = \sqrt{65}$
$\sin \alpha = 4/\sqrt{65} = 4\sqrt{65}/65$,
$\cos \alpha = 7\sqrt{65}/65$, $\tan \alpha = 4/7$,
$\cot \alpha = 7/4$, $\sec \alpha = \sqrt{65}/7$,
$\csc \alpha = \sqrt{65}/4$

23. $a = 8$, $c = 14$, so
$b = \sqrt{132} = 2\sqrt{33}$.
$\sin \alpha = 8/14 = 4/7$,
$\cos \alpha = \sqrt{33}/7$,
$\tan \alpha = 4\sqrt{33}/33$,
$\cot \alpha = \sqrt{33}/4$,
$\sec \alpha = 7\sqrt{33}/33$,
$\csc \alpha = 7/4$

25. $b = 1/4$, $c = 1$, so
$a = \sqrt{1 - (1/4)^2} = \sqrt{15}/4$
$\sin \alpha = \sqrt{15}/4$, $\cos \alpha = 1/4$,
$\tan \alpha = \sqrt{15}$, $\cot \alpha = \sqrt{15}/15$,
$\sec \alpha = 4$, $\csc \alpha = 4\sqrt{15}/15$

27. $\sin x = 3/8$
Sketch the corresponding right
triangle and find that the other
leg is $\sqrt{55}$. Thus

$\cos x = \sqrt{55}/8$, $\tan x = 3\sqrt{55}/55$,
$\cot x = \sqrt{55}/3$, $\sec x = 8\sqrt{55}/55$,
$\csc x = 8/3$.

29. $\tan x = 3$
Find that $c = \sqrt{10}$. Thus,
$\sin x = 3\sqrt{10}/10$, $\cos x = \sqrt{10}/10$,
$\cot x = 1/3$, $\sec x = \sqrt{10}$,
$\csc x = \sqrt{10}/3$

31. $\sin(90° - x) = 2/5 = \cos x$
We have $\tan \alpha = \sqrt{21}/2$

33. $\alpha = 30°$, $a = 6$

$\sin \alpha = \dfrac{6}{c}$

$\dfrac{1}{2} = \dfrac{6}{c}$, $c = 12$

$b^2 + 6^2 = 12^2$, $b = \sqrt{108} = 6\sqrt{3} \approx 10.4$
$\beta = 60°$

35. $\beta = 25°10'$, $b = 12$
$\alpha = 90 - 25°10' = 64°50'$

$\sin 25°10' = \dfrac{12}{c}$

$c = \dfrac{12}{0.4252} = 28.2 \approx 28$

$\tan 25°10' = \dfrac{12}{a}$

$a = \dfrac{12}{0.4699} = 25.5 \approx 26$

37. $a = 7$, $b = 24$
$c^2 = 7^2 + 24^2$, $c = \sqrt{625} = 25$

$\sin \alpha = \dfrac{7}{25} = 0.28$, $\alpha \approx 16°16'$

$\beta = 90° - 16°16' = 73°44'$

39. $\alpha = 45°20'$, $c = 10.4$
$\beta = 44°40'$

$\sin \alpha = \dfrac{a}{10.4}$

$a = (10.4)(0.7112) \approx 7.4$

$\sin \beta = \dfrac{b}{10.4}$

$b = (10.4)(0.7030) \approx 7.3$

41.

$\tan \alpha = \dfrac{5}{120} = 0.0416$

So $\alpha = 2°23'$

43. $\tan 65°20' = \dfrac{h}{18}$

So $h = 39.19$ ft ≈ 39 ft

45.

We have

$\dfrac{h}{1200 + x} = \tan 30 = \dfrac{\sqrt{3}}{3}$

$h = \dfrac{\sqrt{3}}{3}(1200 + x)$

Also

$\dfrac{h}{x} = \tan 60° = \sqrt{3}$

$h = \sqrt{3}\, x$

Thus

$\sqrt{3}\, x = \dfrac{\sqrt{3}}{3}(1200 + x)$

$x = 400 + \dfrac{x}{3}$

$\dfrac{2x}{3} = 400$

$x = 600$ m and
$h = 600\sqrt{3}$ m
≈ 1039 m

47.

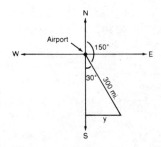

$$\frac{x}{300} = \cos 30° = \frac{\sqrt{3}}{2}$$

$$x = 150\sqrt{3} \text{ mi}$$

$$\frac{y}{300} = \sin 30° = \frac{1}{2}$$

$$y = 150 \text{ mi}$$

The plane will be $150\sqrt{3}$ mi $\simeq 260$ mi south and 150 mi east of the airport.

49. Angle: 30°16',
    altitude: 10,500 ft

$$\frac{x}{10500} = \cot 30°16' = 1.7136$$

$$x \simeq 17,993 \text{ ft}$$

Chapter 7 Test

1. Find the angle α, $0° \le \alpha < 360°$, that is coterminal with the angle of measure
    a) 420°, b) -210°.

2. Find the angle β, $0 \le \beta < 2\pi$, that is coterminal with the angle of measure
    a) $17\pi/6$, b) $-5\pi/4$.

3. In what quadrant does the terminal side of the following angles lie?
    a) 475°   b) $-8\pi/3$

4. Convert each of the following degree measures to radians. Leave your answers in simplified form and as multiple of π.
    a) -330°   b)150°

5. Convert each of the following radian measures to degrees.
    a) $11\pi/9$ b) $-7\pi/18$

6. On a circle of radius 8 cm, a central angle θ is subtended by and arc whose length is 36 cm. What is the radian measure of θ?

7. What are the radian and degree measures of a central angle subtended by an arc $\pi/6$ meters long if the radius of the circle measures a) 1 meter, b) 3 meters?

8. Convert each of the following angles to decimal degrees. Round off your answers to two decimal places.
    a) 2 radians   b) $2\pi/11$ radians

9. A 12-inch diameter record is rotating at a rate of 33 revolutions per minute. Find a) the angular speed of the record, b) the angle swept in 5 seconds, c) the linear speed of a point on the rim.

10. The cities of Quito (Equador) and Macapa (Brazil) are approximately on the equator. Quito is at 78° of longitude west and Macapa is at 51° of longitude west. Find the approximate distance (in miles) between these two cities knowing that the radius of the Earth is 4,000 miles.

11. A right triangle has sides $a = 21$ cm, $b = 20$ cm, and $c = 29$ cm. Find the trigonometric function values of β, the angle opposite to side b.

12. The hypotenuse of a right triangle is 12 inches long and one of its legs is 4 inches long. Find the trigonometric function values of the angle α adjacent to the leg 4 inches long.

13. Let α be one of the acute angles in a right triangle such that $\cos \alpha = 3/8$. Find the remaining five trigonometric function values of α. Give your answer in exact form. Do not approximate.

14. Suppose that csc β = 4, with β an acute angle in a right triangle. Find the remaining five trigonometric function values of β. Give your answer in exact form.

15. Suppose that cos $(90° - \alpha)$ = 0.40. Find sin α and cos α.

16. Let tan $(90° - \alpha)$ = 2/3, with α an acute angle in a right triangle. Find sin α and cos α.

17. In a 45° – 45° – 90° triangle, one leg measures 6 inches. Find the length of the hypotenuse.

18. The hypotenuse of a 30° – 60° – 90° triangle is 12 feet long. Find the lengths of the two legs of the triangle.

19. Use Table 4 in the textbook to approximate the following numbers.
    a) sin 46°30'   b) tan 20°40'

20. Approximate each of the numbers to four decimal places
    a) cos 25.40°   b) cot 42.16°

In problems 21-24, two elements of a right triangle are given. Use Table 4 in the textbook or a calculator to approximate the other elements.

21. α = 30°, a = 8   22. β = 45°, a = 6

23. α = 40°10', b = 3.4   24. α = 15°30', a = 5.6

25. What is the perimeter of a square inscribed in a circle of radius 2 m?

26. Find the area of a pentagon inscribed in a circle of radius 4 ft.

27. The angle of elevation of a kite is 62°. How high is the kite if the string is taut and measures 55 ft?

28. How long is the shadow of a tower 250 ft high when the angle of elevation of the sun is 35°?

29. At a point 180 meters from the base of a monument, the angle of elevation to the top of the monument is 32°. How tall is the monument?

30. From a point 2 miles away, Kristen is watching the launching of the space shuttle. At a certain instant, she observes that the angle of elevation is 8°. A few seconds later, the angle of elevation is 15°. How far did the shuttle rise between the two observations?

Chapter 8
Trigonometric or Circular Functions

Exercises 8.1

1. $\theta = 5\pi/4$, $P(-\sqrt{2}/2, -\sqrt{2}/2)$
   $\sin \theta = -\sqrt{2}/2$,
   $\cos \theta = -\sqrt{2}/2$, $\tan \theta = 1$,
   $\cot \theta = 1$, $\sec \theta = -\sqrt{2}$,
   $\csc \theta = -\sqrt{2}$

3. $\alpha = 255°$, $P(-\sqrt{2}/2, -\sqrt{2}/2)$
   $\sin \alpha = -\sqrt{2}/2$,
   $\cos \alpha = -\sqrt{2}/2$, $\tan \alpha = 1$,
   $\cot \alpha = 1$, $\sec \alpha = -\sqrt{2}$,
   $\csc \alpha = -\sqrt{2}$

5. $\alpha = 5\pi/2$, $P(0,1)$
   $\sin \alpha = 1$, $\cos \alpha = 0$
   $\tan \alpha$ - not defined,
   $\cot \alpha = 0$, $\sec \alpha$ - not defined,
   $\csc \alpha = 1$

7. $\theta = -135°$, $P(-\sqrt{2}/2, -\sqrt{2}/2)$
   $\sin \theta = -\sqrt{2}/2$, $\cos \theta = -\sqrt{2}/2$,
   $\tan \theta = 1$, $\cot \theta = 1$,
   $\sec \theta = -\sqrt{2}$, $\csc \theta = -\sqrt{2}$

9. $\theta = 3\pi$, $P(-1,0)$
   $\sin \theta = 0$, $\cos \theta = -1$,
   $\tan \theta = 0$, $\cot \theta$ - not defined,

   $\sec \theta = -1$, $\csc \theta$ - not defined

11. $\theta = 100\pi$, $P(1,0)$
    $\sin \theta = 0$, $\cos \theta = 1$,
    $\tan \theta = 0$, $\cot \theta$ - not defined,
    $\sec \theta = 1$, $\csc \theta$ - not defined

13. $\theta = -180°$, $P(-1,0)$
    $\sin \theta = 0$, $\cos \theta = -1$,
    $\tan \theta = 0$, $\cot \theta$ - not defined,
    $\sec \theta = -1$, $\csc \theta$ - not defined

15. $\sin \theta = 2/3$, $\cos \theta = \sqrt{5}/3$

    $\tan \theta = \dfrac{\sin \theta}{\cos \theta} = \dfrac{2/3}{\sqrt{5}/3}$

    $= 2/\sqrt{5} = 2\sqrt{5}/5$

    $\cot \theta = \dfrac{\cos \theta}{\sin \theta} = \dfrac{\sqrt{5}/3}{2/3} = \sqrt{5}/2$

    $\sec \theta = \dfrac{1}{\cos \theta} = \dfrac{1}{\sqrt{5}/3}$

    $= 3/\sqrt{5} = 3\sqrt{5}/5$

    $\csc \theta = \dfrac{1}{\sin \theta} = \dfrac{1}{2/3} = 3/2$

167

17. $\sin \theta = 1/4$, $\sec \theta = -4\sqrt{15}/15$

$$\cos \theta = \frac{1}{\sec \theta} = \frac{1}{-4\sqrt{15}/15}$$

$$= -\sqrt{15}/4$$

$$\tan \theta = \sin \theta \cdot \sec \theta$$
$$= (1/4)(-4\sqrt{15}/15)$$
$$= -\sqrt{15}/15$$

$$\cot \theta = \frac{1}{\tan \theta} = -15/\sqrt{15}$$

$$= -\sqrt{15}$$

$$\csc \theta = \frac{1}{\sin \theta} = \frac{1}{1/4} = 4$$

19. $\tan \theta = \sqrt{2}/4$, $\sin \theta = -1/3$
$\cot \theta = 4/\sqrt{2} = 4\sqrt{2}/2 = 2\sqrt{2}$
$\cos \theta = \cot \theta \cdot \sin \theta$
$\qquad = 2\sqrt{2}(-1/3) = -2\sqrt{2}/3$

$$\sec \theta = \frac{1}{-2\sqrt{2}/3} = -3\sqrt{2}/4$$

$$\csc \theta = \frac{1}{-1/3} = -3$$

21. $\sin \theta = -3/4$, $\sec \theta = 4\sqrt{7}/7$

$$\cos \theta = \frac{1}{4\sqrt{7}/7} = 7/4\sqrt{7} = \sqrt{7}/4$$

$$\tan \theta = \frac{-3/4}{\sqrt{7}/4} = -3/\sqrt{7} = -3\sqrt{7}/7$$

$\cot \theta = -\sqrt{7}/3$
$\csc \theta = -4/3$

23. $\sin \theta = 3/4$ and $\cos \theta < 0$

$\cos \theta = \pm\sqrt{1 - (3/4)^2} = \pm\sqrt{7}/4$
Since $\cos \theta < 0$, we have
$\cos \theta = -\sqrt{7}/4$. Then
$\tan \theta = -3\sqrt{7}/7$, $\cot \theta = -\sqrt{7}/3$,
$\sec \theta = -4\sqrt{7}/7$, $\csc \theta = 4/3$

25. $\tan \theta = 3$ and $\sin \theta > 0$.
Since $\tan \theta = 3 > 0$, $\sin \theta > 0$,
it follows that $\cos \theta > 0$ and also
$\sec \theta > 0$. Now

$$\sec \theta = \sqrt{1 + \tan^2\theta} = \sqrt{1 + 9}$$

$$= \sqrt{10}.$$

Next
$\cos \theta = 1/\sqrt{10} = \sqrt{10}/10$
$\sin \theta = \tan \theta \cdot \cos \theta$
$\qquad = 3\sqrt{10}/10$
$\cot \theta = 1/3$
$\csc \theta = \sqrt{10}/3$

27. $\cos \theta = 1/3$ and θ is a fourth-quadrant angle. Since θ is a fourth-quadrant angle, it follows that $\sin \theta < 0$, so

$$\sin \theta = -\sqrt{1 - (1/3)^2} = -2\sqrt{2}/3.$$

Next,
$$\tan \theta = \frac{-2\sqrt{2}/3}{1/3} = -2\sqrt{2}$$

$\cot \theta = -\sqrt{2}/4$
$\sec \theta = 3$
$\csc \theta = -3\sqrt{2}/4$

29. $\tan \theta = -1$ and θ is a second-quadrant angle. If θ is a second-quadrant angle, then $\sin \theta > 0$ and $\cos \theta < 0$, so $\sec \theta < 0$. Now

$$\sec \theta = -\sqrt{1 + \tan^2\theta}$$

$$= -\sqrt{1 + (-1)^2} = -\sqrt{2}.$$

Next
$\cos \theta = -\sqrt{2}/2$,
$\sin \theta = \sqrt{2}/2$,
$\cot \theta = -1$, $\csc \theta = \sqrt{2}$.

31. $\sin \theta = 0.8192$ and $\cos \theta < 0$.

$$\cos \theta = -\sqrt{1 - (0.8192)^2}$$
$$= -0.5735$$
$$\tan \theta = -1.4284,$$
$$\cot \theta = -0.7001,$$
$$\sec \theta = -1.7437,$$
$$\csc \theta = 1.2207$$

33. $\tan \theta = 1.1925$ and $\cos \theta < 0$. If $\cos \theta < 0$, then $\sec \theta < 0$, so

$$\sec \theta = -\sqrt{1 + (1.1925)^2}$$
$$= -1.5563$$
$$\sin \theta = -0.7662,$$
$$\cos \theta = -0.6426,$$
$$\sec \theta = -1.5563,$$
$$\csc \theta = -1.3051$$

35. Be formulas (8.4) in the text we have

$$\sec(-\theta) = \frac{1}{\cos(-\theta)}$$
$$= \frac{1}{\cos \theta} = \sec \theta$$

and

$$\csc(-\theta) = \frac{1}{\sin(-\theta)} = \frac{1}{-\sin \theta}$$
$$= -\frac{1}{\sin \theta} = -\csc \theta.$$

37. By formulas (8.5) and (8.6) in the text, we have

$$\cot \theta = \frac{\cos \theta}{\sin \theta} = \frac{1}{\dfrac{\sin \theta}{\cos \theta}}$$
$$= \frac{1}{\tan \theta}.$$

39. The right triangles OAP and OPM are similar, so

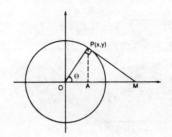

$$\frac{\overline{OM}}{\overline{OP}} = \frac{\overline{OP}}{\overline{OA}}. \text{ But } \overline{OP} = 1$$

(radius of the unit circle) and

$$\overline{OA} = x. \text{ Thus } \overline{OM} = \frac{1}{x} = \frac{1}{\cos \theta}$$
$$= \sec \theta.$$

Exercises 8.2

1. If r denotes the distance from P to the origin, then

$$r = \sqrt{2^2 + 5^2} = \sqrt{29}$$

We have
$\sin \theta = 5/\sqrt{29} = 5\sqrt{29}/29,$
$\cos \theta = 2\sqrt{29}/29,$ $\tan \theta = 5/2$
$\cot \theta = 2/5,$ $\sec \theta = \sqrt{29}/2,$
$\csc \theta = \sqrt{29}/5$

3. $r = \sqrt{(-2)^2 + 6^2} = \sqrt{40} = 2\sqrt{10}$
$\sin \theta = 6/2\sqrt{10} = 3\sqrt{10}/10,$
$\cos \theta = -2/2\sqrt{10} = -\sqrt{10}/10,$
$\tan \theta = 6/-2 = -3,$ $\cot \theta = -1/3$
$\sec \theta = -\sqrt{10},$ $\csc \theta = \sqrt{10}/3$

5. $r = \sqrt{(-3)^2 + (-7)^2} = \sqrt{58}$
$\sin \theta = -7\sqrt{58}/58$
$\cos \theta = -3\sqrt{58}/58,$ $\tan \theta = 7/3,$
$\cot \theta = 3/7,$ $\sec \theta = -\sqrt{58}/3,$
$\csc \theta = -\sqrt{58}/7$

7. $r = \sqrt{2^2 + (-4)^2} = \sqrt{20} = 2\sqrt{5}$
   $\sin \theta = -4/2\sqrt{5} = -2\sqrt{5}/5$,
   $\cos \theta = 2/2\sqrt{5} = \sqrt{5}/5$,
   $\tan \theta = -2$, $\cot \theta = -1/2$,
   $\sec \theta = \sqrt{5}$, $\csc \theta = -\sqrt{5}/2$

9. $r = \sqrt{4^2 + (-5)^2} = \sqrt{41}$
   $\sin \theta = -5\sqrt{41}/41$,
   $\cos \theta = 4\sqrt{41}/41$,
   $\tan \theta = -5/4$, $\cot \theta = -4/5$
   $\sec \theta = \sqrt{41}/4$, $\csc \theta = -\sqrt{41}/5$

11. First-quadrant; $y = 2x$
    The point $P(1,2)$ is on the terminal side of the angle so

    $r = \sqrt{1^2 + 2^2} = \sqrt{5}$

    $\sin \theta = 2\sqrt{5}/5$, $\cos \theta = \sqrt{5}/5$,
    $\tan \theta = 2$, $\cot \theta = 1/2$,
    $\sec \theta = \sqrt{5}$, $\csc \theta = \sqrt{5}/2$

13. Second-quadrant; $y = -3x$
    $P(-1,3)$ lies on the terminal side of the angle and $r = \sqrt{10}$.
    $\sin \theta = 3\sqrt{10}/10$,
    $\cos \theta = -\sqrt{10}/10$, $\tan \theta = -3$,
    $\cot \theta = -1/3$, $\sec \theta = -\sqrt{10}$
    $\csc \theta = \sqrt{10}/3$

15. Third quadrant; $x - y = 0$.
    $P(-1,-1)$, $r = \sqrt{2}$
    $\sin \theta = -\sqrt{2}/2$, $\cos \theta = -\sqrt{2}/2$,
    $\tan \theta = 1$, $\cot \theta = 1$,
    $\sec \theta = -\sqrt{2}$, $\csc \theta = -\sqrt{2}$

17. First-quadrant; $2x - 3y = 0$
    $P(3,2)$, $r = \sqrt{13}$
    $\sin \theta = 2\sqrt{13}/13$,
    $\cos \theta = 3\sqrt{13}/13$, $\tan \theta = 2/3$,
    $\cot \theta = 3/2$, $\sec \theta = \sqrt{13}/3$,
    $\csc \theta = \sqrt{13}/2$

19. Third-quadrant; $3x - 2y = 0$
    $P(-2,-3)$, $r = \sqrt{13}$
    $\sin \theta = -3\sqrt{13}/13$,
    $\cos \theta = -2\sqrt{13}/13$,

$\tan \theta = 3/2$, $\cot \theta = 2/3$,
$\sec \theta = -\sqrt{13}/2$, $\csc \theta = -\sqrt{13}/3$

21. If θ is a first-quadrant angle and $\tan \theta = 3/5$, then $P(5,3)$ lies on the terminal side of the angle and

    $r = \sqrt{5^2 + 3^2} = \sqrt{34}$

    Thus
    $\sin \theta = 3\sqrt{34}/34$,
    $\cos \theta = 5\sqrt{34}/34$,
    $\tan \theta = 3/5$, $\cot \theta = 5/3$,
    $\sec \theta = \sqrt{34}/5$, $\csc \theta = \sqrt{34}/3$

23. If $\cot \alpha = -2$ and $\sin \alpha > 0$, then α is a second-quadrant angle and the point $P(-2,1)$ lies on the terminal side of α. We have

    $r = \sqrt{(-2)^2 + 1^2} = \sqrt{5}$

    $\sin \alpha = \sqrt{5}/5$, $\cos \alpha = -2\sqrt{5}/5$,
    $\tan \alpha = -1/2$, $\cot \alpha = -2$,
    $\sec \alpha = -\sqrt{5}/2$, $\csc \alpha = \sqrt{5}$

25. $\sin \theta = 1/3$ and θ is a second-quadrant angle.
    <u>First method.</u> Since θ is a second-quadrant angle, it follows that $\cos \theta < 0$, so

    $\cos \theta = -\sqrt{1 - (1/3)^2}$

    $\qquad = -2\sqrt{2}/3.$

    Next,
    $\tan \theta = -\sqrt{2}/4$, $\cot \theta = -2\sqrt{2}$
    $\sec \theta = -3\sqrt{2}/4$, $\csc \theta = 3$
    <u>Second method.</u> From the triangle

    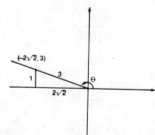

    we can read off the values of the trigonometric functions of θ.

27. $\cos \theta = -2/5$ and θ is a third-quadrant angle. It follows that $\sin \theta < 0$, so

$$\sin \theta = -\sqrt{1 - (-2/5)^2}$$

$$= -\sqrt{21}/5.$$

Next,
$$\tan \theta = \sqrt{21}/2,$$
$$\cot \theta = 2\sqrt{21}/21,$$
$$\sec \theta = -5/2,$$
$$\csc \theta = -5\sqrt{21}/21.$$

29. The point of intersection has coordinates $(\cos \theta, \sin \theta)$. Since

$$r = \sqrt{4^2 + (-3)^2} = 5,$$
it follows that
$$\cos \theta = 4/5 \text{ and } \sin \theta = -3/5$$

Execises 8.3

1. $60°$

3. $40°$

5. $\pi/3$

7. $\pi/4$

9. $40°$

11. $\pi/7$

13. $\sin \dfrac{2\pi}{3} = \sin \dfrac{\pi}{3} = \sqrt{3}/2$

15. $\sec 210° = -\sec 30°$
$$= -2\sqrt{3}/3$$

17. $\cot(-\dfrac{3\pi}{4}) = \cot \dfrac{\pi}{4} = 1$

19. $\cos \dfrac{11\pi}{6} = \cos \dfrac{\pi}{6} = \sqrt{3}/2$

21. $\sin(-390°) = -\sin 30° = -1/2$

23. $\cos(-300°) = \cos 60° = 1/2$

25. $\tan 15°30' = 0.2773$

27. $\sec(-56°40') = 1.820$

29. $\sin 105°20' = \sin 74°40'$
$$= 0.9644$$

31. $\sin 0.1074 = 0.1045$

33. $\tan 0.3927 = 0.4142$

35. $\tan (-0.8639) = -1.171$

37. $\sin 22.6° = 0.3843$

39. $\sin 23°15' = 0.3947$

41. $\csc(-26°32') = -2.2385$

43. $\sin 42°30'16" = 0.6756$

45. $\cot (-\dfrac{2\pi}{7}) = -0.7975$

47. $\cos \dfrac{\sqrt{2}\pi}{5} = 0.6305$

49. $\sec(-12.38) = 1.0176$

Exercises 8.4

Find the graphs in the answer section of your text.

11. $y = \cos 4x$
Amplitude: $|A| = |1| = 1$
Period: $2\pi/B = 2\pi/4 = \pi/2$
Phase shift: $-C/B = 0$

13. $y = \sin (x/3)$
Amplitude: 1

Period: $\dfrac{2\pi}{1/3} = 6\pi$

Phase shift $= 0$

15. $y = 3 \sin (-2x) = -3 \sin 2x$
Amplitude: $|A| = |-3| = 3$
Period: $2\pi/2 = \pi$
Phase shift: 0

17. $y = 3 \cos (\pi x/2)$
Amplitude: 3

Period: $\dfrac{2\pi}{\pi/2} = 4$

Phase shift: 0

19. $y = \sin (x + \dfrac{\pi}{3})$

Amplitude: 1
Period: 2π
Phase shift: $-\pi/3$.
(Set $x + \pi/3 = 0$ and solve for x.)

21. $y = \sin \pi(x - 1) = \sin(\pi x - \pi)$
Amplitude: 1
Period: $2\pi/\pi = 2$
Phase shift: 1
(Set $\pi x - \pi = 0$ and solve for x.)

23. $y = 2 \cos(x - \dfrac{\pi}{3})$

Amplitude: 2
Period: 2π
Phase shift: $\pi/3$

25. $y = 3 \sin(2x - \pi)$
Amplitude: 3
Period: $2\pi/2 = \pi$
Phase shift: $\pi/2$

27. $y = \tan 3x$
Period: $\pi/3$
Phase shift: 0

29. $y = \tan (- \dfrac{x}{2}) = -\tan \dfrac{x}{2}$

Period: $\dfrac{\pi}{1/2} = 2\pi$

Phase shift: 0

37. $f(-x) = \sin |-x| = \sin x = f(x)$.
So f is even.

39. $f(-x) = \tan (-(-x)) = \tan x$
$= - \tan (-x)$. So f is odd.

Exercises 8.5

1. $x = 5 \cos(t - \dfrac{\pi}{4})$
Amplitude: 5
Angular speed: 1
Period: 2π
Frequency: $1/2\pi$
Phase angle: $-\pi/4$

3. $x = 4 \cos(\pi t - 2)$
Amplitude: 4
Angular speed: π
Period: 2
Frequency: $1/2$
Phase angle: -2

5. $x = \dfrac{3}{4} \cos(2\pi t + \dfrac{\pi}{4})$

Amplitude: 3/4
Angular speed: 2π
Period: 1
Frequency: 1
Phase angle: $\pi/4$

7. $x = 2 \cos(\dfrac{\pi}{2}t - \dfrac{\pi}{3})$

Amplitude: 2
Angular speed: $\pi/2$
Period: 4
Frequency: $1/4$
Phase angle: $-\pi/3$

9. $x = 6 \cos(\pi t + \dfrac{\pi}{2}) + 1$

Amplitude: 6
Angular speed: π
Period: 2
Frequency: $1/2$
Phase angle: $\pi/2$

11. $v = 30$ cycles per second, $r = 4$.
The angular speed is $\omega = 2\pi v$
$= 60\pi$ rad/s. If the initial
position is $P_o(0,4)$, then the
phase angle is $\theta_o = \pi/2$. Thus

$$x = 4\,\cos(60\pi t + \frac{\pi}{2}).$$

13. $m = 2$ lb, $y = 0.25\,\cos(2.5t)$

a) The amplitude is 0.25 ft

b) The frequency is

$$v = \frac{\omega}{2\pi} = \frac{2.5}{2\pi} \simeq 0.4 \text{ Hz}$$

c) Since $\omega = \sqrt{k/m}$, it follows that
$k = \omega^2 m = (2.5)^2(2) = 12.5$ lb/ft

15. $x = 6\,\cos(20\pi t + \frac{\pi}{2})$

Amplitude: 6

Period: $T = \dfrac{2\pi}{\omega} = \dfrac{2\pi}{20\pi}$

$$= \frac{1}{10}\text{ s}$$

Frequency: $v = \dfrac{1}{T} = 10$ Hz

After 1/120 seconds the
position is

$$x = 6\,\cos(\frac{20\pi}{120} + \frac{\pi}{2})$$

$$= 6\,\cos\frac{2\pi}{3}$$

$$= 6(-\frac{1}{2}) = -3$$

17. $x = a\,\cos(\omega t + \theta_o)$

a) $\theta_o = 0$,

b) $\theta_o = \pi/2$,

c) $\theta_o = -\pi/3$

19. $I = 20\,\sin(120\pi t)$

a) $T = \dfrac{2\pi}{\omega} = \dfrac{2\pi}{120\pi} = \dfrac{1}{60}\text{ s}$;

$$v = \frac{1}{T} = 60 \text{ Hz}$$

b) Maximum intensity:
$I = 20$ amperes.

21. Length of pendulum: 1.5 ft;
arc displacement: 0.05 ft

$$\omega = \sqrt{g/\ell} = = \sqrt{32/1.5} = 8\sqrt{3}/3$$

$$y = 0.05\,\cos(\frac{8\sqrt{3}}{3}\,t)$$

$$T = \frac{2\pi}{8\sqrt{3}/3} = \frac{\sqrt{3}\pi}{4}\text{ s} \simeq 1.36 \text{ s}$$

23. $T = \dfrac{2\pi}{\omega}$, where $\omega = \sqrt{k/m}$.

If the mass is quadrupled, then
the new angular frequency is

$$\omega_1 = \sqrt{k/4m} = \sqrt{k/m}/2 = \omega/2.$$

The new period is

$$T_1 = \frac{2\pi}{\omega_1} = \frac{2\pi}{\omega/2} = \frac{4\pi}{\omega}$$

$$= 2(\frac{2\pi}{\omega}) = 2T.$$

Thus, the period is doubled.

25. $C = 1500 - 250\,\sin(\dfrac{\pi t}{4})$

The maximum population is reached

when $\sin(\dfrac{\pi t}{4}) = -1$, that is, when

$\dfrac{\pi t}{4} = \dfrac{3\pi}{2}$, that is, $t = 6$ years.

The maximum is 1750.

27. $T = 2\pi\sqrt{\ell/g}$

$\quad = 2\pi\sqrt{3/32/6}$

$\quad = 2\pi(\dfrac{3}{4}) \approx 4.7$ s

29. $T = 2\pi\sqrt{\ell/g}$
When the temperature is 68°F, the period of the pendulum is 2 seconds, so

(*) $2 = 2\pi\sqrt{\ell/g}$

Due to an increase in temperature the length of the pendulum becomes $\ell + 10^{-4}\ell = 1.0001\ \ell$. Thus the new period is

(**) $T = 2\pi\sqrt{1.0001\ \ell/g}$.

Dividing (**) by (*) and simplifying, we get

$\quad T = 2\sqrt{1.0001}$
$\quad\quad \approx 2(1.00005)$
$\quad\quad = 2.0001$ seconds.

Thus, every 2 seconds, the clock loses 0.0001 s. Since a day has 86,400 s, in one day the clock loses

$$\dfrac{86,400(0.0001)}{2} \approx 4.32$$

seconds.

Exercises 8.6

1. arc $\sin(\sqrt{3}/2) = y$ if and only if $\sin y = \sqrt{3}/2$, $-\pi/2 \le y \le \pi/2$. Thus $y = \pi/3$.

3. $\cos^{-1}0 = y$ if and only if $\cos y = 0$, $0 \le y \le \pi$. Thus $y = \pi/2$.

5. $\tan^{-1}(-\sqrt{3}/3) = y$ if and only if $\tan y = -\sqrt{3}/3$, $-\pi/2 < y < \pi/2$. So $y = -\pi/6$.

7. arc $\cos(\sqrt{2}/2) = \pi/4$

9. arc $\sin(\sqrt{2}/2) = \pi/4$

11. arc $\tan(\sqrt{3}/3) = \pi/6$

13. arc $\sin(0.1994) = 11°30'$

15. $\cos^{-1}(0.3907) = 67°$

17. arc $\tan(0.9490) = 43°30'$

19. $\tan^{-1}(0.9380) = 43°10'$

21. Set $y = \sin^{-1}(\sqrt{2}/2)$. Then $\sin y = \sqrt{2}/2$, $-\pi/2 \le y \le \pi/2$, which implies that $y = \pi/4$. Thus $\cos(\sin^{-1}(\sqrt{2}/2)) = \cos y = \cos \pi/4 = \sqrt{2}/2$.

23. Set $\cos^{-1}(\sqrt{3}/2) = y$. Then $\cos y = \sqrt{3}/2$, $0 \le y \le \pi$, which implies that $y = \pi/6$. So $\tan(\cos^{-1}(\sqrt{3}/2)) = \tan \pi/6 = \sqrt{3}/3$.

25. $\cot(\sin^{-1}(\sqrt{3}/2)) = \sqrt{3}/3$

27. $\csc(\cos^{-1}(-1/2)) = 2\sqrt{3}/3$

29. $\cos(\tan^{-1}(\sqrt{3}/3) = \sqrt{3}/2$

31. Let $y = \sin^{-1}(\sqrt{2}/3)$, so that $\sin y = \sqrt{2}/3$, $-\pi/2 \le y \le \pi/2$.

Next, $\cos y = \sqrt{1 - (\sqrt{2}/3)^2} = \sqrt{7}/3$. (We have taken the positive square root because $\cos y \ge 0$ when $-\pi/2 \le y \le \pi/2$.)

33. If you set $y = \cos^{-1}(\sqrt{3}/5)$, you obtain $\cos y = \sqrt{3}/5$, $0 \le y \le \pi/2$. Next sketch a right triangle where one leg measures $\sqrt{3}$ and the hypotenuse measures 5. The other leg (facing the angle y) will measure $\sqrt{22}$. Thus $\tan(\cos^{-1}(\sqrt{3}/5)) = \tan y = \sqrt{22}/\sqrt{3} = \sqrt{66}/3$.

35. $\cos(\tan^{-1}(-2)) = \sqrt{5}/5$

37. $\sin(\tan^{-1}\sqrt{5}) = \sqrt{30}/6$

39. $\csc(\tan^{-1}(-2/3)) = -\sqrt{13}/2$

41. $\arcsin(-0.4563) - -27°9'$

43. $\tan^{-1}(-5.3641) = -79°26'$

45. $\sin[\cos^{-1}(-0.6324)] = 0.7746$

47. $\tan[\sin^{-1}(0.3535)] = 0.3779$

49. $\sin[\tan^{-1}(15.3214)] = 0.9979$

51. $x = \dfrac{1}{2} \arcsin y$

53. $x = \dfrac{1}{3} \sin^{-1}(y/2)$

55. $x = \arccos(3y) + 2$

57. $x = 2\cos(y/3)$

59. $x = \dfrac{1}{2} \tan(3y/2)$

Review Exercises - Chapter 8

1. $\alpha = 3\pi/4$, $P(-\sqrt{2}/2, \sqrt{2}/2)$
    $\sin\alpha = \sqrt{2}/2$, $\cos\alpha = -\sqrt{2}/2$,
    $\tan\alpha = -1$, $\cot\alpha = -1$,
    $\sec\alpha = -\sqrt{2}$, $\csc\alpha = \sqrt{2}$

3. $P(-1/2, -\sqrt{3}/2)$

    $\sin\gamma = -\sqrt{3}/2$, $\cos\gamma = -1/2$

    $\tan\gamma = \sqrt{3}$, $\cot\gamma = \sqrt{3}/3$

    $\sec\gamma = -2$, $\csc\gamma = -2\sqrt{3}/3$

5. $\cos\alpha = 2\sqrt{2}/3$, $\csc\alpha = 3$
    Since $\csc\alpha = 3$, it follows that
    $\sin\alpha = 1/3$. Next,
    $$\tan\alpha = \frac{1/3}{2\sqrt{2}/3} = 1/2\sqrt{2}$$

$$= \sqrt{2}/4$$

$\cot\alpha = 2\sqrt{2}$, $\sec\alpha = 3\sqrt{2}/4$

7. $\sin\beta = 2/5$, $\tan\beta = -2\sqrt{21}/21$
    $\csc\beta = 5/2$, $\cot\beta = -\sqrt{21}/2$
    $\cos\beta = \cot\beta \cdot \sin\beta$

    $$= -\sqrt{21}/5$$

    $\sec\beta = -5\sqrt{21}/21$

9. $\cos\theta = 2/5$ and θ is a fourth-quadrant angle, we have that $\sin\theta < 0$. So $\sin\theta =$
    $$-\sqrt{1 - (\frac{2}{5})^2} = -\frac{\sqrt{21}}{5}. \text{ Next}$$

    $\tan\theta = -\sqrt{21}/2$, $\cot\theta = -2\sqrt{21}/21$,

    $\sec\theta = 5/2$, $\csc\theta = -5\sqrt{21}/21$

11. $\cot\theta = -3$ and $\cos\theta < 0$
    $\tan\theta = -1/3$

    $\sec\theta = -\sqrt{1 + (-1/3)^2}$

    $$= -\sqrt{10}/3$$

    $\cos\theta = -3\sqrt{10}/10$

    $\sin\theta = \sqrt{10}/10$, $\csc\theta = \sqrt{10}$

13. $\sin\theta = 0.5948$ and $\cos\theta < 0$

    $\cos\theta = -\sqrt{1 - (0.5948)^2}$
    $\qquad = -0.8039$
    $\tan\theta = -0.7399$
    $\cot\theta = -1.3515$
    $\sec\theta = -1.2440$
    $\csc\theta = 1.6812$

15. $P(3,-5)$, $r = \sqrt{3^2 + (-5)^2} = \sqrt{34}$

    $\sin\theta = -5\sqrt{34}/34$, $\cos\theta = 3\sqrt{34}/34$
    $\tan\theta = -5/3$, $\cot\theta = -3/5$
    $\sec\theta = \sqrt{34}/3$, $\csc\theta = -\sqrt{34}/5$

17. $X(-3,6)$

    $$r = \sqrt{(-3)^2 + 6^2} = \sqrt{45}$$

    $$= 3\sqrt{5}$$

    $\sin \theta = 2\sqrt{5}/5$, $\cos \theta = -\sqrt{5}/5$,
    $\tan \theta = -2$, $\cot \theta = -1/2$

    $\sec \theta = -\sqrt{5}$, $\csc \theta = \sqrt{5}/2$

19. $U(-3,-2)$,

    $$r = \sqrt{(-3)^2 + (-2)^2} = \sqrt{13}$$
    $\sin \theta = -2\sqrt{13}/13$, $\cos \theta = -3\sqrt{13}/13$
    $\tan \theta = 2/3$, $\cot \theta = 3/2$,
    $\sec \theta = -\sqrt{13}/3$, $\csc \theta = -\sqrt{13}/2$

21. The point $P(1,2)$ lies on the line
    $y = 2x$ and is in the first
    quadrant. Thus

    $$r = \sqrt{1^2 + 2^2} = \sqrt{5} \text{ and}$$

    $$\sin \theta = 2/\sqrt{5} = 2\sqrt{5}/5,$$

    $\cos \theta = 1/\sqrt{5} = \sqrt{5}/5$,
    $\tan \theta = 2$, $\cot \theta = 1/2$,

    $\sec \theta = \sqrt{5}$, $\csc \theta = \sqrt{5}/2$

23. The point $P(-1,-3)$ lies on the
    line $3x - y = 0$ and is in the
    third quadrant. So

    $$r = \sqrt{(-1)^2 + (-3)^2} = \sqrt{10} \text{ and}$$

    $\sin \theta = -3/\sqrt{10} = -3\sqrt{10}/10$,

    $\cos \theta = -1/\sqrt{10} = -\sqrt{10}/10$,
    $\tan \theta = 3$, $\cot \theta = 1/3$,

    $\sec \theta = -\sqrt{10}$, $\csc \theta = -\sqrt{10}/3$

25. a) $50°$, b) $40°$, c) $\pi/2$
    d) $\pi/6$, e) $30°15'$
    f) $24°30'$

27. a) $\sin 45°20' = 0.7112$
    b) $\cos 15°30' = 0.9636$
    c) $\sec 25°30' = 1.108$
    d) $\csc 56°20' = 1.202$
    e) $\cos 0.1484 = 0.9890$
    f) $\sin 0.4102 = 0.3987$

The graphs for exercises 29-33 are
in the answer section of your
text.

35. $y = 2 \cos(x/5)$
    Amplitude: 2

    Period: $\dfrac{2\pi}{1/5} = 10\pi$

    Phase shift: 0

37. $y = \cos \pi(x - 2)$
    Amplitude: 1
    Period: $2\pi/\pi = 2$
    Phase shift: 2

39. $y = -2 \sin(\pi x - 4)$
    Amplitude: $|-2| = 2$
    Period: $2\pi/\pi = 2$
    Phase shift: $4/\pi$

The graphs for exercises 41-45 are
in the answer section of your
text.

47. $v = \dfrac{\omega}{2\pi} = \dfrac{1}{2\pi}\sqrt{\dfrac{k}{m}}$

If $m = 0.4$ kg and $v = 32$ Hz, we have the relation

(*) $32 = \dfrac{1}{2\pi}\sqrt{\dfrac{k}{0.4}}$

If $m = 0.8$ kg, we have

(**) $v = \dfrac{1}{2\pi}\sqrt{\dfrac{k}{0.8}}$

Dividing (**) by (*) and simplifying, we obtain

$$\dfrac{v}{32} = \sqrt{\dfrac{0.4}{0.8}} = \sqrt{0.5}$$

$v = 32\sqrt{0.5} \simeq 22.63$ Hz.

49. Initial displacement: 20 cm, period: 1.5 s

$$\omega = \dfrac{2\pi}{T} = \dfrac{2\pi}{1.5} = \dfrac{4\pi}{3}$$

$$y = 20\cos(\dfrac{4\pi}{3}\,t)$$

At $t = 1.8$ s,

$$y = 20\cos(\dfrac{4\pi}{3}\,(1.8)) \simeq 6.2\text{ cm}.$$

51. $\ell = 40$ cm, $v = 0.787$ Hz

$$T = 2\pi\sqrt{\ell/g}\ \text{ or }\ v = \dfrac{1}{2\pi}\sqrt{g/\ell}$$

(Can you see why?) From the second equation, we obtain

$$\sqrt{g} = 2\pi v\sqrt{\ell}$$
$$g = 4\pi^2 v^2 \ell$$
$$= 4\pi^2(0.787)^2(40)$$
$$\simeq 978.07\text{ cm/s}^2$$
$$\simeq 9.78\text{ m/s}^2$$

To write the equation giving the position of the pendulum, we need the initial displacement 0.5 cm and the angular frequency

$$\omega = 2\pi v = 2\pi(0.787) \simeq 4.9.$$

Thus

$$y = 0.5\cos(4.9\,t).$$

53. $\sin(\sin^{-1}(\sqrt{2}/2)) = \sqrt{2}/2$

55. $\cos^{-1}(\sin 4\pi/3)$

$$= \cos^{-1}(-\sqrt{3}/2) = 5\pi/6$$

57. Let $\theta = \tan^{-1}4$, $-\pi/2 < \theta < \pi/2$. Then, $\tan\theta = 4$ and $\sec\theta = \sec(\tan^{-1}4)$

$$= \sqrt{1 + 4^2} = \sqrt{17}.$$

59. arc $\sin(0.8312) = 56°13'$

61. $\cos^{-1}(-0.4715) = 118°8'$

63. $\sin^{-1}(0.7071) = 45°$

65. $\cos[\sin^{-1}(-0.7413)]$
    $= 0.6712$

67. $\tan[\cos^{-1}(0.5561)]$
    $= 1.4945$

69. $\cos[\sin^{-1}(0.7071)]$
    $= 0.7071$

Chapter 8 Test

In Problems 1-2, find the coordinates of the point on the unit circle corresponding to the given angle. Evaluate the six trigonometric function values of the angle. Do not use a calculator or table. Use your knowledge of special angles.

1. $\theta = 5\pi/6$

2. $\alpha = -120°$

3. Let $\sin \alpha = 1/3$ and $\cos \alpha = 2\sqrt{2}/3$. Find the exact values of the four remaining trigonometric functions.

4. If $\csc \beta = 4$ and $\cos \beta > 0$, find the other five trigonometric function values of β.

5. Let α be a second-quadrant angle such that $\sec \alpha = -2.1045$. Approximate the other five trigonometric function values of α to four decimal places.

In Problems 6-7, the coordinates of a point on the terminal side of α, an angle in standard position, are given. Find the exact values of the six trigonometric functions of α. Do not approximate. Do not use a calculator.

6. A(2,-6)

7. B(-3,4)

8. A point whose second coordinate is -1 lies on the terminal side of an angle β and on the line $y = -x/2$. Find all trigonometric function values of β.

9. The terminal side of α lies in the first quadrant and on the line $4x - y = 0$. Find all trigonometric function values of α.

10. Find the coordinates of the point where the half-line from the origin to the point (-6,8) intersects the unit circle. What do they represent?

In Problems 11-12, find the reference angle of the given angle.

11. -510°

12. $13\pi/3$

In Problems 13-14, use the notion of reference angles and the special angles table to find the following values.

13. $\cos 210°$

14. $\tan(-5\pi/4)$

15. What is the reference angle of 130°20'? Use Table 4 in the Appendix of the textbook and approximate $\cos 130°20'$.

In Problems 16-19, find the amplitude, period, and phase shift of each function

16. $f(x) = 4 \sin(3x)$

17. $f(x) = -3 \cos(x/4)$

18. $f(x) = -2 \cos(4x - \pi)$

19. $f(x) = -5 \tan\left(x - \dfrac{\pi}{2}\right)$

In Problems 20-21, graph the given functions

20. $f(x) = 2 \cos(x - \dfrac{\pi}{2})$ 21. $f(x) = 3 \sin(x - \dfrac{\pi}{4})$

22. Is the function $f(x) = 3 \sin x^2 + 1$ even odd or neither?

23. In a spring-mass system, a mass of 250 grams attached to the end of the spring bobs up and down according to the equation

$$y = -3 \cos 5t$$

where y is measured in centimeters and t in seconds. Find a) the amplitude of the oscillation, b) the angular frequency and the frequency, c) the stiffness of the spring.

24. A pendulum 2 ft long is pulled to the right through an arc of 1/2 ft and released at time $t = 0$. Write the equation describing the oscillation of the pendulum. What is its period? (Acceleration of gravity: 32 ft/s^2)

25. What is the acceleration of gravity in a location where a pendulum 1 meter long has a period of 2 seconds?

26. In a certain habitat, the field mice population was given by the formula

$$M = 4800 + 600 \sin(\dfrac{\pi}{6}t)$$

where t represents the number of months after April 1, 1984. What was the population on July 1, 1984? What was the minimum population? When was the first time that the minimum population was reached?

In Problems 27-28, find the given numbers without the use of tables or calculators.

27. $\tan(\sin^{-1}(2/3))$ 28. $\cos(\arctan(-2))$

29. Solve for x the equation $y = 5 \cos 2x$, $0 \le x \le \pi/2$.

30. Solve for x the equation $y = 3 \sin(x - \pi)$, $\dfrac{\pi}{2} \le x \le \dfrac{3\pi}{2}$.

Chapter 9
Trigonometric Identities and
Applications of Trigonometry

Exercises 9.1

1. $\sin \theta(\cot \theta + \csc \theta)$

    $= \sin \theta \left(\dfrac{\cos \theta}{\sin \theta} + \dfrac{1}{\sin \theta}\right)$

    $= \cos \theta + 1$

3. $\dfrac{\sin x}{\csc x} + \dfrac{\cos x}{\sec x}$

    $= \dfrac{\sin x}{1/\sin x} + \dfrac{\cos x}{1/\cos x}$

    $= \sin^2 x + \cos^2 x = 1$

5. $\dfrac{\tan \alpha}{\sec \alpha} + \dfrac{\cot \alpha}{\csc \alpha}$

    $= \dfrac{\dfrac{\sin \alpha}{\cancel{\cos \alpha}}}{\dfrac{1}{\cancel{\cos \alpha}}} + \dfrac{\dfrac{\cos \alpha}{\cancel{\sin \alpha}}}{\dfrac{1}{\cancel{\sin \alpha}}}$

    $= \sin \alpha + \cos \alpha$

7. $(\sin u + \cos u)^2$

    $= \sin^2 u + 2 \sin u \cos u + \cos^2 u$

    $= 1 + 2 \sin u \cos u$

9. $\dfrac{1}{1 - \sin \theta} + \dfrac{1}{1 + \sin \theta}$

    $= \dfrac{1 + \cancel{\sin \theta} + 1 - \cancel{\sin \theta}}{(1 - \sin \theta)(1 + \sin \theta)}$

    $= \dfrac{2}{1 - \sin^2 \theta}$

    $= \dfrac{2}{\cos^2 \theta} = 2 \sec^2 \theta$

11. $\sin \theta \csc \theta = 1$

    $\sin \theta \csc \theta = \cancel{\sin \theta} \cdot \dfrac{1}{\cancel{\sin \theta}}$

    $= 1$

13. $\tan \theta \cos \theta = \sin \theta$

    $\tan \theta \cos \theta = \dfrac{\sin \theta}{\cancel{\cos \theta}} \cdot \cancel{\cos \theta}$

    $= \sin \theta$

15. $\tan \theta = \sin \theta \sec \theta$

$$\tan \theta = \frac{\sin \theta}{\cos \theta}$$

$$= \sin \theta \cdot \frac{1}{\cos \theta}$$

$$= \sin \theta \sec \theta$$

17. $\cos^2\theta - \sin^2\theta = 2 \cos^2\theta - 1$

$\cos^2\theta - \sin^2\theta$
$$= \cos^2\theta - (1 - \cos^2\theta)$$
$$= \cos^2\theta - 1 + \cos^2\theta$$
$$= 2 \cos^2\theta - 1$$

19. $(\sin \alpha - \cos \alpha)^2$
$$= 1 - 2 \sin \alpha \cos \alpha$$

$(\sin \alpha - \cos \alpha)^2$
$$= \sin^2\alpha - 2 \sin \alpha \cos \alpha + \cos^2\alpha$$
$$= (\sin^2\alpha + \cos^2\alpha)$$
$$\qquad - 2 \sin \alpha \cos \alpha$$
$$= 1 - 2 \sin \alpha \cos \alpha$$

21. $\tan \alpha + \cot \alpha = \sec \alpha \csc \alpha$

$$\tan \alpha + \cot \alpha = \frac{\sin \alpha}{\cos \alpha} + \frac{\cos \alpha}{\sin \alpha}$$

$$= \frac{\sin^2 \alpha + \cos^2 \alpha}{\cos \alpha \sin \alpha}$$

$$= \frac{1}{\cos \alpha \sin \alpha}$$

$$= \sec \alpha \csc \alpha$$

23. $2 \cos^2x - 1 = 1 - 2 \sin^2x$
$2 \cos^2x - 1 = 2 \cos^2x$
$$\qquad\qquad - (\sin^2x + \cos^2x)$$

$$= 2 \cos^2x - \sin^2x$$
$$\qquad - \cos^2x$$
$$= \cos^2x - \sin^2x$$
$$= (1 - \sin^2x) - \sin^2x$$
$$= 1 - 2 \sin^2x$$

25. $\cot^2x - \cos^2x = \cot^2x \cos^2x$
$\cot^2x - \cos^2x$

$$= \frac{\cos^2x}{\sin^2x} - \cos^2x$$

$$= \frac{\cos^2x - \cos^2x \sin^2x}{\sin^2x}$$

$$= \frac{\cos^2x(1 - \sin^2x)}{\sin^2x}$$

$$= \frac{\cos^2x}{\sin^2x} \cdot \cos^2x$$

$$= \cot^2x \cos^2x$$

27. $\dfrac{1 + \sin u}{\cos u} + \dfrac{\cos u}{1 + \sin u}$

$$= 2 \sec u$$

$$\frac{1 + \sin u}{\cos u} + \frac{\cos u}{1 + \sin u}$$

$$= \frac{(1 + \sin u)^2 + (\cos u)^2}{\cos u(1 + \sin u)}$$

$$= \frac{1 + 2 \sin u + \sin^2u + \cos^2u}{\cos u(1 + \sin u)}$$

$$= \frac{2 + 2 \sin u}{\cos u(1 + \sin u)}$$

$$= \frac{2(\cancel{1 + \sin u})}{\cos u(\cancel{1 + \sin u})}$$

$$= 2 \sec u$$

29. $\sec u = \dfrac{\cos u}{1 + \sin u} + \tan u$

$\dfrac{\cos u}{1 + \sin u} + \tan u$

$= \dfrac{\cos u}{1 + \sin u} + \dfrac{\sin u}{\cos u}$

$= \dfrac{(\cos u)^2 + \sin u(1 + \sin u)}{(1 + \sin u)\cos u}$

$= \dfrac{\cos^2 u + \sin u + \sin^2 u}{(1 + \sin u)\cos u}$

$= \dfrac{\cancel{1 + \sin u}}{(\cancel{1 + \sin u})\cos u}$

$= \sec u$

31. $\dfrac{1 - \sin u}{\cos u} = \dfrac{\cos u}{1 + \sin u}$

$\dfrac{1 - \sin u}{\cos u}$

$= \dfrac{(1 - \sin u)(1 + \sin u)}{\cos u(1 + \sin u)}$

$= \dfrac{1 - \sin^2 u}{\cos u(1 + \sin u)}$

$= \dfrac{\cos^{\cancel{2}} u}{\cancel{\cos} u(1 + \sin u)}$

$= \dfrac{\cos u}{1 + \sin u}$

33. $\dfrac{\cot \theta}{\csc \theta + 1} = \dfrac{\csc \theta - 1}{\cot \theta}$

$\dfrac{\csc \theta - 1}{\cot \theta}$

$= \dfrac{(\csc \theta - 1)(\csc \theta + 1)}{\cot \theta(\csc \theta + 1)}$

$= \dfrac{\csc^2 \theta - 1}{\cot \theta(\csc \theta + 1)}$

$= \dfrac{\cot^2 \theta}{\cot \theta(\csc \theta + 1)}$

$= \dfrac{\cot \theta}{\csc \theta + 1}$

35. $\dfrac{\tan^2 \theta - 1}{1 - \cot^2 \theta} = \tan^2 \theta$

$\dfrac{\tan^2 \theta - 1}{1 - \cot^2 \theta} = \dfrac{\dfrac{\sin^2 \theta}{\cos^2 \theta} - 1}{1 - \dfrac{\cos^2 \theta}{\sin^2 \theta}}$

$= \dfrac{\dfrac{\sin^2 \theta - \cos^2 \theta}{\cos^2 \theta}}{\dfrac{\sin^2 \theta - \cos^2 \theta}{\sin^2 \theta}}$

$= \dfrac{\sin^2 \theta}{\cos^2 \theta}$

$= \tan^2 \theta$

37. $\dfrac{1 - \sin \alpha}{1 + \sin \alpha} = (\sec \alpha - \tan \alpha)^2$

$\dfrac{1 - \sin \alpha}{1 + \sin \alpha}$

$= \dfrac{(1 - \sin \alpha)(1 - \sin \alpha)}{(1 + \sin \alpha)(1 - \sin \alpha)}$

$= \dfrac{1 - 2\sin \alpha + \sin^2 \alpha}{1 - \sin^2 \alpha}$

$= \dfrac{1 - 2\sin \alpha + \sin^2 \alpha}{\cos^2 \alpha}$

$= \dfrac{1}{\cos^2 \alpha} - 2\dfrac{\sin \alpha}{\cos^2 \alpha}$

$\quad + \dfrac{\sin^2 \alpha}{\cos^2 \alpha}$

$= \sec^2 \alpha - 2\sec \alpha \tan \alpha$
$\quad + \tan^2 \alpha$

$= (\sec \alpha - \tan \alpha)^2$

39. $$\frac{1}{1 + \cos \alpha} + \frac{1}{1 - \cos \alpha}$$

$$= 2 \csc^2\alpha$$

$$\frac{1}{1 + \cos \alpha} + \frac{1}{1 - \cos \alpha}$$

$$= \frac{1 - \cancel{\cos \alpha} + 1 + \cancel{\cos \alpha}}{(1 + \cos \alpha)(1 - \cos \alpha)}$$

$$= \frac{2}{1 - \cos^2\alpha}$$

$$= \frac{2}{\sin^2\alpha}$$

$$= 2 \csc^2\alpha$$

41. $$\frac{\sec x + 1}{\sin x + \tan x} = \csc x$$

$$\frac{\sec x + 1}{\sin x + \tan x} = \frac{\dfrac{1}{\cos x} + 1}{\sin x + \dfrac{\sin x}{\cos x}}$$

$$= \frac{\dfrac{1 + \cos x}{\cancel{\cos x}}}{\dfrac{\sin x \cos x + \sin x}{\cancel{\cos x}}}$$

$$= \frac{\cancel{1 + \cos x}}{\sin x(\cancel{\cos x + 1})}$$

$$= \csc x$$

43. $$\frac{\sec x}{1 + \sec x} - \frac{\sec x}{1 - \sec x}$$

$$= 2 \csc^2x$$

$$\frac{\sec x}{1 + \sec x} - \frac{\sec x}{1 - \sec x}$$

$$= \frac{\dfrac{1}{\cos x}}{1 + \dfrac{1}{\cos x}} - \frac{\dfrac{1}{\cos x}}{1 - \dfrac{1}{\cos x}}$$

$$= \frac{\dfrac{1}{\cancel{\cos x}}}{\dfrac{\cos x + 1}{\cancel{\cos x}}} - \frac{\dfrac{1}{\cancel{\cos x}}}{\dfrac{\cos x - 1}{\cancel{\cos x}}}$$

$$= \frac{1}{\cos x + 1} - \frac{1}{\cos x - 1}$$

$$= \frac{\cancel{\cos x} - 1 - \cancel{\cos x} - 1}{(\cos x + 1)(\cos x - 1)}$$

$$= \frac{-2}{\cos^2x - 1}$$

$$= \frac{-2}{-\sin^2x}$$

$$= 2 \csc^2x$$

45. $$\frac{\sin x}{1 + \cos x} = \csc x - \cot x$$

$$\csc x - \cot x = \frac{1}{\sin x} - \frac{\cos x}{\sin x}$$

$$= \frac{(1 - \cos x)(1 + \cos x)}{\sin x \, (1 + \cos x)}$$

$$= \frac{1 - \cos^2x}{\sin x(1 + \cos x)}$$

$$= \frac{\sin^2x}{\sin x(1 + \cos x)}$$

$$= \frac{\sin x}{1 + \cos x}$$

47. $\dfrac{\tan \beta + \sin \beta}{\tan \beta - \sin \beta} = \dfrac{\sec \beta + 1}{\sec \beta - 1}$

$\dfrac{\tan \beta + \sin \beta}{\tan \beta - \sin \beta}$

$= \dfrac{\dfrac{\sin \beta}{\cos \beta} + \sin \beta}{\dfrac{\sin \beta}{\cos \beta} - \sin \beta}$

$= \dfrac{\cancel{\sin \beta}(\dfrac{1}{\cos \beta} + 1)}{\cancel{\sin \beta}(\dfrac{1}{\cos \beta} - 1)}$

$= \dfrac{\sec \beta + 1}{\sec \beta - 1}$

49. $\dfrac{\sin a \cos b + \cos a \sin b}{\cos a \cos b - \sin a \sin b}$

$= \dfrac{\tan a + \tan b}{1 - \tan a \tan b}$

$\dfrac{\sin a \cos b - \cos a \sin b}{\cos a \cos b - \sin a \sin b}$

$= \dfrac{\dfrac{\sin a \cancel{\cos b}}{\cos a \cancel{\cos b}} + \dfrac{\cancel{\cos a} \sin b}{\cancel{\cos a} \cos b}}{\dfrac{\cancel{\cos a \cos b}}{\cancel{\cos a \cos b}} - \dfrac{\sin a \sin b}{\cos a \cos b}}$

$= \dfrac{\tan a + \tan b}{1 - \tan a \tan b}$

51. $\sin^4 x - \cos^4 x = 2 \sin^2 x - 1$

$\sin^4 x - \cos^4 x$
$= (\sin^2 x + \cos^2 x)(\sin^2 x - \cos^2 x)$
$= \sin^2 x - \cos^2 x$
$= \sin^2 x - (1 - \sin^2 x)$
$= \sin^2 x - 1 + \sin^2 x$
$= 2 \sin^2 x - 1$

53. $\dfrac{\sec^4 x - 1}{\tan^2 x} = 2 + \tan^2 x$

$\dfrac{\sec^4 x - 1}{\tan^2 x}$

$= \dfrac{(\sec^2 x - 1)(\sec^2 x + 1)}{\tan^2 x}$

$[\sec^2 x = 1 + \tan^2 x]$

$= \dfrac{\cancel{\tan^2 x}(\sec^2 x + 1)}{\cancel{\tan^2 x}}$

$= 1 + \tan^2 x + 1$

$= 2 + \tan^2 x$

55. $\dfrac{\sin^2 \theta + 3 \sin \theta + 2}{\cos^2 \theta} = \dfrac{\sin \theta + 2}{1 - \sin \theta}$

$\dfrac{\sin^2 \theta + 3 \sin \theta + 2}{\cos^2 \theta}$

$= \dfrac{(\sin \theta + 2)(\sin \theta + 1)}{1 - \sin^2 \theta}$

$= \dfrac{(\sin \theta + 2)\cancel{(\sin \theta + 1)}}{\cancel{(1 + \sin \theta)}(1 - \sin \theta)}$

$= \dfrac{\sin \theta + 2}{1 - \sin \theta}$

57. $\dfrac{\tan \theta}{\sin \theta + 3 \tan \theta} = \dfrac{1}{\cos \theta + 3}$

$\dfrac{\tan \theta}{\sin \theta + 3 \tan \theta}$

$= \dfrac{\cancel{\tan \theta}}{\cancel{\tan \theta}(\cos \theta + 3)}$

$= \dfrac{1}{\cos \theta + 3}$

59. $\dfrac{\sin^3 \theta - \cos^3 \theta}{\sin \theta - \cos \theta} = 1 + \sin \theta \cos \theta$

First recall that $a^3 - b^3 = (a - b)(a^2 + ab + b^2)$. Next write

$\dfrac{\sin^3 \theta - \cos^3 \theta}{\sin \theta - \cos \theta}$

$= \dfrac{\cancel{(\sin \theta - \cos \theta)}(\sin^2 \theta + \sin \theta \cos \theta + \cos^2 \theta)}{\cancel{\sin \theta - \cos \theta}}$

$= 1 + \sin \theta \cos \theta$

Exercises 9.2

1. $\sin 15° = \sin(45° - 30°)$
$\quad = \sin 45° \cos 30°$
$\qquad - \cos 45° \sin 30°$

$\quad = \sqrt{2}/2 \cdot \sqrt{3}/2 - \sqrt{2}/2 \cdot 1/2$

$\quad = \sqrt{6}/4 - \sqrt{2}/4$

$\quad = (\sqrt{6} - \sqrt{2})/4$

3. $\tan 15° = \tan(45° - 30°)$

$\quad = \dfrac{\tan 45° - \tan 30°}{1 + \tan 45° \ \tan 30°}$

$\quad = \dfrac{1 - 1/\sqrt{3}}{1 + 1/\sqrt{3}}$

$\quad = \dfrac{(\sqrt{3} - 1)(\sqrt{3} - 1)}{(\sqrt{3} + 1)(\sqrt{3} - 1)}$

$\quad = \dfrac{3 - 2\sqrt{3} + 1}{3 - 1} = \dfrac{4 - 2\sqrt{3}}{2}$

$\quad = 2 - \sqrt{3}$

5. $\cos 75° = \cos(45° + 30°)$
$\quad = \cos 45° \cos 30°$
$\qquad - \sin 45° \sin 30°$

$\quad = \sqrt{2}/2 \cdot \sqrt{3}/2 - \sqrt{2}/2 \cdot 1/2$

$\quad = \sqrt{6}/4 - \sqrt{2}/4$

$\quad = \dfrac{\sqrt{6} - \sqrt{2}}{4}$

7. $\sin 105° = \sin(60° + 45°)$
$\quad = \sin 60° \cos 45°$
$\qquad + \cos 60° \sin 45°$

$\quad = \sqrt{3}/2 \cdot \sqrt{2}/2 + 1/2 \cdot \sqrt{2}/2$

$\quad = \sqrt{6}/4 + \sqrt{2}/4$

$\quad = \dfrac{\sqrt{6} + \sqrt{2}}{4}$

9. $\cos 210° = \cos(180° + 30°)$
$\quad = \cos 180° \cos 30°$
$\qquad - \sin 180° \cos 30°$

$\quad = (-1)(\sqrt{3}/2) - (0)(1/2)$

$\quad = -\sqrt{3}/2$

11. $\cos \dfrac{5\pi}{12} = \cos(\dfrac{\pi}{4} + \dfrac{\pi}{6})$

$\quad = \cos \dfrac{\pi}{4} \cos \dfrac{\pi}{6}$

$\qquad - \sin \dfrac{\pi}{4} \sin \dfrac{\pi}{6}$

$\quad = \sqrt{2}/2 \cdot \sqrt{3}/2 - \sqrt{2}/2 \cdot 1/2$

$\quad = \sqrt{6}/4 - \sqrt{2}/4$

$\quad = \dfrac{\sqrt{6} - \sqrt{2}}{4}$

13. $\sec \dfrac{5\pi}{12} = \dfrac{1}{\cos(\dfrac{\pi}{4} + \dfrac{\pi}{6})}$

$\quad = \dfrac{1}{\cos\dfrac{\pi}{4} \cos\dfrac{\pi}{6} - \sin\dfrac{\pi}{4} \sin\dfrac{\pi}{6}}$

$\quad = \dfrac{1}{\sqrt{2}/2 \cdot \sqrt{3}/2 - \sqrt{2}/2 \cdot 1/2}$

$\quad = \dfrac{1}{\dfrac{\sqrt{6} - \sqrt{2}}{4}}$

$\quad = \dfrac{4}{(\sqrt{6} - \sqrt{2})} \dfrac{(\sqrt{6} + \sqrt{2})}{(\sqrt{6} + \sqrt{2})}$

$\quad = \dfrac{4(\sqrt{6} + \sqrt{2})}{6 - 2} = \sqrt{6} + \sqrt{2}$

15. $\cos \dfrac{3\pi}{4} = \cos(\dfrac{\pi}{2} + \dfrac{\pi}{4})$

$\quad\quad = \cos \dfrac{\pi}{2} \cos \dfrac{\pi}{4}$

$\quad\quad\quad - \sin \dfrac{\pi}{2} \sin \dfrac{\pi}{4}$

$\quad\quad = (0)(\sqrt{2}/2) - (1)(\sqrt{2}/2)$

$\quad\quad = -\sqrt{2}/2$

17. $\sin \dfrac{4\pi}{3} = \sin(\pi + \dfrac{\pi}{3})$

$\quad\quad = \sin \pi \cos \dfrac{\pi}{3} + \cos \pi \sin \dfrac{\pi}{3}$

$\quad\quad = (0)(1/2) + (-1)(\sqrt{3}/2)$

$\quad\quad = -\sqrt{3}/2$

19. $\csc \dfrac{7\pi}{6} = \dfrac{1}{\sin(\pi + \dfrac{\pi}{6})}$

$\quad = \dfrac{1}{\sin \pi \cos \dfrac{\pi}{6} + \cos \pi \sin \dfrac{\pi}{6}}$

$\quad = \dfrac{1}{(0)(\sqrt{3}/2) + (-1)(1/2)}$

$\quad = \dfrac{1}{-1/2}$

$\quad = -2$

21. $\sin 32° \cos 28° + \cos 32° \sin 28°$
$\quad\quad = \sin(32° + 28°)$
$\quad\quad = \sin 60°$

23. $\cos 76° \cos(-21°)$
$\quad + \sin 76° \sin(-21°)$
$\quad\quad = \cos(76° - (-21°))$
$\quad\quad = \cos(76° + 21°)$
$\quad\quad = \cos 97°$

25. $\cos \dfrac{\pi}{5} \cos \dfrac{\pi}{4} - \sin \dfrac{\pi}{5} \sin \dfrac{\pi}{4}$

$\quad\quad = \cos(\dfrac{\pi}{5} + \dfrac{\pi}{4})$

$\quad\quad = \cos(\dfrac{4\pi}{20} + \dfrac{5\pi}{20})$

$\quad\quad = \cos \dfrac{9\pi}{20}$

27. $\sin \dfrac{7\pi}{6} \cos(-\dfrac{\pi}{3})$

$\quad - \cos \dfrac{7\pi}{6} \sin(-\dfrac{\pi}{3})$

$\quad\quad = \sin(\dfrac{7\pi}{6} - (-\dfrac{\pi}{3}))$

$\quad\quad = \sin(\dfrac{7\pi}{6} + \dfrac{\pi}{3})$

$\quad\quad = \sin(\dfrac{7\pi}{6} + \dfrac{2\pi}{6})$

$\quad\quad = \sin \dfrac{9\pi}{6} = \sin \dfrac{3\pi}{2}$

29. $\sin 5 \cos 3 - \cos 5 \sin 3$
$\quad\quad = \sin(5 - 3)$
$\quad\quad = \sin 2$

31. $\sin(\cos^{-1}(4/5) + \sin^{-1}(4/5))$

Set $\alpha = \cos^{-1}(4/5)$ and check that
$\cos \alpha = 4/5$ and $\sin \alpha = 3/5$.
Similarly, if $\beta = \sin^{-1}(4/5)$, then
$\sin \beta = 4/5$ and $\cos \beta = 3/5$. Next

$\quad \sin(\alpha + \beta) = \sin \alpha \cos \alpha$
$\quad\quad\quad\quad\quad\quad\quad + \sin \beta \cos \alpha$

$\quad\quad = (\dfrac{3}{5})(\dfrac{3}{5}) + (\dfrac{4}{5})(\dfrac{4}{5})$

$\quad\quad = \dfrac{9 + 16}{25} = 1$

33. $\sin(\tan^{-1}(3/4) - \cos^{-1}(5/13))$.

If $\alpha = \tan^{-1}(3/4)$, then $\sin \alpha = 3/5$ and $\cos \alpha = 4/5$. If $\beta = \cos^{-1}(5/13)$ then $\cos \beta = 5/13$ and $\sin \beta = 12/13$.

$\sin(\alpha - \beta)$
$= \sin \alpha \cos \beta - \cos \alpha \sin \beta$

$= (\frac{3}{5})(\frac{5}{13}) - (\frac{4}{5})(\frac{12}{13})$

$= \frac{15 - 48}{65} = -\frac{33}{65}$

35. $\cos(\tan^{-1}(3) - \sin^{-1}(\sqrt{5}/3))$
$\alpha = \tan^{-1}(3)$ so $\sin \alpha = 3/\sqrt{10}$,
$\cos \alpha = 1/\sqrt{10}$; $\beta = \sin^{-1}(\sqrt{5}/3)$ so
$\sin \beta = \sqrt{5}/3$, $\cos \beta = 2/3$.

$\cos(\alpha - \beta)$
$= \cos \alpha \cos \beta + \sin \alpha \sin \beta$

$= (\frac{1}{\sqrt{10}})(\frac{2}{3}) + (\frac{3}{\sqrt{10}})(\frac{\sqrt{5}}{3})$

$= \frac{2 + 3\sqrt{5}}{3\sqrt{10}} = \frac{2\sqrt{10} + 15\sqrt{2}}{30}$

37. $\sin \alpha = 4/5$, $\cos \beta = 5/13$
$\cos \alpha = 3/5$, $\sin \beta = 12/13$

$\cos(\alpha - \beta)$

$= (\frac{3}{5})(\frac{5}{13}) + (\frac{4}{5})(\frac{12}{13})$

$= \frac{15 + 48}{65} = \frac{63}{65}$

$\sin(\alpha + \beta)$

$= (\frac{4}{5})(\frac{5}{13}) + (\frac{3}{5})(\frac{12}{13})$

$= \frac{20 + 36}{65} = \frac{56}{65}$

39. $\sin \alpha = 5/13$, $\cos \beta = -3/5$
$\cos \alpha = 12/13$, $\sin \beta = -4/5$

$\cos(\alpha + \beta)$
$= (\frac{12}{13})(-\frac{3}{5}) - (\frac{5}{13})(-\frac{4}{5})$

$= \frac{-36 + 20}{65} = -\frac{16}{65}$

$\cos(\alpha - \beta) = \frac{-36 - 20}{65} = -\frac{56}{65}$

41. α in quadrant II $\Rightarrow$
$\sin \alpha > 0$, $\cos \alpha < 0$
$\sin \alpha = 1/3$

$\cos \alpha = -\sqrt{1 - (1/3)^2} = -\frac{2\sqrt{2}}{3}$

β in quadrant I $\Rightarrow$
$\sin \beta > 0$, $\cos \beta > 0$
$\cos \beta = 2/3$

$\sin \beta = \sqrt{1 - (2/3)^2} = \frac{\sqrt{5}}{3}$

$\sin(\alpha + \beta) = \sin \alpha \cos \beta$
$\qquad\qquad + \cos \alpha \sin \beta$

$= (1/3)(2/3) + (-\frac{2\sqrt{2}}{3})(\frac{\sqrt{5}}{3})$

$= 2/9 - 2\sqrt{10}/9$

$= \frac{2 - 2\sqrt{10}}{9}$

$\cos(\alpha + \beta) = \cos \alpha \cos \beta$
$\qquad\qquad - \sin \alpha \sin \beta$

$= (-\frac{2\sqrt{2}}{3})(2/3) - (1/3)(\frac{\sqrt{5}}{3})$

$= \frac{-4\sqrt{2}}{9} - \frac{\sqrt{5}}{9}$

$= \frac{-4\sqrt{2} - \sqrt{5}}{9}$

$\sin(\alpha + \beta) < 0$ and $\cos(\alpha + \beta) < 0$. Therefore $\alpha + \beta$ is in quadrant III.

43. α acute $\Rightarrow \sin \alpha > 0, \cos \alpha > 0$
$\sin \alpha = 4/5$
$\cos \alpha = 3/5$
$\tan \alpha = 4/3$

β acute $\Rightarrow \sin \beta > 0, \cos \beta > 0$
$\cos \beta = 2/5$

$\sin \beta = \sqrt{21}/5$

$\tan \beta = \sqrt{21}/2$

$\sin(\alpha - \beta) = \sin \alpha \cos \beta$
$\qquad - \cos \alpha \sin \beta$
$\qquad = (4/5)(2/5)$

$\qquad \quad - (3/5)(\sqrt{21}/5)$

$\qquad = 8/25 - 3\sqrt{21}/25$

$\qquad = \dfrac{8 - 3\sqrt{21}}{25}$

$\tan(\alpha - \beta) = \dfrac{\tan \alpha - \tan \beta}{1 + \tan \alpha \tan \beta}$

$= \dfrac{4/3 - \sqrt{21}/2}{1 + (4/3)(\sqrt{21}/2)} = \dfrac{\dfrac{8 - 3\sqrt{21}}{6}}{\dfrac{6 + 4\sqrt{21}}{6}}$

$= \dfrac{8 - 3\sqrt{21}}{6 + 4\sqrt{21}} = \dfrac{6 - \sqrt{21}}{-6}$

$= \dfrac{\sqrt{21} - 6}{6}$

Sin$(\alpha - \beta) < 0$ and tan$(\alpha - \beta) < 0$.
Therefore $\alpha - \beta$ is in quadrant IV.

45. α in quadrant IV $\Rightarrow$
$\sin \alpha < 0, \cos \alpha > 0$
$\cos \alpha = 3/4$

$\sin \alpha = -\sqrt{7}/4$

$\tan \alpha = -\sqrt{7}/3$

β in quadrant II $\Rightarrow$
$\sin \beta > 0, \cos \beta < 0$.
$\cos \beta = -2/3$
$\sin \beta = \sqrt{5}/3$

$\tan \beta = -\sqrt{5}/2$

$\cos(\alpha - \beta) = \cos \alpha \cos \beta$
$\qquad + \sin \alpha \sin \beta$
$\qquad = (3/4)(-2/3)$

$\qquad \quad + (-\sqrt{7}/4)(\sqrt{5}/3)$

$\qquad = -6/12 - \sqrt{35}/12$

$\qquad = \dfrac{-6 - \sqrt{35}}{12}$

$\qquad = -\dfrac{(6 + \sqrt{35})}{12}$

$\tan(\alpha - \beta) = \dfrac{\tan \alpha - \tan \beta}{1 + \tan \alpha \tan \beta}$

$= \dfrac{-\sqrt{7}/3 - (-\sqrt{5}/2)}{1 + (-\sqrt{7}/3)(-\sqrt{5}/2)}$

$= \dfrac{\dfrac{-2\sqrt{7} + 3\sqrt{5}}{6}}{\dfrac{6 + \sqrt{35}}{6}}$

$= \dfrac{-2\sqrt{7} + 3\sqrt{5}}{6 + \sqrt{35}}$

$= 32\sqrt{5} - 27\sqrt{7}$

Cos$(\alpha - \beta) < 0$ and tan$(\alpha - \beta) > 0$.
Therefore $\alpha - \beta$ is in
quadrant III.

47. $2 \sin 6\theta \cos 2\theta$
$\quad = \sin(6\theta + 2\theta) + \sin(6\theta + 2\theta)$
$\quad = \sin 8\theta + \sin 4\theta$

49. $\sin \theta \cos 3\theta$

$$= \frac{1}{2}(2 \sin \theta \cos 3\theta)$$

$$= \frac{1}{2}[\sin(\theta + 3\theta) + \sin(\theta - 3\theta)]$$

$$= \frac{1}{2}[\sin 4\theta + \sin(-2\theta)]$$

$$= \frac{\sin 4\theta - \sin 2\theta}{2}$$

51. $2 \cos 5x \cos 2x$
$$= \cos(5x - 2x) + \cos(5x + 2x)$$
$$= \cos 3x + \cos 7x$$

53. $\sin 3x \sin 8x$

$$= \frac{1}{2}[2 \sin 3x \sin 8x]$$

$$= \frac{1}{2}[\cos(3x-8x) - \cos(3x+8x)]$$

$$= \frac{1}{2}[\cos(-5x) - \cos 11x]$$

$$= \frac{1}{2}[\cos 5x - \cos 11x]$$

$$= \frac{\cos 5x - \cos 11x}{2}$$

55. $4 \sin 8a \cos 3b$
$$= 2[2 \sin 8a \cos 3b]$$
$$= 2[\sin(8a + 3b) + \sin(8a-3b)]$$
$$= 2 \sin(8a + 3b) + 2 \sin(8a-3b)$$

57. $3 \sin 5a \sin a$
$$= 3/2[2 \sin 5a \sin a]$$
$$= 3/2[\cos(5a - a) - \cos(5a+a)]$$
$$= 3/2[\cos 4a - \cos 6a]$$

$$= \frac{3 \cos 4a - 3 \cos 6a}{2}$$

59. $\sin(a - b) = \sin a \cos b$
$$- \cos a \sin b$$

$$\sin(a - b) = \sin[a + (-b)]$$
$$= \sin a \cos(-b)$$
$$+ \cos a \sin(-b)$$

$$= \sin a \cos b$$
$$- \cos a \sin b$$

61. $\tan a - \tan b = \dfrac{\sin(a - b)}{\cos a \cos b}$

$$\frac{\sin(a - b)}{\cos a \cos b}$$

$$= \frac{\sin a \cos b - \cos a \sin b}{\cos a \cos b}$$

$$= \frac{\sin a \; \cancel{\cos b}}{\cos a \; \cancel{\cos b}} - \frac{\cancel{\cos a} \sin b}{\cancel{\cos a} \cos b}$$

$$= \tan a - \tan b$$

63. $\cot a - \tan b = \dfrac{\cos(a + b)}{\sin a \cos b}$

$$\frac{\cos(a + b)}{\sin a \cos b}$$

$$= \frac{\cos a \cos b - \sin a \sin b}{\sin a \cos b}$$

$$= \frac{\cos a \; \cancel{\cos b}}{\sin a \; \cancel{\cos b}} - \frac{\cancel{\sin a} \sin b}{\cancel{\sin a} \cos b}$$

$$= \cot a - \tan b$$

65. $\cot(a + b) = \dfrac{\cot a \cot b - 1}{\cot a + \cot b}$

$\cot(a + b) = \dfrac{\cos(a + b)}{\sin(a + b)}$

$= \dfrac{\cos a \cos b - \sin a \sin b}{\sin a \cos b + \cos a \sin b}$

[Divide numerator and
denominator by
$\sin a \sin b$]

$= \dfrac{\dfrac{\cos a \cos b}{\sin a \sin b} - \dfrac{\sin a \sin b}{\sin a \sin b}}{\dfrac{\sin a \cos b}{\sin a \sin b} + \dfrac{\cos a \sin b}{\sin a \sin b}}$

$= \dfrac{\cot a \cot b - 1}{\cot b + \cot a}$

67. $\sin a + \sin b$

$= 2 \sin \dfrac{a + b}{2} \cos \dfrac{a - b}{2}$

According to formula (9.10),
we have

$\sin(\alpha + \beta) + \sin(\alpha - \beta)$
$= 2 \sin \alpha \cos \beta.$

If we set $\alpha = \dfrac{a + b}{2}$ and

$\beta = \dfrac{a - b}{2}$, then $\alpha + \beta = a$ and

$\alpha - \beta = b$. Replacing above,
we obtain

$\sin a + \sin b$

$= 2 \sin \dfrac{a + b}{2} \cos \dfrac{a - b}{2}.$

69. $\cos a + \cos b$

$= 2 \cos \dfrac{a + b}{2} \cos \dfrac{a - b}{2}$

According to formula (9.12) we
have $\cos(\alpha - \beta) + \cos(\alpha + \beta) =$
$2 \cos \alpha \cos \beta.$ If we let

$\alpha = \dfrac{a + b}{2}$ and $\beta = \dfrac{a - b}{2}$, we

get $\alpha + \beta = a$ and $\alpha - \beta = b.$
Thus,

$\cos a + \cos b$
$= \cos(\alpha + \beta) + \cos(\alpha - \beta)$
$= 2 \cos \alpha \cos \beta$

$= 2 \cos \dfrac{a + b}{2} \cos \dfrac{a - b}{2}.$

Exercises 9.3

1. $\cos \alpha = 3/5, \ 0 < \alpha < \pi/2$

$\sin \alpha = \sqrt{1 - (3/5)^2} = 4/5$

$\sin 2\alpha = 2 \sin \alpha \cos \alpha$
$= 2(4/5)(3/5)$
$= 24/25$

$\cos 2\alpha = \cos^2\alpha - \sin^2\alpha$
$= (3/5)^2 - (4/5)^2$
$= 9/25 - 16/25$
$= -7/25$

$\tan 2\alpha = \sin 2\alpha/\cos 2\alpha$

$= \dfrac{24/25}{-7/25}$

$= -24/7$

3. $\sin \alpha = -3/4$, $180° < \alpha < 270°$

$\cos \alpha = -\sqrt{7}/4$

$\sin 2\alpha = 2(-3/4)(-\sqrt{7}/4)$

$\qquad = 3\sqrt{7}/8$

$\cos 2\alpha = (-\sqrt{7}/4)^2 - (-3/4)^2$
$\qquad = -1/8$

$\tan 2\alpha = \dfrac{\dfrac{3\sqrt{7}}{8}}{-\dfrac{1}{8}} = -3\sqrt{7}$

5. $\tan \alpha = 4/3$, α acute
$\sin \alpha = 4/5$
$\cos \alpha = 3/5$
$\sin 2\alpha = 2(4/5)(3/5) = 24/25$

$\cos 2\alpha = (3/5)^2 - (4/5)^2$
$\qquad = -7/25$

$\tan 2\alpha = \dfrac{24/25}{-7/25} = \dfrac{-24}{7}$

7. $\sec \alpha = 5/2$, $3\pi/2 < \alpha < 2\pi$

$\sin \alpha = -\sqrt{21}/5$
$\cos \alpha = 2/5$

$\sin 2\alpha = 2(-\sqrt{21}/5)(2/5)$

$\qquad = \dfrac{-4\sqrt{21}}{25}$

$\cos 2\alpha = (2/5)^2 - (-\sqrt{21}/5)^2$
$\qquad = -17/25$

$\tan 2\alpha = \dfrac{4\sqrt{21}}{17}$

9. $\sin \alpha = -1/2$, $\pi < \alpha < 3\pi/2$

$\cos \alpha = -\sqrt{3}/2$

$\pi/2 < \alpha/2 < 3\pi/4 \Rightarrow$
$\quad \sin \alpha/2 > 0$, $\tan \alpha/2 < 0$.

$\sin \alpha/2 = \sqrt{\dfrac{1 - \cos \alpha}{2}}$

$\qquad = \sqrt{\dfrac{1 + \sqrt{3}/2}{2}}$

$\qquad = \sqrt{\dfrac{2 + \sqrt{3}}{4}}$

$\qquad = \dfrac{\sqrt{2 + \sqrt{3}}}{2}$

$\tan \alpha/2 = -\sqrt{\dfrac{1 - \cos \alpha}{1 + \cos \alpha}}$

$\qquad = -\sqrt{\dfrac{1 + \sqrt{3}/2}{1 - \sqrt{3}/2}}$

$\qquad = -\sqrt{\dfrac{2 + \sqrt{3}}{2 - \sqrt{3}}}$

$\qquad = -\sqrt{(2 + \sqrt{3})^2}$

$\qquad = -(2 + \sqrt{3})$

11. $\cos \alpha = 1/4$, $3\pi/2 < \alpha < 2\pi$

$3\pi/4 < \alpha/2 < \pi \Rightarrow$
$\quad \sin \alpha/2 > 0$, $\tan \alpha/2 < 0$

$\sin \alpha/2 = \sqrt{\dfrac{1 - 1/4}{2}} = \sqrt{\dfrac{3/4}{2}}$

$\qquad = \sqrt{3/8} = \dfrac{\sqrt{6}}{4}$

$\tan \alpha/2 = -\sqrt{\dfrac{1 - 1/4}{1 + 1/4}}$

$\qquad = -\sqrt{\dfrac{3/4}{5/4}} = -\sqrt{3/5}$

$\qquad = -\dfrac{\sqrt{15}}{5}$

13. $\tan \alpha = 3$,
α acute $\Rightarrow \alpha/2$ acute
$\cos \alpha = \sqrt{10}/10$

$$\sin \alpha/2 = \sqrt{\frac{1 - \sqrt{10}/10}{2}}$$

$$= \sqrt{\frac{10 - \sqrt{10}}{20}}$$

$$\tan \alpha/2 = \sqrt{\frac{1 - \sqrt{10}/10}{1 + \sqrt{10}/10}}$$

$$= \sqrt{\frac{10 - \sqrt{10}}{10 + \sqrt{10}}}$$

$$= \sqrt{\frac{(10 - \sqrt{10})^2}{90}}$$

$$= \frac{1}{3} \frac{10 - \sqrt{10}}{\sqrt{10}}$$

$$= \frac{1}{3} (\sqrt{10} - 1)$$

15. $\csc \alpha = -4$, $180° < \alpha < 270°$

$\cos \alpha = -\sqrt{15}/4$
$90° < \alpha/2 < 135° \Rightarrow$
$\sin \alpha/2 > 0$, $\tan \alpha/2 < 0$

$$\sin \alpha/2 = \sqrt{\frac{1 + \sqrt{15}/4}{2}}$$

$$= \sqrt{\frac{4 + \sqrt{15}}{8}}$$

$$= \frac{\sqrt{8 + 2\sqrt{15}}}{4}$$

$$\tan \alpha/2 = -\sqrt{\frac{1 + \sqrt{15}/4}{1 - \sqrt{15}/4}}$$

$$= -\sqrt{\frac{4 + \sqrt{15}}{4 - \sqrt{15}}}$$

$$= -(4 + \sqrt{15})$$

17. $\sin \alpha = 12/13$, $\pi/2 < \alpha < \pi \Rightarrow$
$\cos \alpha = -5/13$

Since $\pi/4 < \alpha/2 < \pi/2$, it follows
that $\cos \alpha/2 > 0$. Thus

$$\cos \frac{\alpha}{2} = \sqrt{\frac{1 - 5/13}{2}}$$

$$= \sqrt{4/13} = 2\sqrt{13}/13$$

19. $\sin 2\alpha = 24/25$, $\pi/4 < \alpha < \pi/2 \Rightarrow$
$\cos 2\alpha = -7/25$

$$\cos \alpha = \sqrt{\frac{1 - 7/25}{2}}$$

$$= \sqrt{18/50} = 3/5$$

21. $\sin \dfrac{\pi}{8} = \sin \dfrac{\pi/4}{2}$, $\alpha = \pi/4$

$$= \sqrt{\frac{1 - \cos \pi/4}{2}}$$

$$= \sqrt{\frac{1 - \sqrt{2}/2}{2}}$$

$$= \sqrt{\frac{2 - \sqrt{2}}{4}}$$

$$= \frac{\sqrt{2 - \sqrt{2}}}{2}$$

23. $\cos \dfrac{\pi}{12} = \cos \dfrac{\pi/6}{2}$, $\alpha = \pi/6$

$$= \sqrt{\frac{1 + \cos \pi/6}{2}}$$

$$= \sqrt{\frac{1 + \sqrt{3}/2}{2}}$$

$$= \sqrt{\frac{2 + \sqrt{3}}{4}} = \frac{\sqrt{2 + \sqrt{3}}}{2}$$

25. $\sin 75° = \sin \dfrac{150°}{2}$, $\alpha = 150°$

$$= \sqrt{\frac{1 - \cos 150°}{2}}$$

$$= \sqrt{\frac{1 + \sqrt{3}/2}{2}}$$

$$= \sqrt{\frac{2 + \sqrt{3}}{4}} = \frac{\sqrt{2 + \sqrt{3}}}{2}$$

27. $\cos 22°30' = \cos \dfrac{45°}{2}$

$= \sqrt{\dfrac{1 + \cos 45°}{2}}$

$= \sqrt{\dfrac{1 + \sqrt{2}/2}{2}}$

$= \sqrt{\dfrac{2 + \sqrt{2}}{4}} = \dfrac{\sqrt{2 + \sqrt{2}}}{2}$

29. Set $x = \cos^{-1}(4/5)$ and get
$\cos x = 4/5$ and $\sin x = 3/5$. Now

$\sin(2 \cos^{-1}(4/5)) = \sin 2x$

$= 2 \sin x \cos x$

$= 2(\dfrac{3}{5})(\dfrac{4}{5}) = \dfrac{24}{25}$

31. Set $x = \tan^{-1}(5/12)$. Then
$\tan x = -5/12$, $\sin x = -5/13$,
$\cos x = 12/13$.

$\cos(2\tan^{-1}(5/12)) = \cos 2x$
$= \cos^2 x - \sin^2 x$

$= (\dfrac{12}{13})^2 - (-\dfrac{5}{13})^2$

$= \dfrac{119}{169}$

33. Set $x = \sin^{-1}(3/5)$ and get
$\sin x = 3/5$, $\cos x = 4/5$. Now

$\sin(\dfrac{1}{2}\sin^{-1}(3/5)) = \sin \dfrac{x}{2}$

$= \sqrt{\dfrac{1 - 4/5}{2}}$

$= \sqrt{1/10}$

$= \sqrt{10}/10.$

35. $(\sin \theta - \cos \theta)^2 = 1 - \sin 2\theta$

$(\sin \theta - \cos \theta)^2$
$= \sin^2\theta - 2 \sin \theta \cos \theta + \cos^2\theta$
$= 1 - 2 \sin \theta \cos \theta$
$= 1 - \sin 2\theta$

37. $1 + \cot^2\theta = 2 \cot \theta \csc 2\theta$

$2 \cot \theta \csc 2\theta = 2 \dfrac{\cot \theta}{\sin 2\theta}$

$= \dfrac{2 \cot \theta}{2 \sin \theta \cos \theta}$

$= \dfrac{1}{\sin^2\theta}$

$= \csc^2\theta$

$= 1 + \cot^2\theta$

39. $\cot 2\theta = \dfrac{\cot^2\theta - 1}{2 \cot \theta}$

$\dfrac{\cot^2\theta - 1}{2 \cot \theta} = \dfrac{\dfrac{\cos^2\theta}{\sin^2\theta} - 1}{2\dfrac{\cos \theta}{\sin \theta}}$

$= \dfrac{\dfrac{\cos^2\theta - \sin^2\theta}{\sin^2\theta}}{2\dfrac{\cos \theta}{\sin \theta}}$

$= \dfrac{\cos^2\theta - \sin^2\theta}{2 \sin \theta \cos \theta}$

$= \dfrac{\cos 2\theta}{\sin 2\theta}$

$= \cot 2\theta$

41. $\csc 2\theta = \dfrac{1}{2}(\tan\theta + \cot\theta)$

$\dfrac{1}{2}(\tan\theta + \cot\theta)$

$= \dfrac{1}{2}\left(\dfrac{\sin\theta}{\cos\theta} + \dfrac{\cos\theta}{\sin\theta}\right)$

$= \dfrac{\sin^2\theta + \cos^2\theta}{2\sin\theta\cos\theta}$

$= \dfrac{1}{\sin 2\theta}$

$= \csc 2\theta$

43. $\tan\dfrac{\theta}{2} = \dfrac{1 - \cos\theta}{\sin\theta}$

$\dfrac{1 - \cos\theta}{\sin\theta} = \dfrac{1 - \cos\left(\dfrac{\theta}{2} + \dfrac{\theta}{2}\right)}{\sin\left(\dfrac{\theta}{2} + \dfrac{\theta}{2}\right)}$

$= \dfrac{1 - \cos^2\dfrac{\theta}{2} + \sin^2\dfrac{\theta}{2}}{2\sin\dfrac{\theta}{2}\cos\dfrac{\theta}{2}}$

$= \dfrac{2\,\sin^2\dfrac{\theta}{2}}{2\,\sin\dfrac{\theta}{2}\cos\dfrac{\theta}{2}}$

$= \tan\dfrac{\theta}{2}$

45. $\cot\theta - \tan\theta = \dfrac{2\cos 2\theta}{\sin 2\theta}$

$\dfrac{2\cos 2\theta}{\sin 2\theta} = \dfrac{2(\cos^2\theta - \sin^2\theta)}{2\sin\theta\cos\theta}$

$= \dfrac{\cos^2\theta}{\sin\theta\cos\theta}$

$- \dfrac{\sin^2\theta}{\sin\theta\cos\theta}$

$= \cot\theta - \tan\theta$

47. If $\cos\alpha = -0.5736$ and $\pi/2 < \alpha < \pi$ then $\sin\alpha = \sqrt{1 - (-0.5736)^2} = 0.8191$. Next,

$$\begin{aligned}\sin 2\alpha &= 2\sin\alpha\cos\alpha \\ &= 2(0.8191)(-0.5736) \\ &\simeq -0.9397\end{aligned}$$

49. If $\cos\theta = 0.8660$ and $3\pi/2 < \theta < 2\pi$, then

$$\cos\dfrac{\theta}{2} = -\sqrt{\dfrac{1 + \cos\theta}{2}}$$

$$= -\sqrt{\dfrac{1 + 0.8660}{2}}$$

$$\simeq -0.9659$$

Exercises 9.4

1. $2\sin\alpha - 1 = 0$, $[-\pi/2, \pi/2]$
$2\sin\alpha = 1$
$\sin\alpha = 1/2$
$\alpha = \pi/6$

3. $\sqrt{3}\tan\alpha - 1 = 0$, $[-\pi/2, \pi/2]$

$\sqrt{3}\tan\alpha = 1$
$\tan\alpha = 1/\sqrt{3}$
$\alpha = \pi/6$

5. $2\cos x + 5 = 4$, $[0°, 360°)$
$2\cos x = -1$
$\cos x = -1/2$
$x = 120°,\ 240°$

7. $3\csc x + 2 = 2\csc x + 4$,
$[0°, 180°)$
$\csc x = 2$
$\sin x = 1/2$
$x = 30°,\ 150°$

9. $\sin\beta + \cos\beta = 0$, $[0, 2\pi)$
$\sin\beta = -\cos\beta$
$\tan\beta = -1$
$\beta = 3\pi/4,\ 7\pi/4$

11. $(\sin x - 1)(\tan x - 1) = 0$,
    $[0°, 360°)$
    $\sin x - 1 = 0$
    $\sin x = 1 \Rightarrow x = 90°$
    $\tan x - 1 = 0$
    $\tan x = 1 \Rightarrow x = 45°, 225°$

13. $(\sqrt{2} \sin x - 1)(\sec x + 2) = 0$,
    $[0, \pi)$

    $\sqrt{2} \sin x - 1 = 0$

    $\sin x = \dfrac{1}{\sqrt{2}} = \dfrac{\sqrt{2}}{2} \Rightarrow$

    $x = \pi/4, 3\pi/4$

    $\sec x + 2 = 0$
    $\sec x = -2$

    $\cos x = -\dfrac{1}{2} \Rightarrow x = 2\pi/3$

15. $(2 \sin x - \sqrt{3})(\tan x + \sqrt{3}) = 0$,
    $[0°, 180°)$

    $2 \sin x - \sqrt{3} = 0$

    $\sin x = \dfrac{\sqrt{3}}{2} \Rightarrow x = 60°, 120°$

    $\tan x + \sqrt{3} = 0$

    $\tan x = -\sqrt{3} \Rightarrow x = 120°$

17. $\sin \theta \tan \theta - \sin \theta = 0$,
    $[0, \pi/2]$
    $\sin \theta (\tan \theta - 1) = 0$
    $\sin \theta = 0 \Rightarrow \theta = 0$
    $\tan \theta - 1 = 0$
    $\tan \theta = 1 \Rightarrow \theta = \pi/4$

19. $2 \cos^2 x + 3 \cos x + 1 = 0$,
    $[0, 2\pi)$
    $(2 \cos x + 1)(\cos x + 1) = 0$
    $2 \cos x + 1 = 0$
    $\cos x = -1/2 \Rightarrow x = 2\pi/3, 4\pi/3$
    $\cos x + 1 = 0$
    $\cos x = -1 \Rightarrow x = \pi.$

21. $\dfrac{2 \sin x}{\sin^2 x - 3} = 1$, $[0°, 360°)$

    $2 \sin x = \sin^2 x - 3$
    $\sin^2 x - 2 \sin x - 3 = 0$
    $(\sin x + 1)(\sin x - 3) = 0$
    $\sin x + 1 = 0$
    $\sin x = -1 \Rightarrow x = 270°$
    $\sin x - 3 = 0$
    $\sin x = 3.$ No solution

23. $\sin \theta + \cos \theta = 1$, $[-\pi/2, \pi/2]$
    $(\sin \theta + \cos \theta)^2 = 1^2$
    $\sin^2 \theta + 2 \sin \theta \cos \theta + \cos^2 \theta = 1$
    $1 + 2 \sin \theta \cos \theta = 1$
    $\quad 2 \sin \theta \cos \theta = 0$
    $\sin \theta = 0 \Rightarrow \theta = 0$
    $\cos \theta = 0 \Rightarrow \theta = \pi/2, -\pi/2$

    Check that $-\pi/2$ is an extraneous
    solution.

25. $\sin 2\alpha = 0$, $[0°, 360°)$
    $\sin 2\alpha = 2 \sin \alpha \cos \alpha = 0$
    $\sin \alpha = 0 \Rightarrow \alpha = 0, 180°$
    $\cos \alpha = 0 \Rightarrow \alpha = 90°, 270°$

27. $\tan 2x = \sqrt{3}$, $[0°, 360°)$

    $\tan 2x = \dfrac{2 \tan x}{1 - \tan^2 x} = \sqrt{3}$

    $2 \tan x = \sqrt{3} - \sqrt{3} \tan^2 x$

    $\sqrt{3} \tan^2 x + 2 \tan x - \sqrt{3} = 0$

    $(\sqrt{3} \tan x - 1)(\tan x + \sqrt{3}) = 0$

    $\sqrt{3} \tan x - 1 = 0$

    $\tan x = \dfrac{\sqrt{3}}{3} \Rightarrow x = 30°, 210°$

    $\tan x + \sqrt{3} = 0$
    $\tan x = -\sqrt{3} \Rightarrow x = 120°, 300°$

29. $2 \cos 2\beta - 1$, $[0, \pi/2]$
    $2 \cos 2\beta = 1$
    $\cos 2\beta = 1/2$
    $2\beta = \pi/3 \Rightarrow \beta = \pi/6$

31. $\sin 2\theta - \sin \theta = 0$, $[0, \pi]$
$2 \sin \theta \cos \theta - \sin \theta = 0$
$\sin \theta (2 \cos \theta - 1) = 0$
$\sin \theta = 0 \Rightarrow \theta = 0, \pi$
$2 \cos \theta - 1 = 0$

$\cos \theta = \dfrac{1}{2} \Rightarrow \theta = \pi/3$

33. $\cos 2x = \sin 2x - 1$, $[0, \pi]$
$2 \cos^2 x - 1 = 2 \sin x \cos x - 1$
$2 \cos^2 x - 2 \sin x \cos x = 0$
$2 \cos x (\cos x - \sin x) = 0$
$\cos x = 0 \Rightarrow x = \pi/2$
$\cos x - \sin x = 0$
$\cos x = \sin x$
$\cot x = 1 \Rightarrow x = \pi/4$

35. $\sin \dfrac{\theta}{2} = \cos \theta$, $[0°, 360°)$

By the double angle formula

$\cos \theta = 1 - 2 \sin^2 \dfrac{\theta}{2}$

so

$\sin \dfrac{\theta}{2} = 1 - 2 \sin^2 \dfrac{\theta}{2}$

$2 \sin^2 \dfrac{\theta}{2} + \sin \dfrac{\theta}{2} - 1 = 0$

$(2 \sin \dfrac{\theta}{2} - 1)(\sin \dfrac{\theta}{2} + 1) = 0$

$2 \sin \dfrac{\theta}{2} - 1 = 0$

$\sin \dfrac{\theta}{2} = \dfrac{1}{2} \Rightarrow \dfrac{\theta}{2} = 30°, 150°$

$\theta = 60°, 300°$

$\sin \dfrac{\theta}{2} + 1 = 0$

$\sin \dfrac{\theta}{2} = -1 \Rightarrow \dfrac{\theta}{2} = 270°$

$\theta = 540°$ not in the given interval.

37. $\cos \theta = \cos \dfrac{\theta}{2} - 1$, $[0, 2\pi)$

If $\theta \in [0, 2\pi)$, then $\dfrac{\theta}{2} \in [0, \pi)$

On the other hand,

$\cos \theta = \cos 2(\dfrac{\theta}{2}) = 2 \cos^2 \dfrac{\theta}{2} - 1$

so

$2 \cos^2 \dfrac{\theta}{2} - 1 = \cos \dfrac{\theta}{2} - 1$

$2 \cos^2 \dfrac{\theta}{2} = \cos \dfrac{\theta}{2}$

$(2 \cos \dfrac{\theta}{2} - 1) \cos \dfrac{\theta}{2} = 0$

$2 \cos \dfrac{\theta}{2} - 1 = 0 \Rightarrow \dfrac{\theta}{2} = \pi/3$

$\theta = 2\pi/3$

$\cos \dfrac{\theta}{2} = 0 \Rightarrow \dfrac{\theta}{2} = \dfrac{\pi}{2}$

$\theta = \pi$

39. $2\sqrt{3} \cos \theta - 3 = 0$

$2\sqrt{3} \cos \theta = 3$

$2 \cos \theta = \sqrt{3}$

$\cos \theta = \dfrac{\sqrt{3}}{2} \Rightarrow \theta = \pm \dfrac{\pi}{6} + 2k\pi$

41. $2 \sin \alpha - 1 = 0$
$2 \sin \alpha = 1$
$\sin \alpha = 1/2$

$\alpha = \dfrac{\pi}{6} + 2k\pi, \dfrac{5\pi}{6} + 2k\pi$

43. $\tan \alpha + 1 = 0$
$\tan \alpha = -1$

$\alpha = \dfrac{3\pi}{4} + k\pi$

45. $\sec^2 u = 2$

$\sec u = \pm\sqrt{2}$

$u = \dfrac{\pi}{4} + \dfrac{k\pi}{2}$

47. $(1 - \sin t)(1 + \cos t) = 0$
$1 - \sin t = 0$

$\sin t = 1 \Rightarrow t = \dfrac{\pi}{2} + 2k\pi$

$1 + \cos t = 0$

2   2   2 + 2k\pi

49. $\sin 2u + \cos u = 0$
$2 \sin u \cos u + \cos u = 0$
$\cos u(2 \sin u + 1) = 0$

$\cos u = 0 \Rightarrow u = \dfrac{\pi}{2} + k\pi$

$\sin u = -\dfrac{1}{2}$

$u = -\dfrac{\pi}{6} + 2k\pi, \quad \dfrac{7\pi}{6} + 2k\pi$

51. $\cos^2 u - \cos u - 6 = 0$
$(\cos u - 3)(\cos u + 2) = 0$
$\cos u = 3$, no solution
$\cos u = -2$, no solution

53. $\sec^2 x = \tan x + 1$
$\tan^2 x + 1 = \tan x + 1$
$\tan^2 x = \tan x$
$\tan x(\tan x - 1) = 0$
$\tan x = 0 \Rightarrow x = k\pi$

$\tan x = 1 \Rightarrow x = \dfrac{\pi}{4} + k\pi$

55. $2 \sin^2 x + 2 \cos x - 1 = 0$, $[0, \pi]$
$2(1 - \cos^2 x) + 2 \cos x - 1 = 0$
$2 \cos^2 x - 2 \cos x - 1 = 0$

$\cos x = \dfrac{2 \pm \sqrt{4 + 8}}{4} = \dfrac{1 \pm \sqrt{3}}{2}$

$\cos x = \dfrac{1 + \sqrt{3}}{2} \approx 1.3660,$

no solution

$\cos x = \dfrac{1 - \sqrt{3}}{2} \approx -0.3660$

$x \approx 1.946$ rad

57. $2 \tan^2 x - 2 \tan x - 1 = 0$,
$[-\pi/2, \ \pi/2]$

$\tan x = \dfrac{2 \pm \sqrt{4 + 8}}{4} = \dfrac{1 \pm \sqrt{3}}{2}$

$\tan x = \dfrac{1 + \sqrt{3}}{2} \approx 1.3660$

$x \approx 0.9389$ rad

$\tan x = \dfrac{1 - \sqrt{3}}{2} \approx -0.3660$

$x = -0.3509$ rad

59. $\cos^2 u - 2 \sin^2 u + 1 = 0$, $[0, \pi]$
$(1 - \sin^2 u) - 2 \sin^2 u + 1 = 0$
$2 - 3 \sin^2 u = 0$
$\sin^2 u = 2/3$

$\sin u = \pm\sqrt{2}/3 = \pm\sqrt{6}/3$
$u = 0.9553$ rad, 2.1863 rad

Exercises 9.5

1. $\alpha = 38°$, $\beta = 75°$, $a = 12$
$\gamma = 180° - (75° + 38°) = 67°$

$b = \dfrac{12 \sin 75°}{\sin 38°} = 18.83 \approx 19$

$c = \dfrac{12 \sin 67°}{\sin 38°} = 17.94 \approx 18$

3. $\alpha = 105°$, $\beta = 15°$, $c = 9$
$\gamma = 180° - (105° + 15°) = 60°$

$a = \dfrac{9 \sin 105°}{\sin 60°} = 10.04 \approx 10$

$b = \dfrac{9 \sin 15°}{\sin 60°} = 2.69 \approx 3$

5. $\beta = 60°$, $\gamma = 45°$, $a = 8$
   $\alpha = 180 - (60° + 45°) = 75°$

   $b = \dfrac{8 \sin 60°}{\sin 75°} = 7.17 \approx 7$

   $c = \dfrac{8 \sin 45°}{\sin 75°} = 5.86 \approx 6$

7. $\beta = 42°10'$, $\gamma = 50°30'$, $a = 6$
   $\alpha = 180° - (42°10' + 50°30')$
   $\quad = 87°20'$

   $b = \dfrac{6 \sin 42°10'}{\sin 87°20'} = 4.03 \approx 4$

   $c = \dfrac{6 \sin 50°30'}{\sin 87°20'} = 4.63 \approx 5$

9. $\alpha = 30°$, $a = 4$, $b = 8$

   $\dfrac{4}{\sin 30°} = \dfrac{8}{\sin \beta}$

   $\sin \beta = \dfrac{8 \sin 30°}{4} = 1$

   $\beta = 90°$
   (Right triangle), $\gamma = 60°$, $c = 4\sqrt{3}$

11. $\beta = 60°$, $a = 4$, $b = 2$

    $\dfrac{4}{\sin \alpha} = \dfrac{2}{\sin 60°}$

    $\sin \alpha = \dfrac{4 \sin 60°}{2} = \sqrt{3} > 1$

    No solution

13. $\alpha = 61°20'$, $\gamma = 38°30'$, $b = 5$
    $\beta = 180° - (61°20' + 38°30')$
    $\quad = 80°10'$

$a = \dfrac{5 \sin 61°20'}{\sin 80°10'} \approx 4.5$

$c = \dfrac{5 \sin 38°30'}{\sin 80°10'} \approx 3.2$

15. $\alpha = 35°$, $a = 6$, $b = 8$

    $\sin \beta = \dfrac{8 \sin 35°}{6} \approx 0.7648$

    $\beta = 49°53'$, $\gamma = 95°7'$, $c = 10.4$
    or
    $\beta = 130°7'$, $\gamma = 14°53$, $c = 2.7$

17. $\alpha = 30°$, $a = 10$, $b = 8$

    $\dfrac{10}{\sin 30°} = \dfrac{8}{\sin \beta}$

    $\sin \beta = \dfrac{8 \sin 30°}{10} = \dfrac{2}{5}$

    $\beta \approx 23°35'$
    $\gamma \approx 126°25'$
    $c \approx 16.1$

19. $\gamma = 65°$, $b = 9.2$, $c = 7.5$

    $\dfrac{7.5}{\sin 65°} = \dfrac{9.2}{\sin \beta}$

    $\sin \beta = \dfrac{9.2 \sin 65°}{7.5} \approx 1.11$

    No solution

21. $\beta = 35°15'$, $\gamma = 42°30'$, $a = 5.2$
    $\alpha = 102°15'$

    $b = \dfrac{5.2 \sin 35°15'}{\sin 102°15'} \approx 3.1$

    $c = \dfrac{5.2 \sin 42°30'}{\sin 102°15'} \approx 3.6$

23. $\gamma = 40°20'$, $b = 10$, $c = 8$

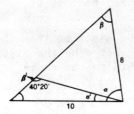

$$\sin \beta = \frac{10 \sin 40°20'}{8} \simeq 0.8090$$

$\beta = 54°$, $\alpha = 85°40'$, $a = 12.3$
or
$\beta = 126°$, $\alpha = 13°40'$, $a = 2.9$

25. $\alpha = 42°$, $a = 12$, $b = 16$;
$0° < \beta < 90°$

$$\sin \beta = \frac{16 \sin 42°}{12} \simeq 0.8922$$

$\beta = 63.15°$
$\gamma = 74.85°$

$$c = \frac{12 \sin 74.85°}{\sin 42°} \simeq 17.31$$

27. $\alpha = 28°40'$, $a = 18$, $b = 24$;
$90° < \beta < 180°$

$$\sin \beta = \frac{24 \sin 28°40'}{18}$$

$$= 0.6396$$
$\beta = 140.24°$
$\gamma = 11.09°$

$$c = \frac{24 \sin 11.09°}{\sin 140.24°} \simeq 7.22$$

29. $\alpha = 30°$, $\beta = 86°$, $a = 5.75$
$\gamma = 64°$

$$b = \frac{5.75 \sin 86°}{\sin 30°} \simeq 11.47$$

$$c = \frac{5.75 \sin 64°}{\sin 30°} \simeq 10.34$$

31. $\alpha = 35.5°$, $\beta = 62.25°$,
$a = 13.5$, $\gamma = 82.25°$

$$b = \frac{13.5 \sin 62.25°}{\sin 35.5°}$$

$$\simeq 20.57$$

$$c = \frac{13.5 \sin 82.25°}{\sin 35.5°}$$

$$\simeq 23.04$$

33. $\alpha = 12.5°$, $\gamma = 105.8°$,
$c = 30.20$, $\beta = 61.7°$

$$a = \frac{30.20 \sin 12.5°}{\sin 105.8°}$$

$$\simeq 6.79$$

$$b = \frac{30.20 \sin 61.7°}{\sin 105.8°}$$

$$\simeq 27.63$$

35. $\beta = 30°$, $a = 2.5$ m, $c = 12.8$ m
$h = 12.8 \sin 30° = 6.4$

$$A = \frac{2.5 \times 6.4}{2} = 8 \text{ m}^2$$

37. $\gamma = 38.5°$, $a = 40.6$ cm,
$b = 12.8$ cm,
$h = 12.8 \sin 38.5° \simeq 7.9682$

$$A = \frac{40.6 \times h}{2} \simeq 161.8 \text{ cm}^2$$

39. Let b be the distance from A to C.

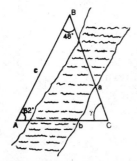

Angle BCA measures 70°

$$b = \frac{320 \sin 48°}{\sin 70°}$$

$$= \frac{(320)(0.7431)}{0.9397}$$

$$\approx 253$$

The distance from A to C is 253 yds.

41. Let x be the distance from A and let y be the distance from B. Angle ACB = 84°

$$x = \frac{24 \sin 35°40'}{\sin 84°} \approx 14.07$$

$$y = \frac{24 \sin 60°20'}{\sin 84°} \approx 20.97$$

The fire is 14 miles from A and 21 miles from B.

43. Let ℓ be the length of the shadow.

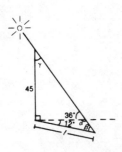

$\alpha = 180° - 36° = 144°$
$\beta = 180° - (144° + 12°) = 24°$
$\gamma = 180° - (36° + 90°) = 54°$

$$\ell = \frac{45 \sin 54°}{\sin 24°} \approx 89.5$$

The shadow is about 90 ft.

45. Let x be the distance from the ship to A. Let d be the distance from the ship to the shore.

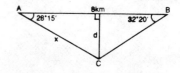

Angle ACB is 119°25'.

$$x = \frac{8 \sin 32°20'}{\sin 119°25'} \approx 4.91$$

$$d = 4.91 \sin 28°15' \approx 2.32$$

The ship is about 5 km from A and 2 km from shore.

47. Let d be distance of plane at 2:00 p.m.

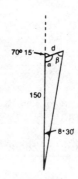

$\alpha = 180° - 70°15' = 109°45'$
$\beta = 180° - (109°45' + 8°30')$
$\quad = 61°45'$

$$d = \frac{150 \sin 8°30'}{\sin 61°45'} \approx 25.17 \text{ miles}$$

The airplane is about 25 miles from the tower.

49.

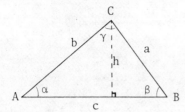

$h = a \sin \beta$ and $h = b \sin \alpha$.
Therefore,
$$a \sin \beta = b \sin \alpha$$

$$\frac{a}{\sin \alpha} = \frac{b}{\sin \beta}$$

Similarly, $\dfrac{a}{\sin \alpha} = \dfrac{c}{\sin \gamma}$.

Thus

$$\frac{a}{\sin \alpha} = \frac{b}{\sin \beta} = \frac{c}{\sin \gamma}.$$

Exercises 9.6

1. $a = 5$, $b = 12$, $c = 13$
    $5^2 + 12^2 = 13^2$, so we have a right triangle
    $\gamma = 90°$

    $\sin \beta = \dfrac{12}{13}$, $\beta = 67°20'$

    $\alpha = 22°40'$

3. $a = 5$, $b = 7$, $c = 9$
    $a^2 = b^2 + c^2 - 2 bc \cos \alpha$
    $25 = 49 + 81 - 126 \cos \alpha$
    $-105 = -126 \cos \alpha$

    $\cos \alpha = \dfrac{105}{126} \simeq 0.8333$

    $\alpha = 33°30'$
    $b^2 = a^2 + c^2 - 2ac \cos \beta$
    $49 = 25 + 81 - 90 \cos \beta$
    $-57 = -90 \cos \beta$

$\cos \beta = \dfrac{57}{90} = 0.6333$

$\beta = 50°40'$
$\gamma = 95°50'$

5. $\beta = 45°$, $a = 3$, $c = 8$
    $b^2 = a^2 + c^2 - 2ac \cos \beta$
    $\quad = 3^2 + 8^2 - 2(3)(8) \cos 45°$
    $\quad = 9 + 64 - (48)(0.7071)$
    $\quad = 39.0589$
    $b = 6.24971 \simeq 6$
    $c^2 = a^2 + b^2 - 2ab \cos \gamma$
    $64 = 9 + 39.0589$
    $\qquad - 2(3)(6.24971) \cos \gamma$
    $\cos \gamma = -0.4251$
    $\qquad \gamma \simeq 115°10'$
    $\qquad \alpha \simeq 19°50'$

7. $\alpha = 110°$, $b = 12$, $c = 18$
    $a^2 = b^2 + c^2 - 2bc \cos \alpha$
    $\quad = 12^2 + 18^2 - 2(12)(18) \cos 110°$
    $\quad = 144 + 324 - 432(-0.3420)$
    $\quad = 615.753$
    $a = 24.814365 \simeq 25$

    $\sin \beta = \dfrac{12 \sin 110°}{24.814365} \simeq 0.4544$

    $\beta = 27°$
    $\gamma \simeq 43°$

9. $a = 5.2$, $b = 8.3$, $c = 10.5$
    $c^2 = a^2 + b^2 - 2ab \cos \gamma$
    $110.25 = 27.04 + 68.89$
    $\qquad - 2(5.2)(8.3) \cos \gamma$
    $14.32 = -86.32 \cos \gamma$
    $\cos \gamma = -0.1659$
    $\gamma = 99°33'$

    $\sin \beta = \dfrac{8.3 \sin 99°33'}{10.5}$

    $\qquad \simeq 0.7795$

    $\beta = 51°13'$
    $\alpha = 29°14'$

11. $a = 10.25, \ b = 5.60, \ c = 8.75$

$a^2 = b^2 + c^2 - 2bc \cos \alpha$

$105.0625 = 31.3600 + 76.5625$
$$- 98 \cos \alpha$$

$-2.8600 = -98 \cos \alpha$

$\cos \alpha = 0.0292$

$\alpha = 88°20'$

$\sin \beta = \dfrac{5.60 \sin 88°20'}{10.25}$

$\simeq 0.5461$

$\beta = 33°6'$

$\gamma = 58°34'$

13. $\alpha = 56°18', \ b = 5.62, \ c = 10.25$

$a^2 = b^2 + c^2 - 2bc \cos \alpha$

$= 31.5844 + 105.0625$
$$-(115.21)(0.5548)$$

$= 136.6469 - 63.9185$

$= 72.7284$

$a \simeq 8.53$

$\sin \beta = \dfrac{5.62 \sin 56°18'}{8.53}$

$\simeq 0.5481$

$\beta = 33°14'$

$\gamma = 90°28'$

15. $\beta = 105°25', \ a = 4.75, \ c = 16.05$

$b^2 = a^2 + c^2 - 2ac \cos \beta$

$= 22.5625 + 257.6025$
$$- (152.475)(-0.2659)$$

$= 280.165 + 40.5431$

$= 320.7281$

$b \simeq 17.91$

$\sin \alpha = \dfrac{4.75 \sin 105°25'}{17.91}$

$\simeq 0.2557$

$\alpha = 14°49'$

$\gamma = 59°46'$

17. $a = 12.5, \ b = 30, \ c = 32.5$

$(12.5)^2 + (30)^2 = (32.5)^2$ so we have a right triangle.

$\gamma = 90°$

$\sin \beta = \dfrac{30}{32.5} = 0.9231$

$\beta = 67.38°$

$\alpha = 22.62°$

19. $a = 4.25, \ b = 3.75, \ c = 5.18$

$\cos \gamma = \dfrac{a^2 + b^2 - c^2}{2ab}$

$= 0.1660423$

$\gamma = 80.44°$

$\beta = 45.55°$

$\alpha = 54.01°$

21. $\beta = 42.5°, \ a = 3.10, \ c = 8.25$

$b^2 = a^2 + c^2 - 2ac \cos \beta$

$= 39.96$

$b \simeq 6.32$

$\alpha = 19.35°$

$\gamma = 118.15°$

23. $\gamma = 75.12°, \ a = 10.5, \ b = 8.25$

$c^2 = a^2 + b^2 - 2ab \cos \gamma$

$= 133.82$

$c = 11.57$

$\beta = 43.57°$

$\alpha = 61.31°$

25. $\alpha = 117.20°, \ b = 6.25, \ c = 8.45$

$a^2 = b^2 + c^2 - 2bc \cos \alpha$

$= 158.75$

$a = 12.60$

$\beta = 26.18°$

$\gamma = 36.62°$

27. $x^2 = 5^2 + 8^2 - 2(5)(8) \cos 60°$

$= 25 + 64 - 80(1/2)$

$= 49$

$x = 7 \text{ in}$

29. $\ell^2 = 15^2 + 15^2 - 2(15)(15) \cos 120°$

$= 675$

$\ell = 15\sqrt{3} \text{ cm} \simeq 25.98 \text{ cm}$

31. $\cos \beta = \dfrac{a^2 + c^2 - b^2}{2ac}$

   $= \dfrac{(4.75)^2 + (12.25)^2 - (10.75)^2}{2(4.75)(12.25)}$

   $= 0.4903329$

   $\beta = 60.64°$

33. Let x be the distance from P to Q.
   $x^2 = 360^2 + 480^2$
   $\quad - 2(360)(480) \cos 38°40'$
   $\quad = 90157.584$
   $\quad x = 300.263 \simeq 300$ yd.

35. Let a be the length of one diagonal and let b be the length of the other diagonal.

   $a^2 = 5^2 + 8^2 - 2(5)(8) \cos 60°$
   $\quad = 25 + 64 - 40$
   $\quad = 49$
   $\quad a = 7$

   $b^2 = 5^2 + 8^2 - 2(5)(8) \cos 120°$
   $\quad = 25 + 64 + 40$
   $\quad = 129$
   $\quad b = 11.36$

   The diagonals measure 7 and 11.36 ft

37. Let d_1 be the length of the diagonal facing the angle of 58°30' and let d_2 be the length of the diagonal facing the angle of 121°30'.

   $d_1^2 = 18^2 + 24^2$
   $\quad - 2(18)(24) \cos 58°30'$
   $\quad = 448.56$
   $\quad d_1 = 21.179 \simeq 21$ ft

   $d_2^2 = 18^2 + 24^2$
   $\quad - 2(18)(24) \cos 121°30'$
   $\quad = 1351.44$
   $\quad d_2 = 36.762 \simeq 37$ ft

39. Let x be the distance apart after 2 hours.

   $x^2 = 16^2 + 20^2 - 2(16)(20) \cos 24°$
   $\quad = 256 + 400 - 640(0.9135)$
   $\quad = 656 - 584.67$
   $\quad = 71.33$
   $\quad x = 8.45$

   They will be 8.45 miles apart.

41. $x^2 = 2^2 + 6^2 - 2(2)(6) \cos 5.4°$
   $\quad \simeq 16.1$
   $\quad x \simeq 4.0$ mi

43. Let x be the distance from the airport.

   $x^2 = 95^2 + 185^2$
   $\quad - 2(95)(185) \cos 105°$
   $\quad = 9025 + 34225$
   $\quad - 35150(-0.2588)$
   $\quad = 52347.48$

   $x = 229$

   The plane is 229 miles from the airport.

45. Let d be the distance between the boats

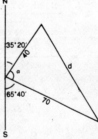

   $\alpha = 180° - (35°20' + 65°40')$
   $\quad = 79°$

   $d^2 = 40^2 + 70^2$
   $\quad - 2(40)(70) \cos 79°$
   $\quad = 1600 + 4900 - 5600(0.1908)$
   $\quad = 5431.47$
   $\quad d = 73.6985$ km $\simeq 73.7$ km

47. $a = 9$ m, $b = 40$ m, $c = 41$ m

$$s = \frac{1}{2}(9 + 40 + 41) = 45$$

$$A = \sqrt{45(36)(5)(4)} = 180 \text{ m}^2$$

49. $a = 15$ in., $b = 9$ in., $c = 18$ in.

$$s = \frac{1}{2}(15 + 9 + 18) = 21$$

$$A = \sqrt{21(6)(12)(3)} \approx 67.3 \text{ in}^2$$

51. $a = 12.5$ yd, $b = 8.9$ yd,
$c = 13.4$ yd

$$s = \frac{1}{2}(12.5 + 8.9 + 13.4) = 17.4$$

$$A = \sqrt{17.4(4.9)(8.5)(4)} \approx 53.8 \text{ yd}^2$$

53. 75 m by 40 m by 85 m

$$s = \frac{1}{2}(75 + 40 + 85) = 100$$

$$A = \sqrt{100(25)(60)(15)} = 1500 \text{ m}^2$$

55. $\cos \alpha = \dfrac{b^2 + c^2 - a^2}{2bc}$

$$1 + \cos \alpha = \frac{b^2 + c^2 - a^2}{2bc} + 1$$

$$= \frac{b^2 + c^2 - a^2 + 2bc}{2bc}$$

$$= \frac{(b + c)^2 - a^2}{2bc}$$

$$= \frac{(b + c + a)(b + c - a)}{2bc}$$

57. $\cos^2 \dfrac{\alpha}{2} = \dfrac{1 + \cos \alpha}{2}$

$$\cos \frac{\alpha}{2} = \sqrt{\frac{1 + \cos \alpha}{2}}$$

$$= \sqrt{\frac{(b + c + a)(b + c - a)}{4bc}}$$

(from exercise 55). On the other hand,

$$s = \frac{a + b + c}{2}$$

$$s - a = \frac{a}{2} + \frac{b}{2} + \frac{c}{2} - a$$

$$= \frac{b + c - a}{2}$$

Thus

$$\cos \frac{\alpha}{2}$$

$$= \sqrt{\frac{b + c + a}{2} \cdot \frac{b + c - a}{2} \cdot \frac{1}{bc}}$$

$$= \sqrt{\frac{s(s - a)}{bc}}$$

59. $A = \dfrac{1}{2} bc \sin \alpha$

$$= \frac{1}{2} bc \sqrt{1 - \cos^2\alpha}$$

$$= \frac{1}{2} bc \sqrt{(1 + \cos \alpha)(1 - \cos \alpha)}$$

Exercises 9.7

1. $A(2,3)$, $B(4,-1)$

$$\overrightarrow{AB} = \langle 4 - 2, \ -1 - 3 \rangle = \langle 2, \ -4 \rangle$$

3. $A(-1,-2)$, $B(3,5)$

$$\overrightarrow{AB} = \langle 3 - (-1), \ 5 - (-2) \rangle$$
$$= \langle 4, 7 \rangle$$

5. $A(-1,1)$, $\vec{v} = (-2,5)$
Let $B(x,y)$ be the terminal point. We have

$$x = -1 + (-2) = -3$$
$$y = 1 + 5 = 6$$

7. $U(2,-6)$, $\vec{u} <-3,3>$
Let $V(x,y)$ be the terminal point.
We have

$$x = 2 + (-3) = -1$$
$$y = -6 + 3 = -3$$

9. $\vec{a} = <-4,1>$, $A(-2,2)$

If $P(x,y)$ is the initial point, then

$$-2 = x + (-4) \text{ and } 2 = y + 1$$
$$x = 2 \quad \text{ and } \quad y = 1$$

11. $\vec{b} = <-3, 5>$, $O(0, 0)$

If $A(x,y)$ is the initial point then

$$0 = x + (-3), \quad 0 = y + 5$$
$$x = 3, \qquad\qquad y = -5$$

13. $A(4,2)$, $B(1,-3)$, $C(-1,5)$

$$\overrightarrow{AB} = <-3,-5>, \overrightarrow{AC} = <-5,3>$$

$$\overrightarrow{AB} + \overrightarrow{AC} = <-3,-5> + <-5,3>$$
$$= <-8,-2>$$

15. $|\vec{u}| = 4$, $|\vec{v}| = 6$, angle $= 150°$

$$|\vec{u} + \vec{v}|^2 = |\vec{u}|^2 + |\vec{v}|^2$$
$$- 2|\vec{u}||\vec{v}| \cos 30°$$

$$= 16 + 36 - 2(4)(6)\frac{\sqrt{3}}{2}$$

$$= 52 - 24\sqrt{3}$$

$$|\vec{u} + \vec{v}| = \sqrt{52 - 24\sqrt{3}} \approx 3.23$$

$$|\vec{u} - \vec{v}|^2 = |\vec{u}|^2 + |\vec{v}|^2$$
$$- 2|\vec{u}||\vec{v}| \cos 150°$$

$$= 16 + 36 - 2(4)(6)(-\frac{\sqrt{3}}{2})$$

$$= 52 + 24\sqrt{3}$$

$$|\vec{u} - \vec{v}| = \sqrt{52 + 24\sqrt{3}} \approx 9.67$$

17. $\overrightarrow{OA} = <-3,5>$

$$|\overrightarrow{OA}| = \sqrt{(-3)^2 + 5^2}$$

$$= \sqrt{9 + 25} = \sqrt{34}$$

19. $\vec{v} = 3\vec{i} - 5\vec{j}$

$$|\vec{v}| = \sqrt{3^2 + (-5)^2} = \sqrt{34}$$

21. $\vec{a} = <-2,3>$, $\vec{b} = <3,-1>$

$$3\vec{a} - \vec{b} = 3<-2,3> - <3,-1>$$
$$= <-6,9> - <3,-1>$$
$$= <-6 - 3, 9 - (-1)>$$
$$= <-9,10>$$

23. $\vec{a} = <-2,3>$, $\vec{b} = <3,-1>$

$$2\vec{a} - 5\vec{b} = 2<-2,3> - 5<3,-1>$$
$$= <-4,6> - <15,-5>$$
$$= <-4 - 15, 6 - (-5)>$$
$$= <-19,11>$$

25. $\vec{v} = <-5,12>$

$$|\vec{v}| = \sqrt{(-5)^2 + 12^2} = 13$$

$$\vec{u} = \frac{\vec{v}}{|\vec{v}|} = \frac{<-5,12>}{13}$$

$$= <-\frac{5}{13}, \frac{12}{13}>$$

$$-\vec{u} = <\frac{5}{13}, -\frac{12}{13}>$$

27. $\vec{u} = -2\vec{i} + 5\vec{j}, \; \vec{v} = 3\vec{i} - \vec{j}$

    $3\vec{u} - 4\vec{v}$

    $\quad = 3(-2\vec{i} + 5\vec{j}) - 4(3\vec{i} - \vec{j})$

    $\quad = -6\vec{i} + 15\vec{j} - 12\vec{i} + 4\vec{j}$

    $\quad = -18\vec{i} + 19\vec{j}$

29. $|\vec{v}| = 2$, direction angle $= 60°$

    $\vec{v} = x\vec{i} + y\vec{j}$

    $x = 2 \cos 60° = 2(\frac{1}{2}) = 1$

    $y = 2 \sin 60° = 2(\frac{\sqrt{3}}{2}) = \sqrt{3}$

    $\vec{v} = \vec{i} = \sqrt{3}\vec{j}$

31. $|\vec{a}| = 6$, direction angle $= 150°$

    $\vec{a} = x\vec{i} + y\vec{j}$

    $x = 6 \cos 150° = 6(-\frac{\sqrt{3}}{2}) = -3\sqrt{3}$

    $y = 6 \sin 150° = 6(\frac{1}{2}) = 3$

    $\vec{a} = -3\sqrt{3}\vec{i} + 3\vec{j}$

33. $\vec{v} = 2\vec{i} + 2\vec{j}, \; \theta = $ direction angle

    $|\vec{v}| = \sqrt{2^2 + 2^2} = 2\sqrt{2}$

    $\cos \theta = \frac{2}{2\sqrt{2}} = \frac{1}{\sqrt{2}}$

    $\sin \theta = \frac{2}{2\sqrt{2}} = \frac{1}{\sqrt{2}}$

    $\theta = 45°$

35. $\vec{v} = \vec{i} - \sqrt{3}\vec{j}, \; \theta = $ direction angle

    $|\vec{v}| = \sqrt{1 + 3} = 2$

    $\cos \theta = \frac{1}{2}, \; \sin \theta = -\frac{\sqrt{3}}{2}$

    $\theta = 300°$

37. $\vec{u} = 4\vec{i} - 3\vec{j}, \; \vec{v} = -5\vec{i} + 2\vec{j}$

    $\vec{u} \cdot \vec{v} = (4)(-5) + (-3)(2)$
    $\quad = -20 - 6 = -26$

39. $\vec{p} = <6,0>, \; \vec{q} = <-2,-4>$

    $\vec{p} \cdot \vec{q} = (6)(-2) + (0)(-4)$
    $\quad = -12 + 0 = -12$

41. $\vec{a} = 4\vec{i} - 3\vec{j}, \; \vec{b} = -2\vec{i} + \vec{j}$

    $\vec{a} \cdot \vec{b} = (4)(-2) + (-3)(1) = -11$

    $|\vec{a}| = \sqrt{4^2 + (-3)^2} = 5$

    $|\vec{b}| = \sqrt{(-2)^2 + 1^2} = \sqrt{5}$

    $\cos \theta = \frac{-11}{5\sqrt{5}} \approx -0.9839$

    $\theta \approx 169.70°$

43. $\vec{u} = <3,-7>, \; \vec{v} = <7,3>$

    $\vec{u} \cdot \vec{v} = (3)(7) + (-7)(3) = 0$

    Thus $\vec{u}$ and $\vec{v}$ are orthogonal
    vectors. Angle is $90°$.

45. $\vec{u} = x\vec{i} + 4\vec{j}$ perpendicular to

    $\vec{v} = -6\vec{i} + 2\vec{j}$.

    $0 = \vec{u} \cdot \vec{v} = x(-6) + (4)(2)$
    $\quad = -6x + 8$

    $x = 4/3$

47. $|\vec{f_1}| = 12$, $|\vec{f_2}| = 18$, angle $= 135°$

$$|\vec{f_1} + \vec{f_2}|^2$$

$$= |\vec{f_1}|^2 + |\vec{f_2}|^2 - 2|\vec{f_1}||\vec{f_2}|\cos 45°$$

$$= 12^2 + 18^2 - 2(12)(18)\,\frac{\sqrt{2}}{2}$$

$$= 124 + 324 - 216\sqrt{2}$$

$$= 448 - 216\sqrt{2}$$

$$|\vec{f_1} + \vec{f_2}| = 12.75 \text{ N}$$

49. 350 kg, ramp inclination $= 15°$
$$f = 350 \sin 15°$$
$$\simeq 350(0.2588)$$
$$\simeq 90.6 \text{ kg}$$

51. The magnitude of the second force is obtained by solving the oblique triangle

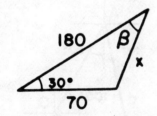

$$x^2 = 70^2 + 180^2$$
$$- 2(70)(180) \cos 30°$$
$$x \simeq 124.4 \text{ lb}$$

Next we use the law of sines to find β, the angle facing the side of length 70.

$$\sin \beta = \frac{70 \sin 30°}{124.4} \simeq 0.2813$$

$$\beta = 16.3°$$

Thus the angle between the two forces is

$$30° + 16.3° = 46.3°$$

53. Let x be the equal magnitude of the two forces making an angle of $60°$. If the magnitude of the resultant is $400\sqrt{3}$, then by the law of cosines, we have

$$(400\sqrt{3})^2 = x^2 + x^2$$
$$- 2(x)(x) \cos 120°$$

$$= 2x^2 - 2x^2(-\frac{1}{2})$$

$$= 3x^2$$

or

$$(400\sqrt{3})^2 = (\sqrt{3}x)^2,$$

so

$$x = 400 \text{ lb.}$$

55. Let $\vec{v} = x\vec{i} + y\vec{j}$, where x and y are, respectively, the easterly and northerly components of $\vec{v}$.

Since $|\vec{v}| = 750$ and the direction is $30°$ west of north, we have

$$y = 750 \cos 30° = 375\sqrt{3}$$
$$x = -750 \cos 60° = -375,$$
thus

$$\vec{v} = -375\vec{i} + 375\sqrt{3}\vec{j}.$$

b) $2 \times 375\sqrt{3} = 750\sqrt{3}$ km

57. The speeds of the boat and the current are shown in the following diagram

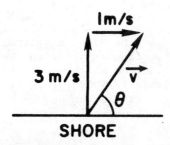

$$|\vec{v}| = \sqrt{3^2 + 1^2} = \sqrt{10} \text{ m/s}$$

$$\theta = \arctan\left(\frac{3}{1}\right) \approx 71.6°$$

59. Let x denote how far downstream Lisa lands. You can find x from the similar triangles shown in the diagram

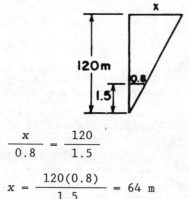

$$\frac{x}{0.8} = \frac{120}{1.5}$$

$$x = \frac{120(0.8)}{1.5} = 64 \text{ m}$$

61. The tensions t_1 and t_2 are obtained from the following diagram

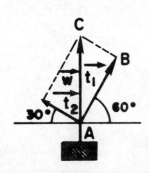

$$|\vec{w}| = 1000$$

Referring to the right triangle ABC, we obtain

$$|\vec{t_1}| = 1000 \cos 30° = 500\sqrt{3} \text{ lb}$$

$$|\vec{t_2}| = 1000 \sin 30° = 500 \text{ lb}$$

Thus a tension of 500 lb is being applied on the 30°-angle cable and a tension of $500\sqrt{3}$ lb on the 60°-angle cable.

63. Referring to the diagram, we have

$$d^2 = (450)^2 + (450)^2 - 2(450)(450) \cos 100°,$$

hence

$$d \approx 689.44 \approx 689 \text{ miles}$$

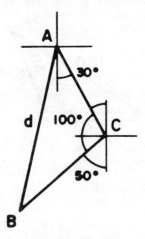

In the triangle ABC, the angles with vertices A and B measure 40° (why?). Thus the bearing relative to A is 10° west of south.

65. Referring to the figure in the text, let a be the magnitude of the force along the horizontal support and let b be the magnitude on the inclined support. We have

$$b = \frac{60}{\cos 50°} \approx 93.3 \text{ lb}$$

and

$$a = 60 \tan 50° \approx 71.5 \text{ lb}$$

67. If $\vec{u} = \langle u_1, u_2 \rangle$, $\vec{v} = \langle v_1, v_2 \rangle$, and $\vec{w} = \langle w_1, w_2 \rangle$, then

$$(\vec{u} + \vec{v}) + \vec{w}$$
$$= \langle (u_1 + v_1) + w_1, (u_2 + v_2) + w_2 \rangle$$
$$= \langle u_1 + (v_1 + w_1), u_2 + (v_2 + w_2) \rangle$$
$$= \vec{u} + (\vec{v} + \vec{w}).$$

$$\vec{u} + \vec{0} = \langle u_1, u_2 \rangle + \langle 0, 0 \rangle$$
$$= \langle u_1 + 0, u_2 + 0 \rangle$$
$$= \langle u_1, u_2 \rangle = \vec{u}$$

$$\vec{u} + (-\vec{u}) = \langle u_1, u_2, \rangle + \langle -u_1, -u_2 \rangle$$
$$= \langle u_1 + (-u_1), u_2 + -(u_2) \rangle$$
$$= \langle 0, 0 \rangle$$
$$= \vec{0}$$

69. $\vec{u} = \langle u_1, u_2 \rangle$, $\vec{v} = \langle v_1, v_2 \rangle$,

$\vec{w} = \langle w_1, w_2 \rangle$

$$\vec{u} \cdot (\vec{v} + \vec{w})$$
$$= u_1(v_1 + w_1) + u_2(v_2 + w_2)$$
$$= u_1 v_1 + u_1 w_1 + u_2 v_2 + u_2 w_2$$

$$= (u_1 v_1 + u_2 v_2) + (u_1 w_1 + u_2 w_2)$$
$$= \vec{u} \cdot \vec{v} + \vec{u} \cdot \vec{w}$$

Exercises 9.8

1. $z = 3 - 3i$

$$|z| = \sqrt{3^2 + (-3)^2} = 3\sqrt{2}$$

$$z = 3\sqrt{2}(1/\sqrt{2} - i/\sqrt{2})$$
$$= 3\sqrt{2}(\cos 7\pi/4 + i \sin 7\pi/4)$$

3. $z = \sqrt{3} - i$

$$|z| = \sqrt{3 + 1} = 2$$

$$z = 2(\sqrt{3}/2 - i/2)$$
$$= 2(\cos 11\pi/6 + i \sin 11\pi/6)$$

5. $z = -2 \quad 2i\sqrt{3}$

$$|z| = \sqrt{4 + 12} = 4$$

$$z = 4(-1/2 - i\sqrt{3}/2)$$
$$= 4(\cos 4\pi/3 + i \sin 4\pi/3)$$

7. $z = 5i$
$$= 5(\cos \pi/2 + i \sin \pi/2)$$

9. $z = -15$
$$= 15(-1)$$
$$= 15(\cos \pi + i \sin \pi)$$

11. $z_1 = 5(\cos \pi/4 + i \sin \pi/4)$
$z_2 = 8(\cos 2\pi/3 + i \sin 2\pi/3)$
$z_1 z_2 = (5)(8)[\cos(\pi/4 + 2\pi/3)$
$\qquad + i \sin (\pi/4 + 2\pi/3)]$
$\qquad = 40(\cos 11\pi/12$
$\qquad\qquad + i \sin 11\pi/12)$
$z_1/z_2 = 5/8[\cos(\pi/4 - 2\pi/3)$
$\qquad\qquad + i \sin(\pi/4 - 2\pi/3)]$
$\qquad = 5/8(\cos 5\pi/12$
$\qquad\qquad - i \sin 5\pi/12)$

13. $z_1 = 6(\cos 240° + i \sin 240°)$
$z_2 = 9(\cos 105° + i \sin 105°)$
$z_1 z_2 = (6)(9)[\cos(240° + 105°)$
$\qquad + i \sin(240° + 105°)]$
$\qquad = 54(\cos 345° + i \sin 345°)$
$z_1/z_2 = 6/9[\cos(240° - 105°)$
$\qquad + i \sin(240° - 105°)]$
$\qquad = 2/3(\cos 135° + i \sin 135°)$

$$= \frac{1}{2}(\cos \frac{\pi}{6} - i \sin \frac{\pi}{6})$$

$$= \frac{1}{2}(\frac{\sqrt{3}}{2} - i \frac{1}{2})$$

$$= \frac{\sqrt{3} - i}{4}$$

15. $z_1 = \sqrt{3} + i, \; z_2 = 2 + 2\sqrt{3}i)$

a) $z_1 z_2 = (\sqrt{3} + i)(2 + 2\sqrt{3}i)$
$\qquad = 2\sqrt{3} - 2\sqrt{3} + 2i + 6i$
$\qquad = 8i$

$$\frac{z_1}{z_2} = \frac{\sqrt{3} + i}{2 + 2\sqrt{3}i}$$

$$= \frac{(\sqrt{3} + i)(2 - 2\sqrt{3}i)}{(2 + 2\sqrt{3}i)(2 - 2\sqrt{3}i)}$$

$$= \frac{2\sqrt{3} + 2\sqrt{3} + 2i - 6i}{4 + 12}$$

$$= \frac{4\sqrt{3} - 4i}{16} = \frac{\sqrt{3} - i}{4}$$

b) $z_1 = \sqrt{3} + i$

$$= 2(\cos \frac{\pi}{6} + i \sin \frac{\pi}{6})$$

$z_2 = 2 + 2\sqrt{3}i$

$$= 4(\cos \frac{\pi}{3} + i \sin \frac{\pi}{3})$$

$$z_1 z_2 = 8[\cos(\frac{\pi}{6} + \frac{\pi}{3})$$

$$+ i \sin(\frac{\pi}{6} + \frac{\pi}{3})]$$

$$= 8(\cos \frac{\pi}{2} + i \sin \frac{\pi}{2})$$

$$= 8i$$

$$\frac{z_1}{z_2} = \frac{2}{4}[\cos(\frac{\pi}{6} - \frac{\pi}{3})$$

$$+ i \sin(\frac{\pi}{6} - \frac{\pi}{3})]$$

17. $z_1 = 5 + 5i, \; z_2 = -3i$

a) $z_1 z_2 = (5 + 5i)(-3i)$
$\qquad = -15i - 15i^2$
$\qquad = 15 - 15i$

$$z_1/z_2 = \frac{5 + 5i}{-3i}$$

$$= \frac{(5 + 5i)(i)}{(-3i)(i)}$$

$$= \frac{5i + 5i^2}{-3i^2}$$

$$= \frac{-5 + 5i}{3}$$

b) $z_1 = 5 + 5i$

$$= 5\sqrt{2}(\cos \pi/4 + i \sin \pi/4)$$

$z_2 = -3$
$$= 3(\cos 3\pi/2 + i \sin 3\pi/2)$$

$$z_1 z_2 = (5\sqrt{2})(3)[\cos(\pi/4 + 3\pi/2)$$
$$+ i \sin(\pi/4 + 3\pi/2)]$$

$$= 15\sqrt{2}(\cos 7\pi/4$$
$$+ i \sin 7\pi/4)$$
$$= 15 - 15i$$

$$z_1/z_2 = 5\sqrt{2}/3[\cos(\pi/4 - 3\pi/2)$$
$$+ i \sin(\pi/4 - 3\pi/2)]$$

$$= 5\sqrt{2}/3[\cos(-5\pi/4)$$
$$+ i \sin(-5\pi/4)]$$

$$= 5\sqrt{2}/3(-1/\sqrt{2} + i/\sqrt{2})$$
$$= (-5 + 5i)/3$$

19. $z = r(\cos \theta + i \sin \theta)$

$z^2 = z \cdot z$

$\quad = r \cdot r[\cos(\theta + \theta)$

$\qquad\qquad + i \sin(\theta + \theta)]$

$\quad = r^2(\cos 2\theta + i \sin 2\theta)$

$z^3 = z^2 \cdot z$

$\quad = r^2 \cdot r[\cos(2\theta + \theta)$

$\qquad\qquad + i \sin(2\theta + \theta)]$

$\quad = r^3(\cos 3\theta + i \sin 3\theta)$

$z^n = z^{n-1} \cdot z$

$\quad = r^{n-1} \cdot r[\cos((n-1)\theta + \theta)$

$\qquad\qquad + i \sin((n-1)\theta + \theta))]$

$\quad = r^n(\cos n\theta + i \sin n\theta)$

Exercises 9.9

1. $(-1 + i)^5$

$$= \left[\sqrt{2}\left(\cos \frac{3\pi}{4} + i \sin \frac{3\pi}{4}\right)\right]^5$$

$$= 4\sqrt{2}\left(\cos \frac{15\pi}{4} + i \sin \frac{15\pi}{4}\right)$$

$$= 4\sqrt{2}\left(\frac{1}{\sqrt{2}} - i \frac{1}{\sqrt{2}}\right)$$

$$= 4 - 4i$$

3. $(-\sqrt{3} - i)^{11}$

$\quad = [2(\cos 7\pi/6 + i \sin 7\pi/6)]^{11}$

$\quad = 2048(\cos 77\pi/6 + i \sin 77\pi/6)$

$\quad = 2048(\cos 5\pi/6 + i \sin 5\pi/6)$

$\quad = 2048(-\sqrt{3}/2 + i/2)$

$\quad = -1024(\sqrt{3} - i)$

5. $[2(\cos 60° + i \sin 60°)]^7$

$\quad = 128(\cos 420° + i \sin 420°)$

$\quad = 128(\cos 60° + i \sin 60°)$

$\quad = 128(1/2 + i\sqrt{3}/2)$

$\quad = 64(1 + i\sqrt{3})$

7. $[\sqrt{3}(\cos 120° - i \sin 120°]^6$

$\quad = 27(\cos 720° - i \sin 120°)$

$\quad = 27$

9. $(-1 + i\sqrt{3})^{-4}$

$\quad = [2(\cos 120° + i \sin 120°)]^{-4}$

$\quad = 1/16[\cos(-480°)$

$\qquad\qquad + i \sin(-480°)]$

$\quad = 1/16(\cos 120° - i \sin 120°)$

$\quad = 1/16(-1/2 - i\sqrt{3}/2)$

$\quad = -1/32 - (\sqrt{3}/32)i$

11. $z = 1 + i\sqrt{3}, \ n = 2$

$$z = 2\left(\cos \frac{\pi}{3} + i \sin \frac{\pi}{3}\right)$$

$$w_k = \sqrt{2}\left[\cos \frac{\dfrac{\pi}{3} + 2k\pi}{2}\right.$$

$$\left. + i \sin \frac{\dfrac{\pi}{3} + 2k\pi}{2}\right],$$

$$k = 0, 1.$$

So

$$w_o = \sqrt{2}\left(\cos \frac{\pi}{6} + i \sin \frac{\pi}{6}\right)$$

$$= \frac{\sqrt{6} + i\sqrt{2}}{2}$$

$$w_1 = \sqrt{2}\left(\cos \frac{7\pi}{6} + i \sin \frac{7\pi}{6}\right)$$

$$= \frac{-\sqrt{6} - i\sqrt{2}}{2}$$

13. $z = -i$, $n = 3$

$z = \cos 3\pi/2 + i \sin 3\pi/2$

$$w_k = \cos \frac{3\pi/2 + 2k\pi}{3}$$

$$+ i \sin \frac{3\pi/2 + 2k\pi}{3};$$

$k = 0, 1, 2$

$w_0 = \cos \pi/2 + i \sin \pi/2 = i$

$w_1 = \cos 7\pi/6 + i \sin 7\pi/6$

$$= \frac{-\sqrt{3} - i}{2}$$

$w_2 = \cos 11\pi/6 + i \sin 11\pi/6$

$$= \frac{\sqrt{3} - i}{2}$$

15. $z = -\sqrt{3} + i$, $n = 5$

$z = 2(\cos 150° + i \sin 150°)$

$$w_k = 2^{1/5}(\cos \frac{150° + 360°k}{5}$$

$$+ i \sin \frac{150° + 360°k}{5});$$

$k = 0, 1, 2, 3, 4$

$w_0 = 2^{1/5}(\cos 30° + i \sin 30°)$

$w_1 = 2^{1/5}(\cos 102° + i \sin 102°)$

$w_2 = 2^{1/5}(\cos 174° + i \sin 174°)$

$w_3 = 2^{1/5}(\cos 246° + i \sin 246°)$

$w_4 = 2^{1/5}(\cos 318° + i \sin 318°)$

17. $z = 64 - 64\sqrt{3}i$; $n = 6$

$z = 128(\cos 300° + i \sin 300°)$

$$w_k = \sqrt[6]{128}(\cos \frac{300° + 360°k}{6}$$

$$+ i \sin \frac{300° + 360°k}{6});$$

$k = 0, 1, 2, 3, 4, 5$

$w_0 = 2^6\sqrt{2}(\cos 50° + i \sin 50°)$

$w_1 = 2^6\sqrt{2}(\cos 110° + i \sin 110°)$

$w_2 = 2^6\sqrt{2}(\cos 170° + i \sin 170°)$

$w_3 = 2^6\sqrt{2}(\cos 230° + i \sin 230°)$

$w_4 = 2^6\sqrt{2}(\cos 290° + i \sin 290°)$

$w_5 = 2^6\sqrt{2}(\cos 350° + i \sin 350°)$

19. $z = 16(\cos 120° + i \sin 120°)$;

$n = 4$

$$w_k = \sqrt[4]{16}(\cos \frac{120° + 360°k}{4}$$

$$+ i \sin \frac{120° + 360°k}{4});$$

$k = 0, 1, 2, 3$

$w_0 = 2(\cos 30° + i \sin 30°)$

$$= \sqrt{3} + i$$

$w_1 = 2(\cos 120° + i \sin 120°)$

$$= -1 + \sqrt{3}i$$

$w_2 = 2(\cos 210° + i \sin 210°)$

$$= -\sqrt{3} - i$$

$w_3 = 2(\cos 300° + i \sin 300°)$

$$= 1 - \sqrt{3}i$$

21. $x^3 + 27 = 0$
    $x^3 = -27 = 27(\cos \pi + i \sin \pi)$

    Now, we find the cube roots of $27(\cos \pi + i \sin \pi)$:

    $w_0 = 3(\cos \dfrac{\pi}{3} + i \sin \dfrac{\pi}{3})$

    $\quad = \dfrac{3}{2} + \dfrac{\sqrt{3}}{2} i$

    $w_1 = 3(\cos \pi + i \sin \pi) = -3$

    $w_2 = 3(\cos \dfrac{5\pi}{3} + i \sin \dfrac{5\pi}{3})$

    $\quad = \dfrac{3}{2} - \dfrac{\sqrt{3}}{2} i$

23. $x^4 - 16 = 0$
    $x^4 = 16 = 16(\cos 0 + i \sin 0)$
    $x_0 = 2(\cos 0 + i \sin 0) = 2$
    $x_1 = 2(\cos \pi/2 + i \sin \pi/2)$
    $\quad = 2i$
    $x_2 = 2(\cos \pi + i \sin \pi) = -2$
    $x_3 = 2(\cos 3\pi/2 + i \sin 3\pi/2)$
    $\quad = -2i$

25. $x^6 + 1 = 0$
    $x^6 = -1 = \cos \pi + i \sin \pi$

    $x_k = \cos \dfrac{\pi + 2k\pi}{6}$

    $\qquad + i \sin \dfrac{\pi + 2k\pi}{6}$

    $k = 0, 1, 2, 3, 4, 5$

    $x_0 = \cos \pi/6 + i \sin \pi/6$

    $\quad = \dfrac{\sqrt{3} + i}{2}$

    $x_1 = \cos \pi/2 + i \sin \pi/2 = i$

$x_2 = \cos 5\pi/6 + i \sin 5\pi/6$

$\quad = \dfrac{-\sqrt{3} + i}{2}$

$x_3 = \cos 7\pi/6 + i \sin 7\pi/6$

$\quad = \dfrac{-\sqrt{3} - i}{2}$

$x_4 = \cos 3\pi/2 + i \sin 3\pi/2$
$\quad = -i$

$x_5 = \cos 11\pi/6 + i \sin 11\pi/6$

$\quad = \dfrac{\sqrt{3} - i}{2}$

27. $x^3 - 8i = 0$
    $x^3 = 8i$
    $\quad = 8(\cos \pi/2 + i \sin \pi/2)$

    $x_0 = 2(\cos \pi/6 + i \sin \pi/6)$

    $\quad = \sqrt{3} + i$

    $x_1 = 2(\cos 5\pi/6 + i \sin 5\pi/6)$

    $\quad = -\sqrt{3} + i$

    $x_2 = 2(\cos 3\pi/2 + i \sin 3\pi/2)$

    $\quad = -2i$

29. Let $x = \cos \dfrac{\pi}{4} - i\sin \dfrac{\pi}{4}$.

    Then

    $\quad x^4 = \cos\pi - i\sin\pi = -1$

    $\quad x^2 = \cos\dfrac{\pi}{2} - i\sin \dfrac{\pi}{2} = -i$

    Thus

    $\quad x^4 - x^2 + 1 = -1 - (-i) + 1 = i.$

Exercises 9.10

Find the corresponding graphs in the answer section of your text.

3. (a) $(1, \sqrt{3})$
    (b) $(-\sqrt{2}, -\sqrt{2})$
    (c) $(-3\sqrt{3}/2, 3/2)$
    (d) $(2\sqrt{2}, -2\sqrt{2})$

5. (a) $(\sqrt{2}, \pi/4)$
    (b) $(2, 5\pi/6)$
    (c) $(5, 3\pi/2)$
    (d) $(3\sqrt{2}, 3\pi/4)$

17. $r = 4$
    $r^2 = 16$
    $x^2 + y^2 = 16$

19. $r = 4 \sin \theta$
    $r^2 = 4r \sin \theta$
    $x^2 + y^2 = 4y$
    $x^2 + y^2 - 4y + 4 = 4$
    $x^2 + (y - 2)^2 = 4$

21. $r = 3 \sec \theta$

    $r = \dfrac{3}{\cos \theta}$

    $r \cos \theta = 3$
    $x = 3$

23. $r = \dfrac{2}{3 \sin \theta - \cos \theta}$

    $r(3 \sin \theta - \cos \theta) = 2$
    $3r \sin \theta - r \cos \theta = 2$
    $3y - x = 2$
    $x - 3y + 2 = 0$

25. $r \cos \theta - 2 = 0$
    $x - 2 = 0$
    $x = 2$

27. $r(2 \sin \theta + \cos \theta) = 1$
    $2r \sin \theta + r \cos \theta = 1$
    $2y + x = 1$

29. $r = \dfrac{2}{1 - \sin \theta}$

    $r(1 - \sin \theta) = 2$
    $r = r \sin \theta + 2$

    $\sqrt{x^2 + y^2} = y + 2$
    $x^2 + y^2 = (y + 2)^2$
    $x^2 + y^2 = y^2 + 4y + 4$
    $x^2 = 4y + 4$

    or

    $4y = x^2 - 4$

31. $x^2 + y^2 = 16$
    $r = 4$

33. $x - 5 = 0$
    $r \cos \theta = 5$

    $r = \dfrac{5}{\cos \theta}$

    $r = 5 \sec \theta$

35. $x + 2y = 4$
    $r \cos \theta + 2 r \sin \theta = 4$
    $r(\cos \theta + 2 \sin \theta) = 4$

    $r = \dfrac{4}{\cos \theta + 2 \sin \theta}$

37. $x^2 + y^2 - 2y = 0$
    $r^2 - 2r \sin \theta = 0$
    $r = 2 \sin \theta$

39. $x^2 = 2y$
    $(r \cos \theta)^2 = 2r \sin \theta$
    $r^2 \cos^2\theta = 2r \sin \theta$
    $r \cos^2\theta = 2 \sin \theta$

    $r = 2 \sin \theta \cdot \dfrac{1}{\cos^2\theta}$

    $r = 2 \sin \theta \sec^2\theta$

Review Exercises - Chapter 9

1. $\cos^2 x(\sec^2 x - 1) = \sin^2 x$
   $\cos^2 x(\sec^2 x - 1)$
   $\quad = \cos^2 x \, \tan^2 x$
   $\quad = \sin^2 x$

3. $\dfrac{1 + \cos u}{\sin u} + \dfrac{\sin u}{1 + \cos u}$

   $\quad = 2 \csc u$

   $\dfrac{1 + \cos u}{\sin u} + \dfrac{\sin u}{1 + \cos u}$

   $\quad = \dfrac{(1 + \cos u)^2 + (\sin u)^2}{\sin u(1 + \cos u)}$

   $\quad = \dfrac{1 + 2 \cos u + \cos^2 u + \sin^2 u}{\sin u(1 + \cos u)}$

   $\quad = \dfrac{2(1 + \cos u)}{\sin u(1 + \cos u)}$

   $\quad = 2 \csc u$

5. $\dfrac{\cos u}{1 - \sin u} + \dfrac{\cos u}{1 + \sin u}$

   $\quad = 2 \sec u$

   $\dfrac{\cos u}{1 - \sin u} + \dfrac{\cos u}{1 + \sin u}$

   $\quad = \dfrac{\cos u(1+\sin u)+\cos u(1-\sin u)}{(1 - \sin u)(1 + \sin u)}$

   $\quad = \dfrac{\cos u+\cos u \sin u+\cos u-\cos u \sin u}{1 - \sin^2 u}$

   $\quad = \dfrac{2 \cancel{\cos u}}{\cos^2 u}$

   $\quad = 2 \sec u$

7. $\dfrac{\tan \theta}{\sec \theta - 1} = \dfrac{\sec \theta + 1}{\tan \theta}$

   $\dfrac{\tan \theta}{\sec \theta - 1}$

   $\quad = \dfrac{\tan \theta(\sec \theta + 1)}{(\sec \theta - 1)(\sec \theta + 1)}$

   $\quad = \dfrac{\tan \theta(\sec \theta + 1)}{\sec^2 \theta - 1}$

   $\quad = \dfrac{\tan \theta(\sec \theta + 1)}{\tan^2 \theta}$

   $\quad = \dfrac{\sec \theta + 1}{\tan \theta}$

9. $\dfrac{\tan a}{\sin a - 2 \tan a} = \dfrac{1}{\cos a - 2}$

   $\dfrac{\tan a}{\sin a - 2 \tan a}$

   $\quad = \dfrac{\cancel{\tan a}}{\cancel{\tan a}(\cos a - 2)}$

   $\quad = \dfrac{1}{\cos a - 2}$

11. $\dfrac{\cot x - \cos x}{\cot x + \cos x} = \dfrac{\csc x + 1}{\csc x + 1}$

    $\dfrac{\cot x - \cos x}{\cot x + \cos x} = \dfrac{\dfrac{\cos x}{\sin x} - \cos x}{\dfrac{\cos x}{\sin x} + \cos x}$

    $\quad = \dfrac{\dfrac{\cos x - \cos x \sin x}{\sin x}}{\dfrac{\cos x + \cos x \sin x}{\sin x}}$

    $\quad = \dfrac{\cancel{\cos x}(1 - \sin x)}{\cancel{\cos x}(1 + \sin x)}$

    [Divide numerator and denominator by sin x]

    $\quad = \dfrac{\dfrac{1}{\sin x} - 1}{\dfrac{1}{\sin x} + 1}$

    $\quad = \dfrac{\csc x - 1}{\csc x + 1}$

13. $\dfrac{\sin a \cos b - \cos a \sin b}{\cos a \cos b + \sin a \sin b}$

$= \dfrac{\tan a - \tan b}{1 + \tan a \tan b}$

$\dfrac{\sin a \cos b - \cos a \sin b}{\cos a \cos b + \sin a \sin b}$

$= \dfrac{\dfrac{\sin a \,\cancel{\cos b}}{\cos a \,\cancel{\cos b}} - \dfrac{\cancel{\cos a}\, \sin b}{\cancel{\cos a}\, \cos b}}{\dfrac{\cancel{\cos a}\,\cancel{\cos b}}{\cancel{\cos a}\,\cancel{\cos b}} + \dfrac{\sin a \sin b}{\cos a \cos b}}$

$= \dfrac{\tan a - \tan b}{1 + \tan a \tan b}$

15. $\cot^2 x + 2 = \dfrac{\csc^4 x - 1}{\cot^2 x}$

$\dfrac{\csc^4 x - 1}{\cot^2 x}$

$= \dfrac{(\csc^2 x - 1)(\csc^2 x + 1)}{\cot^2 x}$

$[1 + \cot^2 x = \csc^2 x]$

$= \dfrac{\cot^2 x(\csc^2 x + 1)}{\cot^2 x}$

$= 1 + \cot^2 x + 1$
$= \cot^2 x + 2$

17. $2 - \csc^2 \theta = \dfrac{1 - \cot^4 \theta}{\csc^2 \theta}$

$\dfrac{1 - \cot^4 \theta}{\csc^2 \theta}$

$= \dfrac{(1 - \cot^2 \theta)(1 + \cot^2 \theta)}{\csc^2 \theta}$

$[1 + \cot^2 \theta = \csc^2 \theta]$

$= \dfrac{(1 - \cot^2 \theta)\cancel{\csc^2 \theta}}{\cancel{\csc^2 \theta}}$

$= 1 + 1 - \csc^2 \theta$
$= 2 - \csc^2 \theta$

19. $\dfrac{\sin^3 x + \cos^3 x}{\sin x + \cos x} = 1 - \sin x \cos x$

Recall that $a^3 + b^3 = (a + b)(a^2 - ab + b^2)$. Thus

$\dfrac{\sin^3 x + \cos^3 x}{\sin x + \cos x} = \sin^2 x$

$\quad - \sin x \cos x + \cos^2 x$

$= 1 - \sin x \cos x$

21. $(\sin \theta + \cos \theta)^2 = 1 + \sin 2\theta$
$(\sin \theta + \cos \theta)^2$
$\quad = \sin^2 \theta + 2 \sin \theta \cos \theta$
$\qquad + \cos^2 \theta$
$\quad = 1 + 2 \sin \theta \cos \theta$
$\quad = 1 + \sin 2\theta$

23. $2 \tan \alpha \csc 2\alpha = 1 + \tan^2 \alpha$

$2 \tan \alpha \csc 2\alpha$

$= 2 \dfrac{\sin \alpha}{\cos \alpha} \dfrac{1}{\sin 2\alpha}$

$= \cancel{2} \dfrac{\cancel{\sin \alpha}}{\cos \alpha} \dfrac{1}{\cancel{2} \,\cancel{\sin \alpha} \cos \alpha}$

$= \dfrac{1}{\cos^2 \alpha}$

$= \sec^2 \alpha$
$= 1 + \tan^2 \alpha$

25. $2 \tan u \cot 2u = 1 - \tan^2 u$
$2 \tan u \cot 2u$

$= 2 \dfrac{\sin u}{\cos u} \dfrac{\cos 2u}{\sin 2u}$

$= \cancel{2} \dfrac{\cancel{\sin u}}{\cos u} \dfrac{\cos^2 u - \sin^2 u}{\cancel{2} \,\cancel{\sin u} \cos u}$

$= \dfrac{\cos^2 u - \sin^2 u}{\cos^2 u}$

$= 1 - \tan^2 u$

27. $\tan \dfrac{\theta}{2} = \dfrac{\sin \theta}{1 + \cos \theta}$

$$\dfrac{\sin \theta}{1+\cos \theta} = \dfrac{2 \sin \dfrac{\theta}{2} \cos \dfrac{\theta}{2}}{1 + \cos^2 \dfrac{\theta}{2} - \sin^2 \dfrac{\theta}{2}}$$

$$= \dfrac{\cancel{2} \sin \dfrac{\theta}{2} \cos \cancel{\dfrac{\theta}{2}}}{\cancel{2} \cos^{\cancel{2}} \dfrac{\theta}{2}}$$

$$= \tan \dfrac{\theta}{2}$$

29. If $\cos \alpha = -0.5736$ and
$\pi/2 < \alpha < \pi$, then
$\sin \alpha = \sqrt{1 - (-0.5736)^2}$
$\quad = 0.8191$ and
$\tan \alpha = -1.4281$. Thus

$$\tan 2\alpha = \dfrac{2 \tan \alpha}{1 - \tan^2 \alpha}$$

$$= \dfrac{2(-1.4281)}{1 - (-1.4281)^2}$$

$$= 2.7480$$

31. $2 \sin \theta + 1 = 0, \ [0, \ 2\pi)$
$\quad\quad 2 \sin \theta = -1$
$\quad\quad\quad \sin \theta = -1/2$
$\quad\quad\quad\quad \theta = 7\pi/6, \ 11\pi/6$

33. $\tan 2\theta + \sqrt{3} = 0, \ [-\pi/4, \ \pi/4]$
$\quad\quad \tan 2\theta = -\sqrt{3}$
$\quad\quad\quad\quad 2\theta = -\pi/3$
$\quad\quad\quad\quad\quad \theta = -\pi/6$

35. $\sin \alpha - \cos \alpha = 1$
$(\sin \alpha - \cos \alpha)^2 = 1$
$\sin^2 \alpha - 2 \sin \alpha \cos \alpha + \cos^2 \alpha = 1$
$\quad -2 \sin \alpha \cos \alpha = 0$
$\sin \alpha = 0$ or $\cos \alpha = 0$
$\quad \alpha = k\pi \quad\quad \alpha = \pi/2 + k\pi$

But the solutions of the given
equation are $(2k + 1)\pi$ or
$\pi/2 + 2k\pi$, k any integer.

37. $4 \sin^2 x - 1 = 0$
$(2 \sin x + 1)(2 \sin x - 1) = 0$
$\sin x = -1/2 \quad \sin x = 1/2$
$x = \pm\pi/6 + k\pi$, k any integer

39. $\csc^2 u = \cot u + 1$
$1 + \cot^2 u = \cot u + 1$
$\cot^2 u - \cot u = 0$
$\cot u(\cot u - 1) = 0$
$\cot u = 0 \quad \cot u = 1$
$u = \pi/2 + k\pi, \ \pi/4 + k\pi,$
k any integer

41. $\alpha = 28°, \ \beta = 65°, \ a = 15$
$\gamma = 180° - (28° + 65°) = 87°$

$$b = \dfrac{15 \sin 65°}{\sin 28°} \approx 29.0$$

$$c = \dfrac{15 \sin 87°}{\sin 28°} \approx 31.9$$

43. $\beta = 120°, \ \gamma = 35°, \ a = 18$
$\alpha = 180° - (120° + 35°) = 25°$

$$b = \dfrac{18 \sin 120°}{\sin 25°} \approx 36.9$$

$$c = \dfrac{18 \sin 35°}{\sin 25°} \approx 24.4$$

45. $a = 9, \ b = 12, \ c = 15$

$$\cos \gamma = \dfrac{9^2 + 12^2 - 15^2}{2(9)(12)} = 0$$

$\gamma = 90°$, right triangle

$$\sin \alpha = \dfrac{9}{15} \Rightarrow \alpha \approx 36°50'$$

$$\sin \beta = \dfrac{12}{15} \Rightarrow \beta \approx 53°10'$$

47. $a = 4$, $b = 5$, $c = 6$

$$\cos \gamma = \frac{4^2 + 5^2 - 6^2}{2(4)(5)} = 0.125$$

$\gamma \simeq 82.82° \simeq 82°50'$

$$\frac{4}{\sin \alpha} = \frac{6}{\sin 82°50'}$$

$\sin \alpha = 0.6614$
$\alpha \simeq 41.41° \simeq 41°20'$
$\beta = 55°50'$

49. Let x be the length of the pole.

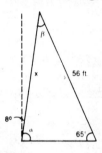

$\alpha = 90° - 8° = 82°$
$\beta = 180° - (82° + 65°) = 33°$

$$x = \frac{56 \sin 65°}{\sin 82°} = 51$$

The pole is 51 ft.

51. Let h be the height of the balloon.

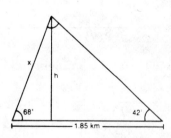

$$\frac{1.85}{\sin 70°} = \frac{x}{\sin 42°}$$

$x = 1.32$
$h = 1.32 \sin 68°$
$\quad = 1.221$

The balloon is 1.22 km above the ground.

53. $a = 10.25$, $b = 8.18$, $c = 6.72$
$a^2 = b^2 + c^2 - 2bc \cos \alpha$
$(10.25)^2 = (8.18)^2 + (6.72)^2$
$\qquad\qquad - 2(8.18)(6.72) \cos \alpha$
$-7.0083 = -109.9392 \cos \alpha$
$\cos \alpha = 0.0637$
$\qquad \alpha = 86°21'$

55. Referring to Figure 5.5 in the text, let us show that the area of the triangle ABC is (1/2) bc sin α. First,

$$\text{Area } \triangle ABC = \frac{ch}{2}$$

Since $h = b \sin \alpha$, we conclude that

$$\text{Area } \triangle ABC = \frac{bc \sin \alpha}{2}$$

Similarly, it can be shown that

$$\text{Area } \triangle ABC = \frac{ac \sin \beta}{2}$$

$$= \frac{ab \sin \gamma}{2}$$

57. a) α is acute
$a^2 = b^2 + c^2 - 2bc \cos \alpha$
$a^2 - b^2 - c^2 = -2bc \cos \alpha$
If α is acute, then $\cos \alpha > 0$ and
$-2bc \cos \alpha < 0$. Thus
$a^2 - b^2 - c^2 < 0$ or $a^2 < b^2 + c^2$.

b) α is a right angle.
If α is a right angle, then
$\cos \alpha = 0$ and $-2bc \cos \alpha = 0$.
Thus $a^2 - b^2 - c^2 = 0$ or
$a^2 = b^2 + c^2$.

c) α is obtuse. If α is obtuse,
then $\cos \alpha < 0$ and $-2bc \cos \alpha > 0$.
Thus $a^2 - b^2 - c^2 > 0$ or
$a^2 > b^2 + c^2$.

59. By the law of Cosines,
$$b^2 = a^2 + c^2 - 2ac \cos \beta$$
$$a^2 = b^2 + c^2 - 2bc \cos \alpha.$$

If we subtract the second relation from the first, we obtain

$$b^2 - a^2 = a^2 - b^2 - 2ac \cos \beta$$
$$+ 2bc \cos \alpha$$
$$2b^2 - 2a^2 = 2c(b \cos \alpha - a \cos \beta)$$
$$b^2 - a^2 = c(b \cos \alpha - a \cos \beta).$$

61. $\vec{v} = <-5, 4>$, $A(-3,4)$

If $B(x, y)$ is the terminal point, then
$$x = -3 + (-5) = -8$$
$$y = 4 + 4 = 8.$$

63. $|\vec{u}| = 5$, $|\vec{v}| = 8$, angle between $\vec{u}$ and $\vec{v}$ measures $120°$.

$$|\vec{u} + \vec{v}|^2 = |\vec{u}|^2 + |\vec{v}|^2$$
$$- 2|\vec{u}||\vec{v}| \cos 60°$$
$$= 25 + 64 - 2(5)(8)(1/2)$$
$$= 49$$

$$|\vec{u} + \vec{v}| = 7$$

$$|\vec{u} - \vec{v}|^2 = |\vec{u}|^2 + |\vec{v}|^2$$
$$- 2|\vec{u}||\vec{v}| \cos 120°$$
$$= 25 + 64 - 2(5)(8)(-1/2)$$
$$= 129$$

$$|\vec{u} - \vec{v}| = \sqrt{129}$$

65. $\vec{v} = <7, -12>$

$$|\vec{v}| = \sqrt{7^2 + (-12)^2} = \sqrt{193}$$

$$\vec{u} = <7/\sqrt{193}, -12/\sqrt{193}>$$

67. $|\vec{u}| = 8$, $\theta = 45°$

$$\vec{u} = xi + yj$$
$$x = 8 \cos 45° = 4\sqrt{2}$$
$$y = 8 \sin 45° = 4\sqrt{2}$$
$$\vec{u} = 4\sqrt{2}\vec{i} + 4\sqrt{2}\vec{j}$$

69. (a) $\vec{a} = 3\vec{i} - 4\vec{j}$, $\vec{b} = -6\vec{i} + 3\vec{j}$

$$\cos \theta = \frac{\vec{a} \cdot \vec{b}}{|\vec{a}||\vec{b}|} = \frac{-30}{5\sqrt{45}} \simeq 0.8944$$

$$\theta \simeq 153.43°$$

(b) $\vec{r} = <8, 3>$, $\vec{s} = <-8, 4>$

$$\cos \theta = \frac{\vec{r} \cdot \vec{s}}{|\vec{r}||\vec{s}|} = \frac{-52}{4\sqrt{5}\sqrt{73}}$$

$$\simeq -0.6805$$
$$\theta \simeq 132.88°$$

71. $4000 - \text{lb}$ vehicle, angle $= 30°$
$$f = 4000 \sin 30°$$
$$= 4000(1/2) = 2000 \text{ lb}$$

73. Let f and $2f$ be the magnitude of the two forces making an angle of $60°$. If the resultant has magnitude 252 lb, then

$$252^2 = f^2 + (2f)^2 - 2f(2f) \cos 120°$$
$$= f^2 + 4f^2 - 4f^2(-1/2)$$
$$252^2 = 7f^2 \Rightarrow f = 252/\sqrt{7} \text{ lb}$$

So
$$f = 36\sqrt{7} \text{ lb and } 2f = 72\sqrt{7} \text{ lb}$$

75. $1 + i = \sqrt{2}(1/\sqrt{2} + i/\sqrt{2})$
$$= \sqrt{2}(\cos \pi/4 + i \sin \pi/4)$$

77. $3\sqrt{3} + 3i = 6(\sqrt{3}/2 + i/2)$
$$= 6(\cos \pi/6 + i \sin \pi/6)$$

79. $-4i = 4(0 - i)$
$$= 4(\cos 3\pi/2 + i \sin 3\pi/2)$$

81. $z_1 = 2(\cos \pi/6 + i \sin \pi/6)$
$z_2 = 3(\cos \pi/3 + i \sin \pi/3)$

$z_1 z_2 = (2)(3)[\cos(\pi/6 + \pi/3)$
$\quad + i \sin (\pi/6 + \pi/3)]$
$\quad = 6(\cos \pi/2 + i \sin \pi/2)$
$\quad = 6i$

$z_1/z_2 = 2/3[\cos(\pi/6 - \pi/3)$
$\quad + i \sin(\pi/6 - \pi/3)]$
$\quad = 2/3(\cos(-\pi/6)$
$\quad + i \sin(-\pi/6))$
$\quad = 2/3(\sqrt{3}/2 - i/2)$
$\quad = (\sqrt{3} - i)/3$

83. $z_1 = \sqrt{2}(\cos 45° + i \sin 45°)$
$z_2 = \sqrt{3}(\cos 285° + i \sin 285°)$
$z_1 z_2 = \sqrt{2} \cdot \sqrt{3}(\cos 330°$
$\quad + i \sin 330°)$
$\quad = \sqrt{6}(\sqrt{3}/2 - i/2)$

$\quad = \dfrac{3\sqrt{2} - i\sqrt{6}}{2}$

$z_1/z_2 = \sqrt{2}/\sqrt{3}(\cos(-240°)$
$\quad + i \sin(-240°))$
$\quad = \sqrt{2}/\sqrt{3}(-1/2 + i\sqrt{3}/2)$

$\quad = \dfrac{-\sqrt{6}}{6} + \dfrac{i\sqrt{2}}{2}$

85. $(1 + i)^4 = [\sqrt{2}(1/\sqrt{2} + i/\sqrt{2})]^4$
$\quad = [\sqrt{2}(\cos 45° + i \sin 45°)]^4$
$\quad = 4(\cos 180° + i \sin 180°)$
$\quad = -4$

87. $(1 - \sqrt{3}i)^6$
$\quad = [2(\cos 300° + i \sin 300°)]^6$
$\quad = 64(\cos 1800° + i \sin 1800°)$
$\quad = 64(\cos 0 + i \sin 0)$
$\quad = 64$

89. $[3(\cos 25° + i \sin 25°)]^4$
$\quad = 81(\cos 100° + i \sin 100°)$

91. $z = 1 - \sqrt{3}i, \quad n = 2$
$z = 2(\cos 5\pi/3 + i \sin 5\pi/3)$

$w_k = \sqrt{2}(\cos \dfrac{5\pi/3 + 2k\pi}{2}$
$\quad + i \sin \dfrac{5\pi/3 + 2k\pi}{2})$;

$k = 0, 1$

$w_0 = \sqrt{2}(\cos 5\pi/6$
$\quad + i \sin 5\pi/6)$

$\quad = \dfrac{-\sqrt{6} + i\sqrt{2}}{2}$

$w_1 = \sqrt{2}(\cos 11\pi/6$
$\quad + i \sin 11\pi/6)$

$\quad = \dfrac{\sqrt{6} - i\sqrt{2}}{2}$

93. $z = -8 - 8i, \quad n = 3$
$z = 8\sqrt{2}(\cos 225°$
$\quad + i \sin 225°)$

$w_k = (8\sqrt{2})^{1/3} (\cos \dfrac{225° + 360°k}{3}$
$\quad + i \sin \dfrac{225° + 360°k}{3})$;

$k = 0, 1, 2$
$w_0 = 2^{7/6} (\cos 75° + i \sin 75°)$
$w_1 = 2^{7/6} (\cos 195° + i \sin 195°)$
$w_2 = 2^{7/6} (\cos 315° + i \sin 315°)$

95. $z = 81(\cos 220° + i \sin 220°)$;
$n = 4$
$w_0 = 3(\cos 55° + i \sin 55°)$
$w_1 = 3(\cos 145° + i \sin 145°)$
$w_2 = 3(\cos 235° + i \sin 235°)$
$w_3 = 3(\cos 325° + i \sin 325°)$

97. $x^3 + 27 = 0$
$x^3 = -27$
$\quad = 27(\cos \pi + i \sin \pi)$

$x_0 = 3(\cos \pi/3 + i \sin \pi/3)$

$\quad = \dfrac{3 + 3\sqrt{3}i}{2}$

$x_1 = 3(\cos \pi + i \sin \pi) = -3$
$x_2 = 3(\cos 5\pi/3 + i \sin 5\pi/3)$

$\quad = \dfrac{3 - 3\sqrt{3}i}{2}$

99. $x^4 - 16i = 0$
$x^4 = 16i$
$\quad = 16(\cos \pi/2 + i \sin \pi/2)$
$x_0 = 2(\cos \pi/8 + i \sin \pi/8)$
$x_1 = 2(\cos 5\pi/8 + i \sin 5\pi/8)$
$x_2 = 2(\cos 9\pi/8 + i \sin 9\pi/8)$
$x_3 = 2(\cos 13\pi/8 + i \sin 13\pi/8)$

101. (a) $(3, \pi/4)$
$x = 3 \cos \pi/4 = 3\sqrt{2}/2$
$y = 3 \sin \pi/4 = 3\sqrt{2}/2$

(b) $(-2, \pi/3)$
$x = -2 \cos \pi/3 = -1$
$y = -2 \sin \pi/3 = -\sqrt{3}$

(c) $(4, -5\pi/6)$
$x = 4 \cos(-5\pi/6) = -2\sqrt{3}$
$y = 4 \sin(-5\pi/6) = -2$

The graphs for exercises 103-107 are in the answer section of your text.

105. $r = 4 \csc \theta = \dfrac{4}{\sin \theta}$

$r \sin \theta = 4$
$\quad y = 4$

107. $r = \dfrac{3}{2 \sin \theta + \cos \theta}$

$2r \sin \theta + r \cos \theta = 3$
$2y + x = 3$

109. Let $x = \dfrac{\sqrt{2}}{2} + \dfrac{\sqrt{2}}{2}i$

$\quad = \cos 45° + i \sin 45°$
Then

$x^4 = \cos 180° + i \sin 180° = -1$
$x^2 = \cos 90° + i \sin 90° = i$

Thus

$x^4 + x^2 + 1 = -1 + i + 1 = i$

Chapter 9 Test

In Problems 1-2, verify each identity

1. $\sin x \tan^2 x = \sec x \tan x - \sin x$

2. $\sin x \cos x + \cos^2 x \cot x + \tan x = \sec x \csc x$

3. Use the addition and subtraction formulas to find exact values for
    a) $\sin(-5\pi/12)$, b) $\cos 255°$.

4. Write $\sin 85° \cos 26° - \cos 85° \sin 26°$ as one trigonometric function of one angle.

5. Write the product $2 \sin 5a \sin 2a$ as a sum or difference of two trigonometric functions.

6. Let α be a first-quadrant angle such that $\cos \alpha = 2/5$ and β be a second-quadrant angle such that $\sin \beta = 1/2$. Find $\sin(\alpha + \beta)$, $\cos(\alpha + \beta)$. In which quadrant does $\alpha + \beta$ lie?

7. Verify the identity $\dfrac{\sin(a + b)}{\cos a \cos b} = \tan a + \tan b$.

8. Let $\sin \alpha = 3/5$ with $90° < \alpha < 180°$. Find $\sin 2\alpha$, $\cos 2\alpha$, and $\tan 2\alpha$.

9. If $\cos \alpha = -5/13$ with $\pi/2 < \alpha < \pi$, find $\sin(\alpha/2)$, $\cos(\alpha/2)$, and $\tan(\alpha/2)$.

10. Use a half-angle formula to find the exact value of $\sin(\pi/12)$.

11. Let α be an acute angle such that $\cos \alpha = 0.5412$. Approximate $\sin 2\alpha$ to four decimal places.

12. Verify the identity $\cot 2x = \dfrac{\cot^2 x - 1}{2 \cot x}$.

13. Find all solutions of the equation $\csc^2 x - 1 = 0$ in the interval $[0, 2\pi)$.

14. Find all solutions of $\sin^2 x - \sin x = 2$ in the interval $[0, 360°)$.

15. Find all solutions of $\tan^2 x + \sec^2 x = 7$ in the interval $[0, \pi)$.

16. Find all solutions of $\sin^2 x = 1/2 \sin x$ in the interval $[0, 360°)$.

17. In a given triangle ABC, $b = 5.40$, $\alpha = 30°$ and $\beta = 50°$. Find the other elements.

18. Solve a triangle ABC knowing that $\alpha = 42.5°$, $\beta = 56.4°$, and $c = 15.60$.

19. The sides of a triangle are $a = 15$, $b = 8$, $c = 17$. Approximate the angles to the nearest tenth of a degree.

20. In a triangle, the adjacent sides of an angle measure 10 cm and 8 cm, and the angle itself measures 30°. Find the length of the side facing the angle.

21. Two surveyors 350 ft apart at points A and B on the bank of a river look at a point C on the opposite bank and determine that the angles CAB and CBA measure 48.5° and 56.5°, respectively. Find the distance from each surveyor to the point C.

22. A cruiser sailing due east at 18 mi/hr records the position of a radio beacon at 45° NE. Two hours later it records the position of the same radio beacon at 75° NE. How close is the cruiser to the radio beacon at the second recording?

23. The hour and minute hands of a clock are 4 and 6 inches long, respectively. At two o'clock how far apart are the outer ends of the hour and minute hands?

24. Two ships leave from the same port at the same time on courses 36° apart, and at speeds of 12 km/h and 15 km/h. How far apart are the ships after one hour?

25. Write $3i$ in trigonometric form.

26. Find $z_1 z_2$ if $z_1 = 6(\cos 83° + i \sin 83°)$ and $z_2 = 5(\cos 79° + i \sin 79°)$.

27. Use De Moivre's formula to compute the number $(\sqrt{3} + i)^4$. Write your answer in rectangular form.

28. If $z_1 = 2(\cos(\pi/3) + i \sin(\pi/3))$ and $z_2 = 3(\cos(\pi/2) + i \sin(\pi/2))$, find z_1/z_2. Give your answer in rectangular form.

29. Find all solutions of the equation $x^3 - 125 = 0$.

30. Show that $(\sqrt{3}/2) - (1/2)i$ is a solution of the equation $x^4 - x^2 + 1 = 0$.

Chapter 10
Systems of Equations

Exercises 10.1

1. $\begin{cases} 4x - 3y = 7 \\ 2y = 6 \end{cases}$

$$y = \frac{6}{2} = 3$$

$$4x - 3(3) = 7$$
$$4x = 16$$
$$x = 4$$

3. $\begin{cases} 3x - 8y = -1 \\ 2x = 5y \end{cases}$

$$x = \frac{5y}{2}$$

$$3\left(\frac{5y}{2}\right) - 8y = -1$$

$$\frac{15y}{2} - 8y = -1$$

$$15y - 16y = -2$$
$$y = 2$$

$$x = \frac{5(2)}{2} = 5$$

5. $\begin{cases} 3x - 2y = 7 \\ x + 4y = 21 \end{cases}$

$$\begin{aligned} 3x - 2y &= 7 \\ -3x - 12y &= -63 \\ \hline -14y &= -56 \\ y &= 4 \end{aligned}$$

$$x + 4(4) = 21$$
$$x + 16 = 21$$
$$x = 5$$

7. $\begin{cases} 4x - 3y = 6 \\ 2x + 5y = 16 \end{cases}$

$$\begin{aligned} 4x - 3y &= 6 \\ -4x - 10y &= -32 \\ \hline -13y &= -26 \\ y &= 2 \end{aligned}$$

$$4x - 3(2) = 6$$
$$4x - 6 = 6$$
$$4x = 12$$
$$x = 3$$

9. $\begin{cases} 5x + 7y = 9 \\ 2x + 6y = -6 \end{cases}$

$$10x + 14y = 18$$
$$-10x - 30y = 30$$
$$\overline{\quad -16y = 48}$$
$$y = -3$$
$$5x + 7(-3) = 9$$
$$5x - 21 = 9$$
$$5x = 30$$
$$x = 6$$

11. $\begin{cases} 8x + 6y = -1 \\ 4x + 10y = -4 \end{cases}$

$$8x + 6y = -1$$
$$-8x - 20y = 8$$
$$\overline{\quad -14y = 7}$$
$$y = -1/2$$
$$8x + 6(-\frac{1}{2}) = -1$$

$$8x - 3 = -1$$
$$8x = 2$$

$$x = \frac{1}{4}$$

13. $\begin{cases} 5x - 6y = 1 \\ 3x - 2y = 7 \end{cases}$

$$5x - 6y = 1$$
$$-9x + 6y = -21$$
$$\overline{\quad -4x = -20}$$
$$x = 5$$
$$3(5) - 2y = 7$$
$$15 - 2y = 7$$
$$-2y = -8$$
$$y = 4$$

15. $\begin{cases} 6x - 4y = 2 \\ 9x - 2y = 5 \end{cases}$

$$6x - 4y = 2$$
$$-18x + 4y = -10$$
$$\overline{\quad -12x = -8}$$

$$x = \frac{2}{3}$$
$$6(\frac{2}{3}) - 4y = 2$$

$$4 - 4y = 2$$
$$-4y = -2$$

$$y = \frac{1}{2}$$

17. $\begin{cases} 0.5x - 0.25y = 3.5 \\ 1.2x - 1.5y = 3 \end{cases}$

$$50x - 25y = 350$$
$$12x - 15y = 30$$
$$150x - 75y = 1050$$
$$-60x + 75y = -150$$
$$\overline{\quad 90x = 900}$$
$$x = 10$$
$$12(10) - 15y = 30$$
$$120 - 15y = 30$$
$$-15y = -90$$
$$y = 6$$

19. $\begin{cases} \dfrac{2}{3}x - \dfrac{1}{2}y = 2 \\[2mm] \dfrac{3}{4}x - \dfrac{2}{3}y = \dfrac{11}{6} \end{cases}$

$$\begin{cases} 4x - 3y = 12 \\ 9x - 8y = 22 \end{cases}$$
$$36x - 27y = 108$$
$$-36x + 32y = -88$$
$$\overline{\quad 5y = 20}$$
$$y = 4$$
$$4x - 12 = 12$$
$$4x = 24$$
$$x = 6$$

21. $\begin{cases} \dfrac{x}{5} + \dfrac{y}{3} = \dfrac{1}{15} \\[2mm] -\dfrac{x}{2} + 2y = \dfrac{11}{2} \end{cases}$

$\begin{cases} 3x + 5y = 1 \\ -x + 4y = 11 \end{cases}$

$\begin{array}{r} 3x + 5y = 1 \\ -3x + 12y = 33 \\ \hline 17y = 34 \\ y = 2 \end{array}$

$-x + 8 = 11$

$-x = 3$

$x = -3$

23. $\begin{cases} \dfrac{5}{x} - \dfrac{1}{y} = 7 \\[2mm] \dfrac{3}{x} + \dfrac{2}{y} = -1 \end{cases}$

If we set $u = 1/x$ and $v = 1/y$, then the given system becomes

$\begin{cases} 5u - v = 7 \\ 3u + 2v = -1 \end{cases}$

$\begin{array}{r} 10u - 2v = 14 \\ 3u + 2v = -1 \\ \hline 13u = 13 \\ u = 1 \end{array}$

$5 - v = 7$

$v = -2$

Thus $x = 1$ and $y = -1/2$.

25. $\begin{cases} \dfrac{4}{x} - \dfrac{3}{y} = 5 \\[2mm] -\dfrac{3}{x} + \dfrac{7}{y} = 1 \end{cases}$

Set $u = 1/x$ and $v = 1/y$:

$\begin{cases} 4u - 3v = 5 \\ -3u + 7v = 1 \end{cases}$

$\begin{array}{r} 12u - 9v = 15 \\ -12u + 28v = 4 \\ \hline 19v = 19 \\ v = 1 \end{array}$

$4u - 3 = 5$

$4u = 8$

$u = 2$

Thus $x = 1/2$ and $y = 1$.

27. $\begin{cases} \dfrac{2}{x - 1} - \dfrac{3}{y - 2} = 3 \\[2mm] \dfrac{1}{x - 1} + \dfrac{4}{y - 2} = 7 \end{cases}$

Set $u = \dfrac{1}{x - 1}$ and $v = \dfrac{1}{y - 2}$

$\begin{cases} 2u - 3v = 3 \\ u + 4v = 7 \end{cases}$

$\begin{array}{r} 2u - 3v = 3 \\ -2u - 8v = -14 \\ \hline -11v = -11 \\ v = 1 \end{array}$

$2u - 3 = 3$

$2u = 6$

$u = 3$

Next $\dfrac{1}{x - 1} = 3$

$x - 1 = \dfrac{1}{3}, \quad x = \dfrac{4}{3}$

$\dfrac{1}{y - 2} = 1$

$y - 2 = 1$

$y = 3$

29. Let x be kilograms of 20% alloy and y be kilograms of 50% alloy.

$\begin{cases} x + y = 30 \\ 20x + 50y = 45(30) \end{cases}$

$\begin{array}{r} -20x - 20y = -600 \\ 20x + 50y = 1350 \\ \hline 30y = 750 \\ y = 25 \end{array}$

$x + 25 = 30$

$x = 5$

5 kg of 20% alloy and 25 kg of 50% alloy.

31. Let x be the speed of the boat and y be the speed of the current.

    Rate = Distance/Time

    upstream $\quad x - y = \dfrac{24}{3} = 8$

    downstream $\quad x + y = \dfrac{24}{1.5} = 16$

    $$\begin{array}{r} 2x = 24 \\ x = 12 \\ y = 4 \end{array}$$

    boat: 12 mph
    current: 4 mph

33. Let x be the number of nickels and y the number of dimes.

    $$\begin{cases} x + y = 30 \\ 5x + 10y = 210 \end{cases}$$

    $$\begin{array}{r} -5x - 5y = -150 \\ 5x + 10y = 210 \\ \hline 5y = 60 \\ y = 12 \\ x + 12 = 30 \\ x = 18 \end{array}$$

    18 nickels and 12 dimes.

35. Let x be the speed of the first car and y be the speed of the second car.

    $$\begin{cases} x - y = 10 \\ 2x + 2y = 200 \end{cases}$$

    $$\begin{array}{r} 2x - 2y = 20 \\ 2x + 2y = 200 \\ \hline 4x = 220 \end{array}$$

    $$\begin{array}{r} x = 55 \text{ mph} \\ y = 45 \text{ mph} \end{array}$$

37. Let x be length and y be width

    $$\begin{cases} 2x - 2y = 140 \\ y = 3x/4 \end{cases}$$

    or

    $$\begin{cases} 2x + 2y = 140 \\ -3x + 4y = 0 \end{cases}$$

$$\begin{array}{r} 6x + 6y = 420 \\ -6x + 8y = 0 \\ \hline 14y = 420 \\ y = 30 \\ x = 40 \end{array}$$

The length is 40 cm and the width is 30 cm.

39. Let x be kilograms of alloy A and y be kilograms of alloy B.

    $$\begin{cases} x + y = 100 \\ 30x + 20y = 26(100) \end{cases}$$

    $$\begin{array}{r} -30x - 30y = -3000 \\ 30x + 20y = 2600 \\ \hline -10y = -400 \\ y = 40 \end{array}$$

    60 kg of alloy A and 40 kg of alloy B.

Exercises 10.2

1. $$\begin{cases} 3x - y - z = 2 \\ 2y - z = 5 \\ 2z = 6 \end{cases}$$

    $$\begin{array}{r} z = 3 \\ 2y - 3 = 5 \\ 2y = 8 \\ y = 4 \\ 3x - 4 - 3 = 2 \\ 3x = 9 \\ x = 3 \end{array}$$

    $x = 3, \; y = 4, \; z = 3$

3. $$\begin{cases} 3x - y + 2z = 1 \\ 2y = 0 \\ 4x - 5y = 4 \end{cases}$$

    $y = 0$
    $4x - 5(0) = 4$
    $4x = 4, \; x = 1$
    $3(1) - 0 + 2z = 1$
    $2z = -2, \; z = -1$

5. $\begin{cases} \quad y + z = 5 \\ x \phantom{{}+y} + z = 4 \\ x + y \phantom{{}+z} = 3 \end{cases}$

$y = 5 - z$

$x = 4 - z$

$4 - z + 5 - z = 3$

$9 - 2z = 3$

$-2z = -6$

$z = 3$

$y = 5 - 3 = 2$

$x = 4 - 3 = 1$

$x = 1, \; y = 2, \; z = 3$

7. $\begin{cases} 2x - y \phantom{{}+ 2z} = 3 \\ \quad y - 3z = -8 \\ x \phantom{{}+y} + 2z = 8 \end{cases}$

$y = 3z - 8$

$x = 8 - 2z$

$2(8 - 2z) - (3z - 8) = 3$

$16 - 4z - 3z + 8 = 3$

$-7z = -21$

$z = 3$

$y = 3(3) - 8 = 1$

$x = 8 - 2(3) = 2$

$x = 2, \; y = 1, \; z = 3$

9. $\begin{cases} x - 2y + z = 0 \\ y + 2z = -2 \\ 2x - 3y + 3z = -1 \end{cases}$

Multiply the first equation by -2
and add it to the third

$\begin{cases} x - 2y + z = 0 \\ y + 2z = -2 \\ y + z = -1 \end{cases}$

Multiply the second equation by -1
and add it to the third

$\begin{cases} x - 2y + z = 0 \\ y + 2z = -2 \\ -z = 1 \end{cases}$

$z = -1$

$y + 2(-1) = -2, \text{ so } y = 0$

$x - 2(0) - 1 = 0, \text{ so } x = 1$

$x = 1, \; y = 0, \; z = -1$

11. $\begin{cases} \quad y - 2z = 5 \\ x \phantom{{}+y} + z = 0 \\ 2x - y \phantom{{}+z} = 3 \end{cases}$

$\begin{cases} x \phantom{{}+y} + z = 0 \\ \quad y - 2z = 5 \\ 2x - y \phantom{{}+z} = 3 \end{cases}$

$\begin{cases} x \phantom{{}+y} + z = 0 \\ \quad y - 2z = 5 \\ -y - 2z = 3 \end{cases}$

$\begin{cases} x + z = 0 \\ y - 2z = 5 \\ -4z = 8 \end{cases}$

$ z = -2$

$y - 2(-2) = 5$

$ y = 1$

$x + (-2) = 0$

$ x = 2$

$x = 2, \; y = 1, \; z = -2$

13. $\begin{cases} 2x + y - 3z = 3 \\ 4x - 2y - 3z = 1 \\ 2x - y + 3z = -1 \end{cases}$

$\begin{cases} 2x + y - 3z = 3 \\ -4y + 3z = -5 \\ \quad 2y - 6z = 4 \end{cases}$

$\begin{cases} 2x + y - 3z = 3 \\ -4y + 3z = -5 \\ -9z = 3 \end{cases}$

$z = -1/3$

$-4y + 3(-1/3) = -5$

$-4y - 1 = -5$

$ 4y = 4$

$ y = 1$

$2x + 1 - 3(-1/3) = 3$

$ 2x + 2 = 3$

$ 2x = 1$

$ x = 1/2$

$x = 1/2, \; y = 1, \; z = -1/3$

15. $\begin{cases} x + 2y + 3z = 6 \\ 2x \quad\;\; + z = -1 \\ 3x + y + z = 0 \end{cases}$

$\begin{cases} x + 2y + 3z = 6 \\ \quad\; 4y + 5z = 13 \\ \quad\; 5y + 8z = 18 \end{cases}$

$\begin{cases} x + 2y + 3z = 6 \\ \quad\; 4y + 5z = 13 \\ \quad\quad\;\; -7z = -7 \end{cases}$

$z = 1$

$4y + 5 = 13$

$4y = 8$

$y = 2$

$x + 4 + 3 = 6$

$x = -1$

$x = -1, \; y = 2, \; z = 1$

17. $\begin{cases} x + y - z - w = 1 \\ 2x - y + 2z + w = 2 \\ x - 2y + z + 3w = 2 \\ 2x + y - z - 2w = 1 \end{cases}$

$\begin{cases} x + y - z - w = 1 \\ \quad\; -3y + 4z + 3w = 0 \\ \quad\; -3y + 2z + 4w = 1 \\ \quad\; -y + z \quad\quad = -1 \end{cases}$

$\begin{cases} x + y - z - w = 1 \\ \quad\; -y + z \quad\quad = -1 \\ \quad\; -3y + 4z + 3w = 0 \\ \quad\; -3y + 2z + 4w = 1 \end{cases}$

$\begin{cases} x + y - z - w = 1 \\ \quad\; -y + z \quad\quad = -1 \\ \quad\quad\quad z + 3w = 3 \\ \quad\quad\; -z + 4w = 4 \end{cases}$

$\begin{cases} x + y - z - w = 1 \\ \quad\; -y + z \quad\quad = -1 \\ \quad\quad\quad z + 3w = 3 \\ \quad\quad\quad\quad 7w = 7 \end{cases}$

$w = 1$

$z + 3 = 3$

$z = 0$

$-y + 0 = -1$

$y = 1$

$x + 1 - 0 - 1 = 1$

$x = 1$

$x = 1, \; y = 1, \; z = 0, \; w = 1$

19. $\begin{bmatrix} 3 & 2 & \vdots & 2 \\ 9 & -4 & \vdots & 1 \end{bmatrix} = \begin{bmatrix} 3 & 2 & \vdots & 2 \\ 0 & -10 & \vdots & -5 \end{bmatrix}$

$= \begin{bmatrix} 3 & 2 & \vdots & 2 \\ 0 & 1 & \vdots & 1/2 \end{bmatrix} = \begin{bmatrix} 3 & 0 & \vdots & 1 \\ 0 & 1 & \vdots & 1/2 \end{bmatrix}$

$= \begin{bmatrix} 3 & 0 & \vdots & 1/3 \\ 0 & 1 & \vdots & 1/2 \end{bmatrix}$

Thus $x = 1/3$ and $y = 1/2$.

21. $\begin{bmatrix} 1 & 2 & \vdots & 2 \\ 3 & -4 & \vdots & 1 \end{bmatrix} = \begin{bmatrix} 1 & 2 & \vdots & 2 \\ 0 & -10 & \vdots & -5 \end{bmatrix}$

$= \begin{bmatrix} 1 & 2 & \vdots & 2 \\ 0 & 1 & \vdots & 1/2 \end{bmatrix} = \begin{bmatrix} 1 & 0 & \vdots & 1 \\ 0 & 1 & \vdots & 1/2 \end{bmatrix}$

Thus $x = 1$ and $y = 1/2$.

23. $\begin{bmatrix} 1 & -2 & -3 & \vdots & 4 \\ 2 & 1 & -2 & \vdots & 4 \\ 3 & 2 & 1 & \vdots & 2 \end{bmatrix} = \begin{bmatrix} 1 & -2 & -3 & \vdots & 4 \\ 0 & 5 & 4 & \vdots & -4 \\ 0 & 8 & 10 & \vdots & -10 \end{bmatrix}$

$= \begin{bmatrix} 1 & -2 & -3 & \vdots & 4 \\ 0 & 5 & 4 & \vdots & -4 \\ 0 & 0 & 18 & \vdots & -18 \end{bmatrix} = \begin{bmatrix} 1 & -2 & -3 & \vdots & 4 \\ 0 & 5 & 4 & \vdots & -4 \\ 0 & 0 & 1 & \vdots & -1 \end{bmatrix}$

$= \begin{bmatrix} 1 & -2 & -3 & \vdots & 4 \\ 0 & 5 & 0 & \vdots & 0 \\ 0 & 0 & 1 & \vdots & -1 \end{bmatrix} = \begin{bmatrix} 1 & -2 & -3 & \vdots & 4 \\ 0 & 1 & 0 & \vdots & 0 \\ 0 & 0 & 1 & \vdots & -1 \end{bmatrix}$

$= \begin{bmatrix} 1 & -2 & 0 & \vdots & 1 \\ 0 & 1 & 0 & \vdots & 0 \\ 0 & 0 & 1 & \vdots & -1 \end{bmatrix} = \begin{bmatrix} 1 & 0 & 0 & \vdots & 1 \\ 0 & 1 & 0 & \vdots & 0 \\ 0 & 0 & 1 & \vdots & -1 \end{bmatrix}$

Thus $x = 1, \; y = 0, \; z = -1$.

25. $\begin{bmatrix} 1 & 2 & 0 & \vdots & -1 \\ 2 & 0 & 1 & \vdots & 8 \\ 0 & 1 & 2 & \vdots & -7 \end{bmatrix} = \begin{bmatrix} 1 & 2 & 0 & \vdots & -1 \\ 0 & -4 & 1 & \vdots & 10 \\ 0 & 1 & 2 & \vdots & -7 \end{bmatrix}$

$= \begin{bmatrix} 1 & 2 & 0 & \vdots & -1 \\ 0 & 1 & 2 & \vdots & -7 \\ 0 & -4 & 1 & \vdots & 10 \end{bmatrix} = \begin{bmatrix} 1 & 2 & 0 & \vdots & -1 \\ 0 & 1 & 2 & \vdots & -7 \\ 0 & 0 & 9 & \vdots & -18 \end{bmatrix}$

$\begin{bmatrix} 1 & 2 & 0 & \vdots & -1 \\ 0 & 1 & 2 & \vdots & -7 \\ 0 & 0 & 1 & \vdots & -2 \end{bmatrix} = \begin{bmatrix} 1 & 2 & 0 & \vdots & -1 \\ 0 & 1 & 0 & \vdots & -3 \\ 0 & 0 & 1 & \vdots & -2 \end{bmatrix}$

$= \begin{bmatrix} 1 & 0 & 0 & \vdots & 5 \\ 0 & 1 & 0 & \vdots & -3 \\ 0 & 0 & 1 & \vdots & -2 \end{bmatrix}$

$x = 5, \ y = -3, \ z = -2$

27. $\begin{cases} 2x - y + 4z = 1 \\ x \quad\ + 3z = 0 \end{cases}$

$x = -3z$

$2(-3z) - y + 4z = 1$

$-6z - y + 4z = 1$

$-2z - y = 1$

$-y = 1 + 2z$

$y = -1 - 2z$

$x = -3z, \ y = -1 - 2z, \ z$ any number.

29. $\begin{cases} x - 2y + 3z = 2 \\ 2x + 3y - z = 4 \end{cases}$

$\begin{cases} x - 2y = 2 - 3z \\ 2x + 3y = 4 + z \end{cases}$

$\begin{cases} x - 2y = 2 - 3z \\ \quad\ 7y = 7z \end{cases}$

$y = z$

$x - 2z = 2 - 3z$

$x = 2 - z$

$x = 2 - z, \ y = z, \ z$ any number.

31. $\begin{cases} x \quad\quad + z = 3 \\ 3x - 2y \quad\ = 1 \end{cases}$

From the first equation, we get
$z = 3 - x$.
From the second equation, we get

$$y = \frac{3x - 1}{2}$$

Thus, the solution is x any number,

$$y = \frac{3x - 1}{2},$$

and $z = 3 - x$.

33. $\begin{cases} x - y + z = 0 \\ 2x + y - 3z = 0 \\ 4x - y - z = 0 \end{cases}$

$\begin{cases} x - y + z = 0 \\ \quad\ 3y - 5z = 0 \\ \quad\ 3y - 5z = 0 \end{cases}$

$3y = 5z$

$$y = \frac{5z}{3}$$

$$x - \frac{5z}{3} + z = 0$$

$$x - \frac{2z}{3} = 0$$

$$x = \frac{2z}{3}$$

$x = 2z/3, \ y = 5z/3, \ z$ any number.

35. $\begin{cases} 2x - 3y = 5 \\ -3x + y = 3 \\ x - y = 1 \end{cases}$

Solve the last two equations

$$-3x + y = 3$$
$$\underline{x - y = 1}$$
$$-2x = 4$$
$$x = -2$$
$$-2 - y = 1$$
$$y = -3$$

Check into the first equation:

$$2(-2) - 3(-3) = 5$$
$$-4 + 9 = 5$$
$$5 = 5$$

The solution of the system is $x = -2$, $y = -3$.

37. $\begin{cases} x + 2y = -1 \\ 5x - y = 6 \\ 2x - y = 0 \end{cases}$

$$y = 2x$$
$$x + 2(2x) = -1$$
$$x + 4x = -1$$
$$5x = -1$$

$$x = -\frac{1}{5}$$

$$5x - (2x) = 6$$
$$3x = 6$$
$$x = 2 \qquad \text{No solution.}$$

39. $\begin{cases} x + y + 2z = 0 \\ y + z = 0 \\ 3x + z = 0 \end{cases}$

$$z = -3x$$
$$y = -z$$
$$= -(-3x)$$
$$= 3x$$
$$x + 3x + 2(-3x) = 0$$
$$x + 3x - 6x = 0$$
$$-2x = 0$$
$$x = 0$$
$$x = y = z = 0$$

41. Let x be the number of $.11 stamps, y the number of $.18 stamps, z the number of $.40 stamps.

$$\begin{cases} 0.11x + 0.18y + 0.40z = 8.86 \\ x + y = z \\ y = x + 4 \end{cases}$$
$$z = x + x + 4$$
$$= 2x + 4$$
$$11x + 18(x + 4) + 40(2x + 4) = 886$$
$$11x + 18x + 72 + 80x + 160 = 886$$
$$109x + 232 = 886$$
$$109x = 654$$
$$x = 6$$

$$y = 6 + 4 = 10, \ z = 6 + 10 = 16$$

Six 11-cent, ten 18-cent, sixteen 40-cent stamps.

43. Let x be kilograms of alloy A, y kilograms of alloy B, and z kilograms of alloy C.

$$\begin{cases} x + y + z = 100 \\ 50x + 60y + 40z = 520(100) \\ 20x + 25y + 30z = 24(100) \end{cases}$$
$$\begin{cases} x + y + z = 100 \\ 5x + 6y + 4z = 520 \\ 4x + 5y + 6z = 480 \end{cases}$$

$$\begin{bmatrix} 1 & 1 & 1 & :100 \\ 5 & 6 & 4 & :520 \\ 4 & 5 & 6 & :480 \end{bmatrix} = \begin{bmatrix} 1 & 1 & 1 & :100 \\ 0 & 1 & -1 & :20 \\ 0 & 1 & 2 & :80 \end{bmatrix}$$

$$= \begin{bmatrix} 1 & 1 & 1 & :100 \\ 0 & 1 & -1 & :20 \\ 0 & 0 & 3 & :60 \end{bmatrix} = \begin{bmatrix} 1 & 1 & 1 & :100 \\ 0 & 1 & -1 & :20 \\ 0 & 0 & 1 & :20 \end{bmatrix}$$

$$= \begin{bmatrix} 1 & 1 & 1 & :100 \\ 0 & 1 & 0 & :40 \\ 0 & 0 & 1 & :20 \end{bmatrix} = \begin{bmatrix} 1 & 1 & 0 & :80 \\ 0 & 1 & 0 & :40 \\ 0 & 0 & 1 & :20 \end{bmatrix}$$

$$= \begin{bmatrix} 1 & 0 & 0 & :40 \\ 0 & 1 & 0 & :40 \\ 0 & 0 & 1 & :20 \end{bmatrix}$$

40 kg of alloy A, 40 kg of alloy B, 20 kg of alloy C.

45. Let x be the shortest side and y be the longest side.

$$\begin{cases} x + y + z = 42 \\ \quad x = y - 6 \\ \quad 2y = z \end{cases}$$

$$y - 6 + y + 2y = 42$$
$$4y = 48$$
$$y = 12$$
$$x = 6, \ z = 24$$

47. Let x be pounds of \$2.60 coffee, y pounds of \$2.80 coffee, and z pounds of \$3.20 coffee.

$$\begin{cases} x + y + z = 150 \\ \quad x = y \\ 26x + 28y + 32z = 30(150) \end{cases}$$

$$z = 150 - 2x$$
$$26x + 28x + 32(150 - 2x) = 4500$$
$$26x + 28x + 4800 - 64x = 4500$$
$$54x - 64x = 4500 - 4800$$
$$-10x = -300$$
$$x = 30$$
$$y = 30$$
$$z = 150 - 2(30)$$
$$= 90$$

30 lb of \$2.60 coffee, 30 lb of \$2.80 coffee, 90 lb of \$3.20 coffee.

49. Let x, y, and z be the amounts invested at 8.5%, 7.5%, and 6.5%, respectively.

$$\begin{cases} x + y + z = 18000 \\ y = x - 2000 \\ 0.085x + 0.075y + 0.065z = 1390 \end{cases}$$

$$\begin{cases} x + (x - 2000) + z = 18000 \\ 0.085x + 0.075(x - 2000) \\ \quad\quad + 0.065z = 1390 \end{cases}$$

$$\begin{cases} 2x + z = 20000 \\ 0.16x + 0.065z = 1540 \end{cases}$$

$$z = 20000 - 2x$$
$$0.16x + 0.065(20000 - 2x) = 1540$$
$$0.03x = 240$$
$$x = 8000$$
$$y = 6000$$
$$z = 4000$$

Exercises 10.3

1. $\begin{bmatrix} 2 & 1 \\ 3 & -2 \end{bmatrix} + \begin{bmatrix} 4 & 0 \\ -5 & 3 \end{bmatrix}$

$\quad = \begin{bmatrix} 6 & 1 \\ -2 & 1 \end{bmatrix}$

3. $\begin{bmatrix} 2 \\ -4 \\ 5 \end{bmatrix} + \begin{bmatrix} -3 \\ 2 \\ -1 \end{bmatrix} = \begin{bmatrix} -1 \\ -2 \\ 4 \end{bmatrix}$

5. $\begin{bmatrix} 3 & 1 & -2 \\ 0 & 1 & 4 \end{bmatrix} - \begin{bmatrix} 1 & -3 & 4 \\ -3 & 1 & -2 \end{bmatrix}$

$\quad = \begin{bmatrix} 2 & 4 & -6 \\ 3 & 0 & 6 \end{bmatrix}$

7. $\begin{bmatrix} 1 & -2 \\ 2 & -3 \\ 0 & -1 \\ 2 & 1 \end{bmatrix} - \begin{bmatrix} 2 & 3 \\ -1 & 0 \\ -2 & 2 \\ 0 & 2 \end{bmatrix}$

$\quad = \begin{bmatrix} -1 & -5 \\ 3 & -3 \\ 2 & -3 \\ 2 & -1 \end{bmatrix}$

9. $x = 3, \ y = -2, \ z = 4$

11. $3r = 6, \ r = 2$

$\quad \dfrac{s}{2} = -2, \ s = -4$

$\quad 2t = 8, \ t = 4$

13. $x - 2 + 4 = 6, \ x = 4$
$\quad y + 2y - 1 = -1, \ y = 0$
$\quad 3z - 2 = 4, \ z = 2$
$\quad 2v + 3 - v = 1, \ v = -2$
$\quad 5 + 7 = 2w, \ w = 6$

15. $A = \begin{bmatrix} 2 & 1 \\ 3 & -2 \end{bmatrix}$ $B = \begin{bmatrix} 1 & -2 \\ 2 & -3 \end{bmatrix}$

$A + B = \begin{bmatrix} 3 & -1 \\ 5 & -5 \end{bmatrix}$ $A - B = \begin{bmatrix} 1 & 3 \\ 1 & 1 \end{bmatrix}$

$3A = \begin{bmatrix} 6 & 3 \\ 9 & -6 \end{bmatrix}$ $2A - 3B = \begin{bmatrix} 1 & 8 \\ 0 & 5 \end{bmatrix}$

17. $A = \begin{bmatrix} 2 & -1 & 5 \\ -1 & 3 & -2 \end{bmatrix}$ $B = \begin{bmatrix} -2 & 1 & -2 \\ 1 & -3 & 4 \end{bmatrix}$

$A + B = \begin{bmatrix} 0 & 0 & 3 \\ 0 & 0 & 2 \end{bmatrix}$

$A - B = \begin{bmatrix} 4 & -2 & 7 \\ -2 & 6 & -6 \end{bmatrix}$

$3A = \begin{bmatrix} 6 & -3 & 15 \\ -3 & 9 & -6 \end{bmatrix}$

$2A - 3B = \begin{bmatrix} 10 & -5 & 16 \\ -5 & 15 & -16 \end{bmatrix}$

19. $A = \begin{bmatrix} 1 \\ -2 \\ 1 \end{bmatrix}$ $B = \begin{bmatrix} -2 \\ 3 \\ 4 \end{bmatrix}$

$A + B = \begin{bmatrix} -1 \\ 1 \\ 5 \end{bmatrix}$ $A - B = \begin{bmatrix} 3 \\ -5 \\ -3 \end{bmatrix}$

$3A = \begin{bmatrix} 3 \\ -6 \\ 3 \end{bmatrix}$ $2A - 3B = \begin{bmatrix} 8 \\ -13 \\ -10 \end{bmatrix}$

21. $A = \begin{bmatrix} 1 \\ 2 \\ 0 \\ 2 \end{bmatrix}$ $B = \begin{bmatrix} 3 \\ 0 \\ 2 \\ 2 \end{bmatrix}$

$A + B = \begin{bmatrix} 4 \\ 2 \\ 2 \\ 4 \end{bmatrix}$ $A - B = \begin{bmatrix} -2 \\ 2 \\ -2 \\ 0 \end{bmatrix}$

$3A = \begin{bmatrix} 3 \\ 6 \\ 0 \\ 6 \end{bmatrix}$ $2A - 3B = \begin{bmatrix} -7 \\ 4 \\ -6 \\ -2 \end{bmatrix}$

23. $A = \begin{bmatrix} 1 & -1 \\ 3 & 0 \end{bmatrix}$ $B = \begin{bmatrix} 2 & 0 \\ 1 & -1 \end{bmatrix}$

$AB = \begin{bmatrix} 2-1 & 0+1 \\ 6+0 & 0+0 \end{bmatrix} = \begin{bmatrix} 1 & 1 \\ 6 & 0 \end{bmatrix}$

$BA = \begin{bmatrix} 2+0 & -2+0 \\ 1-3 & -1+0 \end{bmatrix} = \begin{bmatrix} 2 & -2 \\ -2 & -1 \end{bmatrix}$

25. $A = \begin{bmatrix} 3 \\ -2 \\ 4 \end{bmatrix}$ $B = \begin{bmatrix} 1 & -3 & 2 \end{bmatrix}$

$AB = \begin{bmatrix} 3 & -9 & 6 \\ -2 & 6 & -4 \\ 4 & -12 & 8 \end{bmatrix}$

$BA = [3 + 6 + 8] = [17]$

27. $A = \begin{bmatrix} 2 & -2 & 1 \\ 3 & 0 & 1 \end{bmatrix}$ $B = \begin{bmatrix} 0 & -2 \\ -1 & -1 \\ 3 & 2 \end{bmatrix}$

$AB = \begin{bmatrix} 0+2+3 & -4+2+2 \\ 0+0+3 & -6+0+2 \end{bmatrix}$

$= \begin{bmatrix} 5 & 0 \\ 3 & -4 \end{bmatrix}$

$BA = \begin{bmatrix} 0-6 & 0+0 & 0-2 \\ -2-3 & 2+0 & -1-1 \\ 6+6 & -6+0 & 3+2 \end{bmatrix}$

$= \begin{bmatrix} -6 & 0 & -2 \\ -5 & 2 & -2 \\ 12 & -6 & 5 \end{bmatrix}$

29. $A = \begin{bmatrix} 2 & 0 & -2 \\ 0 & 2 & 2 \\ -2 & 0 & 2 \end{bmatrix}$

$B = \begin{bmatrix} 0 & 1 & 0 \\ 1 & 0 & 1 \\ 0 & 1 & 0 \end{bmatrix}$

$AB = \begin{bmatrix} 0+0+0 & 2+0-2 & 0+0+0 \\ 0+2+0 & 0+0+2 & 0+2+0 \\ 0+0+0 & -2+0+2 & 0+0+0 \end{bmatrix}$

$= \begin{bmatrix} 0 & 0 & 0 \\ 2 & 2 & 2 \\ 0 & 0 & 0 \end{bmatrix}$

$BA = \begin{bmatrix} 0+0+0 & 0+2+0 & 0+2+0 \\ 2+0-2 & 0+0+0 & -2+0+2 \\ 0+0+0 & 0+2+0 & 0+0+0 \end{bmatrix}$

$= \begin{bmatrix} 0 & 2 & 2 \\ 0 & 0 & 0 \\ 0 & 2 & 2 \end{bmatrix}$

31. $A = \begin{bmatrix} 4 \\ 1 \\ 0 \\ 2 \end{bmatrix}$ $B = \begin{bmatrix} 2 & 0 & -1 & 3 \end{bmatrix}$

$BA = [8 + 0 + 0 + 6] = [14]$

$AB = \begin{bmatrix} 8 & 0 & -4 & 12 \\ 2 & 0 & -1 & 3 \\ 0 & 0 & 0 & 0 \\ 4 & 0 & -2 & 6 \end{bmatrix}$

33. $\begin{bmatrix} 1 & 1 \\ 0 & 1 \end{bmatrix} \cdot \begin{bmatrix} a & b \\ c & d \end{bmatrix} = \begin{bmatrix} 1 & 0 \\ 0 & 1 \end{bmatrix}$

$a + c = 1$	$b + d = 0$
$0 + c = 0$	$0 + d = 1$
$c = 0$	$d = 1$
$a = 1$	$b = -1$

The inverse matrix is $\begin{bmatrix} 1 & -1 \\ 0 & 1 \end{bmatrix}$

35. $\begin{bmatrix} 2 & -1 \\ 1 & 1 \end{bmatrix} \begin{bmatrix} a & b \\ c & d \end{bmatrix} = \begin{bmatrix} 1 & 0 \\ 0 & 1 \end{bmatrix}$

$2a - c = 1$	$2b - d = 0$
$a + c = 0$	$b + d = 1$
$3a = 1$	$3b = 1$

$a = \dfrac{1}{3}$ $b = \dfrac{1}{3}$

$c = -\dfrac{1}{3}$ $d = \dfrac{2}{3}$

The inverse matrix is

$\begin{bmatrix} \dfrac{1}{3} & \dfrac{1}{3} \\ -\dfrac{1}{3} & \dfrac{2}{3} \end{bmatrix}$

37. $\begin{bmatrix} 2 & 2 \\ 0 & 2 \end{bmatrix} \begin{bmatrix} a & b \\ c & d \end{bmatrix} = \begin{bmatrix} 1 & 0 \\ 0 & 1 \end{bmatrix}$

$2a + 2c = 1$	$2b + 2d = 0$
$0 + 2c = 0$	$0 + 2d = 1$
$c = 0$	$d = 1/2$
$a = 1/2$	$b = -1/2$

The inverse matrix is

$\begin{bmatrix} \dfrac{1}{2} & -\dfrac{1}{2} \\ 0 & \dfrac{1}{2} \end{bmatrix}$

39. $\begin{bmatrix} 1 & 0 & \vdots & 1 & 0 \\ 1 & 1 & \vdots & 0 & 1 \end{bmatrix}$

$\begin{bmatrix} 1 & 0 & \vdots & 1 & 0 \\ 0 & 1 & \vdots & -1 & 1 \end{bmatrix}$

The inverse is $\begin{bmatrix} 1 & 0 \\ -1 & 1 \end{bmatrix}$

41. $\begin{bmatrix} 1 & 2 & \vdots & 1 & 0 \\ 2 & 4 & \vdots & 0 & 1 \end{bmatrix}$

Multiplying the first row of the augmented matrix by -2 and adding to the second, we get

$\begin{bmatrix} 1 & 2 & \vdots & 1 & 0 \\ 0 & 0 & \vdots & -2 & 1 \end{bmatrix}$

Because of the two zeros in the last row, it will be impossible to transform the last augmented matrix into the form $[I_2 \mid B]$. Thus the given matrix has no inverse.

43. $\begin{bmatrix} 1 & 0 & 1 & \vdots & 1 & 0 & 0 \\ 0 & 1 & 0 & \vdots & 0 & 1 & 0 \\ 0 & 0 & 1 & \vdots & 0 & 0 & 1 \end{bmatrix}$

$= \begin{bmatrix} 1 & 0 & 0 & \vdots & 1 & 0 & -1 \\ 0 & 1 & 0 & \vdots & 0 & 1 & 0 \\ 0 & 0 & 1 & \vdots & 0 & 0 & 1 \end{bmatrix}$

The inverse matrix is

$\begin{bmatrix} 1 & 0 & -1 \\ 0 & 1 & 0 \\ 0 & 0 & 1 \end{bmatrix}$

45. $\begin{bmatrix} 1 & 2 & 3 & \vdots & 1 & 0 & 0 \\ 0 & 1 & 2 & \vdots & 0 & 1 & 0 \\ 0 & 0 & 1 & \vdots & 0 & 0 & 1 \end{bmatrix}$

$= \begin{bmatrix} 1 & 2 & 0 & \vdots & 1 & 0 & -3 \\ 0 & 1 & 0 & \vdots & 0 & 1 & -2 \\ 0 & 0 & 1 & \vdots & 0 & 0 & 1 \end{bmatrix}$

$= \begin{bmatrix} 1 & 0 & 0 & \vdots & 1 & -2 & 1 \\ 0 & 1 & 0 & \vdots & 0 & 1 & -2 \\ 0 & 0 & 1 & \vdots & 0 & 0 & 1 \end{bmatrix}$

The inverse matrix is

$\begin{bmatrix} 1 & -2 & 1 \\ 0 & 1 & -2 \\ 0 & 0 & 1 \end{bmatrix}$

47. $\begin{bmatrix} 2 & 1 & 1 & \vdots & 1 & 0 & 0 \\ 0 & 0 & 1 & \vdots & 0 & 1 & 0 \\ 0 & 0 & 2 & \vdots & 0 & 0 & 1 \end{bmatrix}$

$= \begin{bmatrix} 2 & 1 & 1 & \vdots & 1 & 0 & 0 \\ 0 & 0 & 1 & \vdots & 0 & 1 & 0 \\ 0 & 0 & 1 & \vdots & 0 & 0 & \frac{1}{2} \end{bmatrix}$

$= \begin{bmatrix} 2 & 1 & 0 & \vdots & 1 & -1 & 0 \\ 0 & 0 & 1 & \vdots & 0 & 1 & 0 \\ 0 & 0 & 0 & \vdots & 0 & 1 & -\frac{1}{2} \end{bmatrix}$

No inverse matrix exists.

49. $\begin{bmatrix} 1 & 0 & 0 & \vdots & 1 & 0 & 0 \\ 0 & 2 & 0 & \vdots & 0 & 1 & 0 \\ 0 & 0 & 3 & \vdots & 0 & 0 & 1 \end{bmatrix}$

$= \begin{bmatrix} 1 & 0 & 0 & \vdots & 1 & 0 & 0 \\ 0 & 1 & 0 & \vdots & 0 & \frac{1}{2} & 0 \\ 0 & 0 & 1 & \vdots & 0 & 0 & \frac{1}{3} \end{bmatrix}$

The inverse matrix is

$\begin{bmatrix} 1 & 0 & 0 \\ 0 & \frac{1}{2} & 0 \\ 0 & 0 & \frac{1}{3} \end{bmatrix}$

51. $A = \begin{bmatrix} 2 & -1 \\ 4 & -1 \end{bmatrix}$

$A^{-1} = \begin{bmatrix} -\dfrac{1}{2} & \dfrac{1}{2} \\ -2 & 1 \end{bmatrix}$

$\begin{bmatrix} x \\ y \end{bmatrix} = \begin{bmatrix} -\dfrac{1}{2} & \dfrac{1}{2} \\ -2 & 1 \end{bmatrix} \begin{bmatrix} 2 \\ 1 \end{bmatrix}$

$= \begin{bmatrix} -1 + \dfrac{1}{2} \\ -4 + 1 \end{bmatrix} = \begin{bmatrix} -\dfrac{1}{2} \\ -3 \end{bmatrix}$

Thus $x = -\dfrac{1}{2}$, $y = -3$

53. $A = \begin{bmatrix} 1 & -1 \\ 2 & -3 \end{bmatrix}$ $A^{-1} = \begin{bmatrix} 3 & -1 \\ 2 & -1 \end{bmatrix}$

$\begin{bmatrix} x \\ y \end{bmatrix} = \begin{bmatrix} 3 & -1 \\ 2 & -1 \end{bmatrix} \begin{bmatrix} 5 \\ 12 \end{bmatrix}$

$= \begin{bmatrix} 15 - 12 \\ 10 - 12 \end{bmatrix} = \begin{bmatrix} 3 \\ -2 \end{bmatrix}$

Thus $x = 3$, $y = -2$

55. $A = \begin{bmatrix} 0 & 2 & 0 \\ 0 & 1 & -3 \\ 1 & -2 & -3 \end{bmatrix}$

$A^{-1} = \begin{bmatrix} 3/2 & -1 & 1 \\ 1/2 & 0 & 0 \\ 1/6 & -1/3 & 0 \end{bmatrix}$

$\begin{bmatrix} x \\ y \\ z \end{bmatrix} = \begin{bmatrix} 3/2 & -1 & 1 \\ 1/2 & 0 & 0 \\ 1/6 & -1/3 & 0 \end{bmatrix} \begin{bmatrix} 4 \\ 5 \\ 2 \end{bmatrix}$

$= \begin{bmatrix} 6 - 5 + 2 \\ 2 + 0 + 0 \\ \dfrac{2}{3} - \dfrac{5}{3} + 0 \end{bmatrix} = \begin{bmatrix} 3 \\ 2 \\ -1 \end{bmatrix}$

$x = 3$, $y = 2$, $z = -1$

57. $A = \begin{bmatrix} 1 & 0 & -1 \\ 0 & 2 & 1 \\ 0 & 1 & 1 \end{bmatrix}$ $A^{-1} = \begin{bmatrix} 1 & -1 & 2 \\ 0 & 1 & -1 \\ 0 & -1 & 2 \end{bmatrix}$

$\begin{bmatrix} x \\ y \\ z \end{bmatrix} = \begin{bmatrix} 1 & -1 & 2 \\ 0 & 1 & -1 \\ 0 & -1 & 2 \end{bmatrix} \begin{bmatrix} -2 \\ 13 \\ 9 \end{bmatrix}$

$= \begin{bmatrix} -2 - 13 + 18 \\ 0 + 13 - 9 \\ 0 - 13 + 18 \end{bmatrix} = \begin{bmatrix} 3 \\ 4 \\ 5 \end{bmatrix}$

$x = 3$, $y = 4$, $z = 5$

59. $A = \begin{bmatrix} 2 & -1 & 0 \\ 0 & 2 & 1 \\ 1 & 3 & 2 \end{bmatrix}$

$A^{-1} = \begin{bmatrix} 1 & 2 & -1 \\ 1 & 4 & -2 \\ -2 & -7 & 4 \end{bmatrix}$

$\begin{bmatrix} x \\ y \\ z \end{bmatrix} = \begin{bmatrix} 1 & 2 & -1 \\ 1 & 4 & -2 \\ -2 & -7 & 5 \end{bmatrix} \begin{bmatrix} 7 \\ -10 \\ 15 \end{bmatrix}$

$= \begin{bmatrix} 7 - 20 - 15 \\ 7 - 40 - 30 \\ -14 + 70 + 60 \end{bmatrix} = \begin{bmatrix} -28 \\ -63 \\ 116 \end{bmatrix}$

$x = -28$, $y = -63$, $z = 116$

Exercises 10.4

1. $M_{11} = [-2]$

   $\det M_{11} = -2$

   $A_{11} = -2$

   $M_{21} = [-1]$

   $\det M_{21} = -1$

   $A_{21} = (-1)(-1) = 1$

3. $M_{11} = \begin{bmatrix} -1 & 2 \\ -2 & 1 \end{bmatrix}$

   $\det M_{11} = -1 - (-4)$
   $\qquad\qquad = 3$

   $A_{11} = 3$

   $M_{13} = \begin{bmatrix} 3 & -1 \\ 0 & -2 \end{bmatrix}$

   $\det M_{13} = -6$

   $A_{13} = -6$

   $M_{32} = \begin{bmatrix} 1 & 0 \\ 3 & 2 \end{bmatrix}$

   $\det M_{32} = 2 - 0$
   $\qquad\qquad = 2$

   $A_{32} = (-1)(2)$
   $\qquad\; = -2$

5. $M_{11} = \begin{bmatrix} 1 & 0 & 3 \\ 0 & 0 & -1 \\ 0 & 1 & 2 \end{bmatrix}$

   $\det M_{11} = 1$
   $\qquad A_{11} = 1$

   $M_{13} = \begin{bmatrix} 2 & 1 & 3 \\ -2 & 0 & -1 \\ -3 & 0 & 2 \end{bmatrix}$

   $\det M_{13} = 7$
   $\qquad A_{13} = 7$

   $M_{43} = \begin{bmatrix} 1 & 0 & 2 \\ 2 & 1 & 3 \\ -2 & 0 & -1 \end{bmatrix}$

   $\det M_{43} = 3$
   $\qquad A_{43} = (-1)(3) = -3$

7. $\begin{vmatrix} 2 & -1 \\ 3 & -2 \end{vmatrix} = (2)(-2) - (-1)(3)$
   $\qquad\qquad\quad = -4 - (-3)$
   $\qquad\qquad\quad = -4 + 3$
   $\qquad\qquad\quad = -1$

9. $\begin{vmatrix} 2 & -3 \\ -1 & -2 \end{vmatrix} = (2)(-2) - (-3)(-1)$
   $\qquad\qquad\quad = -4 - 3$
   $\qquad\qquad\quad = -7$

11. $\begin{vmatrix} 0 & 1 \\ 2 & 0 \end{vmatrix} = 0 - (1)(2)$
    $\qquad\qquad = -2$

13. $\begin{vmatrix} 1 & -2 & 3 \\ 3 & -1 & 2 \\ 0 & -2 & 1 \end{vmatrix}$ 3rd row expansion

    $= (-1)(-2)\begin{vmatrix} 1 & 3 \\ 3 & 2 \end{vmatrix} + (1)(1)\begin{vmatrix} 1 & -2 \\ 3 & -1 \end{vmatrix}$

    $\quad = 2(2 - 9) + (-1 + 6)$
    $\quad = 2(-7) + (5)$
    $\quad = -14 + 5 = -9$

15. $\begin{vmatrix} 1 & 2 & 3 \\ 0 & -1 & -2 \\ -3 & -2 & 1 \end{vmatrix}$ 2nd row expansion

    $= (1)(-1)\begin{vmatrix} 1 & 3 \\ -3 & 1 \end{vmatrix}$

    $\quad + (-1)(-2)\begin{vmatrix} 1 & 2 \\ -3 & -2 \end{vmatrix}$

    $= -(1 + 9) + 2(-2 + 6)$
    $= -10 + 2(4)$
    $= -10 + 8 = -2$

17. $\begin{vmatrix} 2 & 1 & 2 \\ 0 & 3 & 1 \\ 0 & 0 & 4 \end{vmatrix}$ 3rd row expansion

    $= (1)(4)\begin{vmatrix} 2 & 1 \\ 0 & 3 \end{vmatrix}$

    $= 4(6) = 24$

19. $\begin{vmatrix} 2 & -1 & 3 \\ 1 & 2 & 1 \\ 4 & -2 & 6 \end{vmatrix}$ 2nd row expansion

$$= (-1)(1) \begin{vmatrix} -1 & 3 \\ -2 & 6 \end{vmatrix}$$

$$+ (1)(2) \begin{vmatrix} 2 & 3 \\ 4 & 6 \end{vmatrix}$$

$$+ (-1)(1) \begin{vmatrix} 2 & -1 \\ 4 & -2 \end{vmatrix}$$

$$= -(-6 + 6) + 2(12 - 12)$$
$$- (-4 + 4) = 0$$

21. $\begin{vmatrix} 1 & 0 & 0 & 2 \\ 2 & 1 & 0 & 0 \\ -2 & 0 & 0 & -1 \\ -3 & 0 & 1 & 2 \end{vmatrix}$ 3rd column expansion

$$= (1)(-1)^{4+3} \begin{vmatrix} 1 & 0 & 2 \\ 2 & 1 & 3 \\ -2 & 0 & -1 \end{vmatrix}$$

2nd column expansion

$$= -(1)(-1)^{2+2} \begin{vmatrix} 1 & 2 \\ -2 & -1 \end{vmatrix}$$

$$= - \begin{vmatrix} 1 & 2 \\ -2 & -1 \end{vmatrix}$$

$$= -[(1)(-1) - 2(-2)]$$
$$= -[-1 + 4] = -3$$

23. $\begin{vmatrix} x & y & 1 \\ 2 & 3 & 1 \\ -1 & -2 & 1 \end{vmatrix} = 0$

25. $\begin{vmatrix} 1 & 2 & -1 \\ 0 & -2 & 1 \\ -1 & 2 & 3 \end{vmatrix}$

$$= (1)(-2)(3) + (2)(1)(-1)$$
$$+ (-1)(2)(0)$$
$$- (-1)(-2)(-1)$$
$$- (0)(2)(3) - (1)(2)(1)$$
$$= -6 - 2 + 2 - 2 = -8$$

27. $\begin{vmatrix} 1 & 2 & 3 \\ 0 & -1 & -2 \\ -3 & -2 & 1 \end{vmatrix}$

$$= (1)(-1)(1) + (2)(-2)(-3)$$
$$+ (3)(-2)(0)$$
$$- (3)(-1)(-3)$$
$$- (2)(0)(1)$$
$$- (1)(-2)(-2)$$
$$= -1 + 12 - 9 - 4 = -2$$

29. $\begin{vmatrix} 1 & 2 & 3 \\ 0 & -1 & -2 \\ 3 & 6 & 9 \end{vmatrix} = 3 \begin{vmatrix} 1 & 2 & 3 \\ 0 & -1 & -2 \\ 1 & 2 & 3 \end{vmatrix}$

by property II.
Rows 1 and 3 are equal, so the
determinant = 0.

31. $\begin{vmatrix} 1 & 2 & -1 \\ 0 & -2 & 1 \\ 2 & 2 & -1 \end{vmatrix}$

[Adding -1 times the second row to
the third one]

$$= \begin{vmatrix} 1 & 2 & -1 \\ 0 & -2 & 1 \\ 2 & 4 & -2 \end{vmatrix}$$

[Factoring 2 from the third rows]

$$= 2 \begin{vmatrix} 1 & 2 & -1 \\ 0 & -2 & 1 \\ 1 & 2 & -1 \end{vmatrix} = 0$$

[First and third rows are equal.]

33.
$$\begin{vmatrix} 2 & 1 & 1 & 3 \\ -1 & -2 & 2 & 1 \\ 1 & 0 & 0 & -1 \\ 5 & 2 & 2 & 5 \end{vmatrix}$$

[Adding -1 times the third row to the fourth one]

$$\begin{vmatrix} 2 & 1 & 1 & 3 \\ -1 & -2 & 2 & 1 \\ 1 & 0 & 0 & -1 \\ 4 & 2 & 2 & 6 \end{vmatrix}$$

[Factoring 2 from the fourth row]

$$= 2 \begin{vmatrix} 2 & 1 & 1 & 3 \\ -1 & -2 & 2 & 1 \\ 1 & 0 & 0 & -1 \\ 2 & 1 & 1 & 3 \end{vmatrix} = 0$$

[First and fourth rows are equal]

35.
$$\begin{vmatrix} 2 & -1 \\ 3 & -2 \end{vmatrix} = (2)(-2) - (3)(-1)$$
$$= -4 + 3$$
$$= -1$$

$$x = \frac{\begin{vmatrix} 4 & -1 \\ 3 & -2 \end{vmatrix}}{-1}$$

$$= \frac{(4)(-2) - (-1)(3)}{-1}$$

$$= \frac{-8 + 3}{-1} = 5$$

$$y = \frac{\begin{vmatrix} 2 & 4 \\ 3 & 3 \end{vmatrix}}{-1}$$

$$= \frac{(2)(3) - (4)(3)}{-1}$$

$$= \frac{6 - 12}{-1} = 6$$

37.
$$\begin{vmatrix} 10 & -5 \\ 9 & 6 \end{vmatrix} = (10)(6) - (-5)(9)$$
$$= 60 + 45 = 105$$

$$x = \frac{\begin{vmatrix} -1 & -5 \\ 3 & 6 \end{vmatrix}}{105}$$

$$= \frac{(-1)(6) - (-5)(3)}{105}$$

$$= \frac{-6 + 15}{105} = \frac{9}{105} = \frac{3}{35}$$

$$y = \frac{\begin{vmatrix} 10 & -1 \\ 9 & 3 \end{vmatrix}}{105}$$

$$= \frac{(10)(3) - (-1)(9)}{105}$$

$$= \frac{30 + 9}{105} = \frac{39}{105} = \frac{13}{35}$$

39. No solution. The determinant of the coefficients is

$$\begin{vmatrix} 3 & -2 \\ 9 & -6 \end{vmatrix} = 0$$

41.
$$\begin{vmatrix} 0 & 1 & -2 \\ 1 & 0 & 1 \\ 2 & -1 & 0 \end{vmatrix} = 4$$

$$x = \frac{\begin{vmatrix} 5 & 1 & -2 \\ 0 & 0 & 1 \\ 3 & -1 & 0 \end{vmatrix}}{4}$$

$$= \frac{8}{4} = 2$$

$$y = \frac{\begin{vmatrix} 0 & 5 & -2 \\ 1 & 0 & 1 \\ 2 & 3 & 0 \end{vmatrix}}{4}$$

$$= \frac{4}{4} = 1$$

$$z = \frac{\begin{vmatrix} 0 & 1 & 5 \\ 1 & 0 & 0 \\ 2 & -1 & 3 \end{vmatrix}}{4}$$

$$= \frac{-8}{4} = -2$$

43. $\begin{vmatrix} 3 & -2 & 1 \\ 1 & 1 & -4 \\ 2 & -1 & -1 \end{vmatrix} = -4$

$$x = \frac{\begin{vmatrix} 1 & -2 & 1 \\ 9 & 1 & -4 \\ 4 & -1 & -1 \end{vmatrix}}{-4} = 1$$

$$y = \frac{\begin{vmatrix} 3 & 1 & 1 \\ 1 & 9 & -4 \\ 2 & 4 & -1 \end{vmatrix}}{-4} = 0$$

$$z = \frac{\begin{vmatrix} 3 & -2 & 1 \\ 1 & 1 & 9 \\ 2 & -1 & 4 \end{vmatrix}}{-4} = -2$$

45. If the points $(1, -1)$ and $(2, -3)$ lie on the line $ax + by + 1 = 0$, then $a(1) + b(-1) + 1 = 0$ and $a(2) + b(-3) + 1 = 0$. Next, solve for a and b the two by two linear system

$$\begin{cases} a - b = -1 \\ 2a - 3b = -1 \end{cases}$$

and get $a = -2$, $b = -1$.

47. Let x be the amount (in grams) of 12-carat gold and y be the amount (in grams) of 20-carat gold.

$$\begin{cases} x + y = 60 \\ \dfrac{12x}{24} + \dfrac{20y}{24} = 60(\dfrac{18}{24}) \end{cases}$$

or

$$\begin{cases} x + y = 60 \\ 12x + 20y = 1080 \end{cases}$$

Solving this system, we obtain $x = 15$ gm, $y = 45$ gm.

49. Let x, y, z be the ages of the father, mother, and daughter.

$$\begin{cases} x + y + z = 112 \\ x = 3z \\ y + 4 = 2(z + 4) \end{cases}$$

or

$$\begin{cases} x + y + z = 112 \\ x - 3z = 0 \\ y - 2z = 4 \end{cases}$$

Solving this system, we obtain $x = 54$, $y = 40$, $z = 18$.

Exercises 10.5

1. $\begin{cases} x = y^2 \\ x - 3y = 0 \end{cases}$

$y^2 - 3y = 0$
$y(y - 3) = 0$

$\begin{array}{c|c} y = 0 & y = 3 \\ x = 0 & x = 9 \end{array}$
$x = 0, y = 0$ or $x = 9, y = 3$

3. $\begin{cases} y = x - 2 \\ x^2 + y = 0 \end{cases}$

$x^2 + x - 2 = 0$
$(x + 2)(x - 1) = 0$

$\begin{array}{c|c} x + 2 = 0 & x - 1 = 0 \\ x = -2 & x = 1 \\ y = -4 & y = -1 \end{array}$
$x = -2, y = -4$ or $x = 1, y = -1$

5. $\begin{cases} x^2 - 4y = 0 \\ 3x - 4y = -4 \end{cases}$

$-4y = -3x - 4$
$x^2 - 3x - 4 = 0$
$(x - 4)(x + 1) = 0$

$\begin{array}{c|c} x - 4 = 0 & x + 1 = 0 \\ x = 4 & x = -1 \\ -4y = -3(4) - 4 & -4y = -3(-1) - 4 \\ 4y = 16 & y = 1/4 \end{array}$
$x = 4, y = 4$ or $x = -1, y = 1/4$

7. $\begin{cases} 3x - y = 0 \\ xy = 3 \end{cases}$

$$y = 3x$$
$$x(3x) = 3$$
$$3x^2 = 3$$
$$x = \pm 1$$
$$y = \pm 3$$
$$x = 1, \, y = 3 \text{ or } x = -1, \, y = -3$$

9. $\begin{cases} 2xy = 1 \\ 2x - 4y = 3 \end{cases}$

$$2x = 3 + 4y$$
$$(3 + 4y)y = 1$$
$$3y + 4y^2 = 1$$
$$4y^2 + 3y - 1 = 0$$
$$(4y - 1)(y + 1) = 0$$

$y = 1/4$	$y = -1$
$2x(1/4) = 1$	$2x(-1) = 1$
$x = 2$	$x = -1/2$

$$x = 2, \, y = 1/4 \text{ or } x = -1/2, \, y = -1$$

11. $\begin{cases} xy = -1 \\ y + x = 1 \end{cases}$

$$y = 1 - x$$
$$x(1 - x) = -1$$
$$x - x^2 = -1$$
$$x^2 - x - 1 = 0$$

$$x = \frac{1 \pm \sqrt{5}}{2}$$

$$y = 1 - \left(\frac{1 \pm \sqrt{5}}{2}\right)$$

$$= \frac{1 \mp \sqrt{5}}{2}$$

$$x = \frac{1 + \sqrt{5}}{2}, \, y = \frac{1 - \sqrt{5}}{2}$$

$$\text{or } x = \frac{1 - \sqrt{5}}{2}, \, y = \frac{1 + \sqrt{5}}{2}$$

13. $\begin{cases} x^2 - 2y = 0 \\ x^2 + y^2 = 8 \end{cases}$

$$x^2 - 2y = 0$$
$$-x^2 - y^2 = -8$$
$$\overline{-2y - y^2 = -8}$$

$$y^2 + 2y - 8 = 0$$
$$(y + 4)(y - 2) = 0$$

$y = -4$	$y = 2$
$x^2 + (-4)^2 = 8$	$x^2 + 2^2 = 8$
$x^2 = -8$	$x^2 = 4$
No real	$x = \pm 2$
solution	

$$x = 2, \, y = 2 \text{ or } x = -2, \, y = 2$$

15. $\begin{cases} y = x^2 - 1 \\ x^2 + y^2 = 13 \end{cases}$

$$x^2 = y + 1$$
$$y + 1 + y^2 = 13$$
$$y^2 + y - 12 = 0$$
$$(y - 3)(y + 4) = 0$$

$y = 3$	$y = -4$
$x^2 = 3 + 1$	$x^2 = -4 + 1$
$x^2 = 4$	$x^2 = -3$
$x = \pm 2$	No real
	solution

$$x = 2, \, y = 3 \text{ or } x = -2, \, y = 3$$

17. $\begin{cases} x^2 + y^2 = 7 \\ x - y^2 = -1 \end{cases}$

$$x^2 + y^2 = 7$$
$$x - y^2 = -1$$
$$\overline{x^2 + x = 6}$$

$$x^2 + x - 6 = 0$$
$$(x + 3)(x - 2) = 0$$

$x = -3$	$x = 2$
$-3 - y^2 = -1$	$2 - y^2 = -1$
$y^2 = -2$	$y^2 = 3$
No real	$y = \pm\sqrt{3}$
solution	

$$x = 2, \, y = \sqrt{3} \text{ or } x = 2, \, y = -\sqrt{3}$$

19. $\begin{cases} 2x^2 + 2y^2 = 1 \\ x^2 + 4y^2 = 1 \end{cases}$

$$\begin{array}{r} 2x^2 + 2y^2 = 1 \\ -2x^2 - 8y^2 = -2 \\ \hline -6y^2 = -1 \end{array}$$

$$y^2 = 1/6$$
$$y = \pm\sqrt{6}/6$$
$$x^2 + 4(\sqrt{6}/6)^2 = 1$$
$$x^2 = 1 - 4(6/36)$$
$$= 1 - 2/3$$
$$= 1/3$$
$$x = \pm\sqrt{3}/3$$
$(\sqrt{3}/3, \sqrt{6}/6)$, $(\sqrt{3}/3, -\sqrt{6}/6)$,
$(-\sqrt{3}/3, \sqrt{6}/6)$, $(-\sqrt{3}/3, -\sqrt{6}/6)$

21. $\begin{cases} x^2 + 4y^2 = 1 \\ 2x^2 + y^2 = 1 \end{cases}$

$$\begin{array}{r} 2x^2 + 8y^2 = 2 \\ -2x^2 - y^2 = -1 \\ \hline 7y^2 = 1 \end{array}$$
$$y^2 = 1/7$$
$$y = \pm\sqrt{7}/7$$
$$x^2 = 1 - 4(1/7)$$
$$= 3/7$$
$$x = \pm\sqrt{21}/7$$
$(\sqrt{21}/7, \sqrt{7}/7)$, $(\sqrt{21}/7, -\sqrt{7}/7)$,
$(-\sqrt{21}/7, \sqrt{7}/7)$, $(-\sqrt{21}/7, -\sqrt{7}/7)$

23. $\begin{cases} x^2 - 2x + y^2 = 0 \\ x^2 + y^2 = 1 \end{cases}$

$$\begin{array}{r} x^2 - 2x + y^2 = 0 \\ -x^2 - y^2 = -1 \\ \hline -2x = -1 \end{array}$$
$$x = 1/2$$
$$(1/2)^2 + y^2 = 1$$
$$y^2 = 1 - 1/4$$
$$= 3/4$$
$$y = \pm\sqrt{3}/2$$
$x = 1/2$, $y = \sqrt{3}/2$ or
$x = 1/2$, $y = -\sqrt{3}/2$

25. $\begin{cases} 2x^2y - xy + x^2 = 8 \\ x^2 + xy - y^2 = 4 \end{cases}$

Add the two equations and obtain
$$3x^2 = 12$$
$$x = \pm 2$$
Substitute 2 for x into the second equation and obtain

$$2^2 + 2y - y^2 = 4$$
$$y^2 - 2y = 0$$
$$y = 0, \ y = 2$$

So (2, 0) and (2, 2) are solutions. In a similar way, check that (-2, 0) and (-2, -2) are also solutions.

27. $\begin{cases} x^2 + 2xy - y^2 = 7 \\ x^2 - y^2 = -5 \end{cases}$

Subtract the second equation from the first one and get

$$2xy = 12$$
$$y = 6/x$$

Substitute into the second equation and obtain

$$x^2 - \frac{36}{x^2} = -5$$

$$x^4 + 5x^2 - 36 = 0$$
$$(x^2 - 4)(x^2 + 9) = 0$$
$$x^2 - 4 = 0, \ x = \pm 2$$
$$y = 6/\pm 2, \ y = \pm 3$$

Next, check that the solutions are (2, 3) and (-2, -3).

29. $\begin{cases} 3^x - y = 0 \\ 9^x - y = 6 \end{cases}$

Rewrite the two equations as follows $3^x = y$ and $9^x - 6 = y$. Thus

$$9^x - 6 = 3^x$$
$$9^x - 3^x - 6 = 0$$

Now, set $u = 3^x$ and obtain the quadratic equation

$$u^2 - u - 6 = 0$$
$$(u + 2)(u - 3) = 0$$
$$u = 3, \quad u = -2$$

Next

$u = 3^x = 3$,
so $x = 1$ and $y = 3$.

This is the only solution because there is no x satisfying $3^x = -2$.

31. $\begin{cases} y - \log_2(x + 2) = 3 \\ y + \log_2(x + 3) = 4 \end{cases}$

Solving for y, get

$$y = \log_2(x + 2) + 3$$
$$y = -\log_2(x + 3) + 4$$

hence

$$\log_2(x + 2) + 3 = -\log_2(x + 3) + 4$$
$$\log_2(x + 2) + \log_2(x + 3) = 1$$
$$\log_2(x + 2)(x + 3) = 1$$
$$(x + 2)(x + 3) = 2$$
$$x^2 + 5x + 6 = 2$$
$$x^2 + 5x + 4 = 0$$
$$(x + 1)(x + 4) = 0$$
$$x = -1, \quad x = -4$$

If $x = -1$, then
$$y = \log_2(-1 + 2) + 3$$
$$= \log_2 1 + 3$$
$$= 0 + 3 = 3$$

If $x = -4$, then $\log_2(x + 2)$ is not defined. Thus the only solution of the system is $(-1, 3)$.

33. $\begin{cases} x - y = 8 \\ xy = 240 \end{cases}$

$$x = 8 + y$$
$$(8 + y)y = 240$$
$$y^2 + 8y - 240 = 0$$
$$(y + 20)(y - 12) = 0$$

$$\begin{array}{l|l} y = -20 & y = 12 \\ x = 8 - 20 & x = 8 + 12 \\ \quad = -12 & \quad = 20 \end{array}$$

The two numbers are 12 and 20.

35. $\begin{cases} xy = 384 \\ \dfrac{x}{y} = \dfrac{2}{3} \end{cases}$

$$x = \frac{2y}{3}$$

$$y\left(\frac{2y}{3}\right) = 384$$

$$y^2 = 576$$
$$y = \pm 24$$
$$x = 2(24)/3 = 16$$

The solution is 16 and 24.

37. $\begin{cases} 2x + 2y = 36 \\ xy = 80 \end{cases}$

$$x + y = 18$$
$$y = 18 - x$$
$$x(18 - x) = 80$$
$$18x - x^2 - 80 = 0$$
$$x^2 - 18x + 80 = 0$$
$$(x - 8)(x - 10) = 0$$

$$\begin{array}{l|l} x = 8 & x = 10 \\ y = 18 - 8 & y = 18 - 10 \\ \quad = 10 & \quad = 8 \end{array}$$

The rectangle is 8 m by 10 m.

39. Let x and y denote the lengths of the two legs

$$\begin{cases} x^2 + y^2 = 625 \\ xy/2 = 150 \end{cases}$$

$$xy = 300$$
$$2xy = 600$$
$$x^2 - 2xy + y^2 = 625 + 600$$
$$(x + y)^2 = 1225$$
$$x + y = 35$$
$$x^2 - 2xy + y^2 = 625 - 600$$
$$(x - y)^2 = 25$$
$$x - y = 5$$
$$x + y = 35$$
$$x - y = 5$$
$$\overline{ 2x = 40}$$
$$x = 20$$

$$20 + y = 35$$
$$y = 15$$

The dimensions of the legs are 15 m and 20 m.

Exercise 10.6

The graphs for this section are found in the answers to odd-numbered exercises in your text.

1. $3x - 2y < 1$
   $(1, -3)$
   $3 - 2(-3) = 9 > 1$
   $(-2, 3)$
   $3(-2) - 2(3) = -12 < 1$
   $(1, -3)$, no; $(-2, 3)$, yes

3. $3x - 4y + 1 \geq 0$
   $(1/3, 1/2)$
   $3(1/3) - 4(1/2) + 1 = 0$
   $(-2, 3)$
   $(1/3, 1/2)$, yes; $(-2, 3)$, no

5. $2x + 1 > 0$
   $2x > -1$
   $x > -1/2$

7. $x + 5y \geq 2$
   $5y \geq -x + 2$

   $$y \geq -\frac{1}{5}x + \frac{2}{5}$$

9. $-2x + 4y - 3 < 0$
   $4y < 2x + 3$

   $$y < \frac{1}{2}x + \frac{3}{4}$$

11. $y \leq \sqrt{x - 4}$, $x \geq 4$

13. $y > x^2 - 1$

15. $\begin{cases} y > x^2 \\ y < x + 2 \end{cases}$
    $x^2 = x + 2$
    $x^2 - x - 2 = 0$
    $(x - 2)(x + 1) = 0$
    $x = 2 \quad x = -1$

17. $\begin{cases} x^2 + y^2 < 1 \\ \quad\quad y \geq x \end{cases}$

19. $\begin{cases} 4x^2 + y^2 < 4 \\ \quad\quad y > 4x \end{cases}$
    $4x^2 + (4x)^2 = 4$
    $4x^2 + 16x^2 = 4$
    $20x^2 = 4$
    $x^2 = 1/5$
    $x = \pm\sqrt{5}/5$
    $y = \pm 4\sqrt{5}/5$

21. $\begin{cases} y \geq x^2 - 1 \\ x^2 + y^2 < 3 \end{cases}$
    $x^2 = y + 1$
    $y + 1 + y^2 = 3$
    $y^2 + y - 2 = 0$
    $(y + 2)(y - 1) = 0$
    $y = -2 \quad y = 1$
    $x = \pm\sqrt{2}$

23. $\begin{cases} x^2 + y^2 < 2 \\ \dfrac{x^2}{9} + y^2 < 1 \end{cases}$

    $y^2 = 2 - x^2$

    $\dfrac{x^2}{9} + 2 - x^2 = 1$

    $x^2 + 18 - 9x^2 = 9$

    $ -8x^2 = -9$

    $x = \pm 3\sqrt{2}/4$

    $y = \pm\sqrt{14}/4$

25. $\begin{cases} y \geq 3x \\ y < -x + 4 \end{cases}$

    $y = 3x$

    $\underline{-y = x - 4}$

    $0 = 4x - 4$

    $x = 1, \ y = 3$

27. $\begin{cases} x - y \geq 2 \\ 3x + y < 0 \end{cases}$

    $x - y = 2$

    $\underline{3x + y = 0}$

    $4x = 2$

    $x = 1/2, \ y = -3/2$

29. $\begin{cases} 3x - 4y - 1 > 0 \\ 5x + 8y + 2 \leq 0 \end{cases}$

    $6x - 8y = 2$

    $\underline{5x + 8y = -2}$

    $11x = 0$

    $x = 0$

    $y = -1/4$

31. $\begin{cases} 3x + 2y \geq 6 \\ 3x + 2y \leq 12 \\ x \geq 0 \\ y \leq 0 \end{cases}$

33. $\begin{cases} 3x + 4y \leq 24 \\ x - 4y \geq -8 \\ x \leq 6 \\ y \geq 0 \end{cases}$

    $3x + 4y = 24$

    $\underline{x - 4y = -8}$

    $4x = 16$

    $x = 4$

    $4 - 4y = -8$

    $-4y = -12$

    $y = 3$

35. $\begin{cases} x + y \leq 4 \\ 2x - y \geq -3 \\ 0 \leq x \leq 3 \\ 0 \leq y \leq 3 \end{cases}$

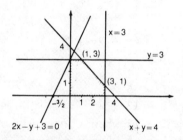

Vertex	L = 2x + 3y − 5
(0, 0)	L = -5
(0, 3)	L = 4
(1, 3)	L = 6
(3, 1)	L = 4
(3, 0)	L = 1

Maximum L = 6 when x = 1, y = 3.

Minimum L = -5 when x = 0, y = 0.

37.
$$\begin{cases} 3x - y \geq 0 \\ x + y \geq 0 \\ 1 \leq x \leq 3 \\ y \geq 0 \end{cases}$$

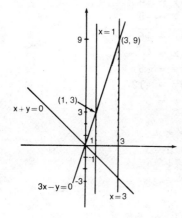

Vertex	$L = 4x - 2y + 1$
(1, 0)	L = 5
(1, 3)	L = -1
(3, 9)	L = -5
(3, 0)	L = 13

Maximum L = 13 when x = 3, y = 0.
Minimum L = -5 when x = 3, y = 9.

39.
$$\begin{cases} 2x + 3y \geq 12 \\ 2x + y \geq 8 \\ x \geq 0 \\ y \geq 0 \end{cases}$$

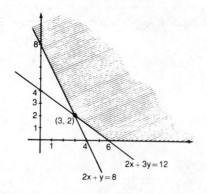

Vertex	$L = 5x + 4y$
(0, 8)	L = 32
(3, 2)	L = 23
(6, 0)	L = 30

Minimum L = 23 when x = 3, y = 2.
No maximum value.

41. Let x be barrels of oil and y be barrels of gasoline.

$$\begin{cases} x + y \leq 2500 \\ x \geq 400 \\ y \geq 300 \\ P = 3x + 4y \end{cases}$$

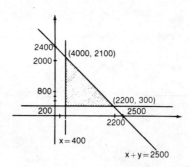

Vertex		P
(400, 300)		P = 2400
(400, 2100)		P = 9600
(2200, 300)		P = 7800

400 barrels of oil, 2100 barrels of gasoline, for maximum profit of $9600.

43. x number of Birdie Custom,
     y number of Bogey De Luxe

$$\begin{cases} 10x + 15y \leq 90 \\ \quad 0 \leq x \leq 6 \\ \quad 0 \leq y \leq 4 \end{cases}$$
$$P = 150x + 240y$$

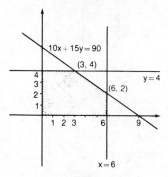

Vertex   P
(0, 0)   P = 0
(0, 4)   P = 960
(3, 4)   P = 1410
(6, 2)   P = 1380
(6, 0)   P = 900

3 Birdie Customs and 4 Bogey De
Luxe for a profit of $1410.

45. x grams of ingredient A,
     y grams of ingredient B

$$\begin{cases} 3x + 2y \geq 180 \\ 3x + 4y \geq 240 \\ \quad\quad x \geq 0 \\ \quad\quad y \geq 0 \end{cases}$$
$$P = 8x + 6y$$

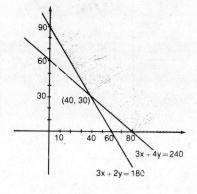

Vertex   P
(0, 90)   P = 540
(40, 30) P = 500
(80, 0)   P = 640
40 grams of A, 30 grams of B.

47. x acres of barley, y acres of corn

$$\begin{cases} x + y \leq 360 \\ \dfrac{x}{2} + \dfrac{3y}{4} \leq 210 \\ \quad x \geq 0 \\ \quad y \geq 0 \end{cases}$$

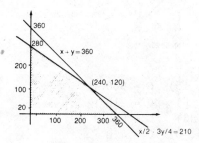

$$P = 35x + 45y$$

Vertex   P
(0, 0)   P = 0
(0, 280)   P = 12600
(240, 120) P = 13800
(360, 0)   P = 12600

Barley: 240 acres, corn: 120
acres, profit: $13800

49. x days center A, y days center B

$$\begin{cases} 360x + 120y \geq 1800 \\ 180x + 540y \geq 2700 \\ 120x + 120y \geq 1080 \\ \qquad\quad x \geq 0 \\ \qquad\quad y \geq 0 \end{cases}$$

$C = 30x + 20y$

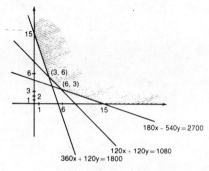

180x – 540y = 2700
120x + 120y = 1080
360x + 120y = 1800

Vertex	C
(0, 15)	C = 300
(3, 6)	C = 210
(6, 3)	C = 240
(15, 0)	C = 450

Center A: 3 days, center B: 6 days.

Review Exercises - Chapter 10

1. $A = \begin{bmatrix} 2 & 1 \\ 2 & -3 \end{bmatrix}$ $B = \begin{bmatrix} 1 & -2 \\ 3 & 2 \end{bmatrix}$

$2A + B = \begin{bmatrix} 5 & 0 \\ 7 & -4 \end{bmatrix}$

$A - 3B = \begin{bmatrix} -1 & 7 \\ -7 & -9 \end{bmatrix}$

$5B = \begin{bmatrix} 5 & -10 \\ 15 & 10 \end{bmatrix}$

3. $A = \begin{bmatrix} 5 & 0 \\ 0 & -5 \end{bmatrix}$ $B = \begin{bmatrix} 0 & 2 \\ -2 & 0 \end{bmatrix}$

$2A + B = \begin{bmatrix} 10 & 2 \\ -2 & -10 \end{bmatrix}$

$A - 3B = \begin{bmatrix} 5 & -6 \\ 6 & -5 \end{bmatrix}$ $5B = \begin{bmatrix} 0 & 10 \\ -10 & 0 \end{bmatrix}$

5. $A = \begin{bmatrix} 0 & 2 & -1 & -2 \end{bmatrix}$, $B = \begin{bmatrix} 2 \\ 1 \\ -3 \\ 2 \end{bmatrix}$

$AB = \begin{bmatrix} 0 + 2 + 3 - 4 \end{bmatrix}$
$\quad = \begin{bmatrix} 1 \end{bmatrix}$

$BA = \begin{bmatrix} 0 & 4 & -2 & -4 \\ 0 & 2 & -1 & -2 \\ 0 & -6 & 3 & 6 \\ 0 & 4 & -2 & -4 \end{bmatrix}$

7. $A = \begin{bmatrix} 1 & -1 & 2 \\ 4 & 0 & 3 \end{bmatrix}$ $B = \begin{bmatrix} 3 & 2 \\ -2 & 0 \\ 1 & -2 \end{bmatrix}$

$AB = \begin{bmatrix} 3+2+2 & 2+0-4 \\ 12+0+3 & 8+0-6 \end{bmatrix} = \begin{bmatrix} 7 & -2 \\ 15 & 2 \end{bmatrix}$

$BA = \begin{bmatrix} 3 + 8 & -3 + 0 & 6 + 6 \\ -2 + 0 & 2 + 0 & -4 + 0 \\ 1 - 8 & -1 + 0 & 2 - 6 \end{bmatrix}$

$\quad = \begin{bmatrix} 11 & -3 & 12 \\ -2 & 2 & -4 \\ -7 & -1 & -4 \end{bmatrix}$

9. $\begin{bmatrix} 2 & -1 \\ 3 & 2 \end{bmatrix} \begin{bmatrix} a & b \\ c & d \end{bmatrix} = \begin{bmatrix} 1 & 0 \\ 0 & 1 \end{bmatrix}$

$\begin{array}{ll} 2a - c = 1 & \qquad 2b - d = 0 \\ 3a + 2c = 0 & \qquad 3b - 2d = 1 \\ 4a - 2c = 2 & \qquad 4b - 2d = 0 \\ 3a + 2c = 0 & \qquad 3b + 2d = 1 \\ \hline \quad 7a = 2 & \qquad\quad 7b = 1 \\ \qquad a = 2/7 & \qquad\quad b = 1/7 \\ 2(2/7) - c = 1 & \quad 2(1/7) - d = 0 \\ 4/7 - c = 1 & \\ \\ \quad -c = 1 - \dfrac{4}{7} & \qquad\quad d = 2/7 \\ \\ \quad c = -3/7 & \end{array}$

The inverse is

$$\begin{bmatrix} \dfrac{2}{7} & \dfrac{1}{7} \\ -\dfrac{3}{7} & \dfrac{2}{7} \end{bmatrix}$$

11. $\begin{bmatrix} 1 & 2 & 3 & \vdots & 1 & 0 & 0 \\ 0 & 2 & 3 & \vdots & 0 & 1 & 0 \\ 0 & 0 & 3 & \vdots & 0 & 0 & 1 \end{bmatrix}$

$= \begin{bmatrix} 1 & 0 & 0 & \vdots & 1 & -1 & 0 \\ 0 & 2 & 0 & \vdots & 0 & 1 & -1 \\ 0 & 0 & 3 & \vdots & 0 & 0 & 1 \end{bmatrix}$

$= \begin{bmatrix} 1 & 0 & 0 & \vdots & 1 & -1 & 0 \\ 0 & 1 & 0 & \vdots & 0 & \frac{1}{2} & -\frac{1}{2} \\ 0 & 0 & 1 & \vdots & 0 & 0 & \frac{1}{3} \end{bmatrix}$

The inverse is

$\begin{bmatrix} 1 & -1 & 0 \\ 0 & 1/2 & -1/2 \\ 0 & 0 & 1/3 \end{bmatrix}$

13. $\begin{bmatrix} 1 & 0 & 0 & 0 & \vdots & 1 & 0 & 0 & 0 \\ 2 & 1 & 0 & 0 & \vdots & 0 & 1 & 0 & 0 \\ 3 & 3 & 1 & 0 & \vdots & 0 & 0 & 1 & 0 \\ 4 & 4 & 4 & 1 & \vdots & 0 & 0 & 0 & 1 \end{bmatrix}$

$= \begin{bmatrix} 1 & 0 & 0 & 0 & \vdots & 1 & 0 & 0 & 0 \\ 0 & 1 & 0 & 0 & \vdots & -2 & 1 & 0 & 0 \\ 0 & 3 & 1 & 0 & \vdots & -3 & 0 & 1 & 0 \\ 0 & 4 & 4 & 1 & \vdots & -4 & 0 & 0 & 1 \end{bmatrix}$

$= \begin{bmatrix} 1 & 0 & 0 & 0 & \vdots & 1 & 0 & 0 & 0 \\ 0 & 1 & 0 & 0 & \vdots & -2 & 1 & 0 & 0 \\ 0 & 0 & 1 & 0 & \vdots & 3 & -3 & 1 & 0 \\ 0 & 0 & 4 & 1 & \vdots & 4 & -4 & 0 & 1 \end{bmatrix}$

$= \begin{bmatrix} 1 & 0 & 0 & 0 & \vdots & 1 & 0 & 0 & 0 \\ 0 & 1 & 0 & 0 & \vdots & -2 & 1 & 0 & 0 \\ 0 & 0 & 1 & 0 & \vdots & 3 & -3 & 1 & 0 \\ 0 & 0 & 0 & 1 & \vdots & -8 & 8 & -4 & 1 \end{bmatrix}$

The inverse is

$\begin{bmatrix} 1 & 0 & 0 & 0 \\ -2 & 1 & 0 & 0 \\ 3 & -3 & 1 & 0 \\ -8 & 8 & -4 & 1 \end{bmatrix}$

15. $\begin{vmatrix} 0 & 2 & 2 \\ 3 & 0 & 2 \\ 3 & 3 & 0 \end{vmatrix}$

$= 0 + (2)(2)(3) + (2)(3)(3)$
$\quad - (3)(0)(2) - (3)(2)(0)$
$\quad - (0)(3)(2)$
$= 12 + 18 = 30$

17. $\begin{vmatrix} 3 & 2 & 0 \\ 1 & 0 & 2 \\ 0 & 1 & 2 \end{vmatrix}$

$= (3)(0)(2) + (2)(2)(0)$
$\quad + (0)(1)(1) - 0 - (2)(1)(2)$
$\quad - (3)(1)(2)$
$= -4 - 6 = -10$

19. $\begin{vmatrix} 1 & 2 & 0 & 3 \\ -1 & -2 & 0 & 1 \\ -1 & 2 & 1 & 1 \\ -3 & 2 & 1 & 2 \end{vmatrix}$ 3rd column expansion

$= (1)(1) \begin{vmatrix} 1 & 2 & 3 \\ -1 & -2 & 1 \\ -3 & 2 & 2 \end{vmatrix}$

$+ (1)(-1) \begin{vmatrix} 1 & 2 & 3 \\ -1 & -2 & 1 \\ -1 & 2 & 1 \end{vmatrix}$

$= -32 - (-16) = -16$

21. $\begin{bmatrix} a & 0 & \vdots & 1 & 0 \\ 0 & b & \vdots & 0 & 1 \end{bmatrix}$

$= \begin{bmatrix} 1 & 0 & \vdots & 1/a & 0 \\ 0 & 1 & \vdots & 0 & 1/b \end{bmatrix}$

The inverse is

$\begin{bmatrix} 1/a & 0 \\ 0 & 1/b \end{bmatrix}$

23. $\begin{cases} x - 3y = 7 \\ 2x - y = -4 \end{cases}$

$\quad\quad -2x + 6y = -14$
$\quad\quad \underline{2x - y = -4}$
$\quad\quad\quad\quad\quad 5y = -18$
$\quad\quad\quad\quad\quady = -18/5$

$$x - 3\left(-\frac{18}{5}\right) = 7$$

$$x + \frac{54}{5} = 7$$

$$x = 7 - \frac{54}{5}$$

$$x = -19/5$$

$$x = -19/5, \quad y = -18/5$$

25. $\begin{cases} x + y - z = 0 \\ 2x - 3y - 4z = 6 \\ 2x - y - z = 8 \end{cases}$

$$\begin{bmatrix} 1 & 1 & -1 & \vdots & 0 \\ 2 & -3 & -4 & \vdots & 6 \\ 2 & -1 & -1 & \vdots & 8 \end{bmatrix}$$

$$= \begin{bmatrix} 1 & 1 & -1 & \vdots & 0 \\ 0 & -5 & -2 & \vdots & 6 \\ 0 & -3 & 1 & \vdots & 8 \end{bmatrix}$$

$$= \begin{bmatrix} 1 & 1 & -1 & \vdots & 0 \\ 0 & -5 & -2 & \vdots & 6 \\ 0 & -15 & 5 & \vdots & 40 \end{bmatrix}$$

$$= \begin{bmatrix} 1 & 1 & -1 & \vdots & 0 \\ 0 & -5 & -2 & \vdots & 6 \\ 0 & 0 & 11 & \vdots & 22 \end{bmatrix}$$

$$= \begin{bmatrix} 1 & 1 & -1 & \vdots & 0 \\ 0 & -5 & -2 & \vdots & 6 \\ 0 & 0 & 1 & \vdots & 2 \end{bmatrix}$$

$$= \begin{bmatrix} 1 & 1 & 0 & \vdots & 2 \\ 0 & -5 & 0 & \vdots & 10 \\ 0 & 0 & 1 & \vdots & 2 \end{bmatrix}$$

$$= \begin{bmatrix} 1 & 1 & 0 & \vdots & 2 \\ 0 & 1 & 0 & \vdots & -2 \\ 0 & 0 & 1 & \vdots & 2 \end{bmatrix}$$

$$= \begin{bmatrix} 1 & 0 & 0 & \vdots & 4 \\ 0 & 1 & 0 & \vdots & -2 \\ 0 & 0 & 1 & \vdots & 2 \end{bmatrix}$$

$$x = 4, \quad y = -2, \quad z = 2$$

27. $\begin{cases} x + y = 1 \\ 2y + z = 0 \\ 4x - 3z = 0 \end{cases}$

$\begin{cases} x + y = 1 \\ 2y + z = 0 \\ -4y - 3z = -4 \end{cases}$

$\begin{cases} x + y = 1 \\ 2y + z = 0 \\ -z = -4 \end{cases}$

$$z = 4$$
$$2y + 4 = 0$$
$$y = -2$$

$$x + (-2) = 1$$
$$x = 3$$

$$x = 3, \quad y = -2, \quad z = 4$$

29. $\begin{vmatrix} x & 2 \\ -3 & -2 \end{vmatrix} = 0$

$$(x)(-2) - (2)(-3) = 0$$
$$-2x + 6 = 0$$
$$-2x = -6$$
$$x = 3$$

31. $\begin{vmatrix} 1 & x & -1 \\ 3 & 0 & 2 \\ 5 & 0 & 4 \end{vmatrix} = -8$

Expanding the determinant by the elements of the second column, obtain

$$(-1)^{1+2}(x) \begin{vmatrix} 3 & 2 \\ 5 & 4 \end{vmatrix} = -8$$

$$-x(12 - 10) = -8$$
$$-2x = -8$$
$$x = 4$$

33. $\begin{vmatrix} x & y & 1 \\ 1 & 2 & 1 \\ -2 & -3 & 1 \end{vmatrix} = 0$

$x\begin{vmatrix} 2 & 1 \\ -3 & 1 \end{vmatrix} - y\begin{vmatrix} 1 & 1 \\ -2 & 1 \end{vmatrix}$

$+ \begin{vmatrix} 1 & 2 \\ -2 & -3 \end{vmatrix} = 0$

$a = \begin{vmatrix} 2 & 1 \\ -3 & 1 \end{vmatrix} = 2 + 3 = 5$

$b = -\begin{vmatrix} 1 & 1 \\ -2 & 1 \end{vmatrix} = -(1 + 2) = -3$

35. $\begin{cases} 3a - 2b + 1 = 0 \\ 3b + 2a - 8 = 0 \end{cases}$

$3a - 2b = -1$
$2a + 3b = 8$

$9a - 6b = -3$
$4a + 6b = 16$
$\overline{13a = 13}$
$a = 1$
$3(1) - 2b = -1$
$-2b = -4$
$b = 2$

37. x amount invested at 8%,
    y amount invested at 7.5%
$\begin{cases} y + 500 = x \\ .08x + .075y = 425 \end{cases}$

$80x + 75(x - 500) = 425000$
$80x + 75x - 37500 = 425000$
$155x = 462500$
$x = \$2983.87$

$\$2983.87$ at 8%, $\$2483.87$ at 7.5%

39. x dimes, y nickels
$\begin{cases} 4x = y \\ 10x + 5y = 480 \end{cases}$

$10x + 5(4x) = 480$
$30x = 480$
$x = 16$

16 dimes and 64 nickels

41. Let x be the rate on $\$15,000$ and
    let y be the rate over $\$15,000$.
$\begin{cases} 15000x + 12000y = 1908 \\ 15000x + 21500y = 2611 \end{cases}$

$15000x + 12000y = 1908$
$-15000x - 21500y = -2611$
$\overline{-9500y = -703}$
$y = 0.074$
$x = 0.068$

43. $\begin{cases} P + \dfrac{1}{2}Q = 16 \\[2mm] \dfrac{3}{4}P + \dfrac{1}{4}Q = 11 \end{cases}$

$2P + Q = 32$
$-3P - Q = -44$
$\overline{-P = -12}$
$P = 12$
$2(12) + Q = 32$
$Q = 8$

12 units of P and 8 units of Q

45. $\begin{cases} c = 1 \\ a + b + c = 0 \\ 4a + 2b + c = 3 \end{cases}$

$\begin{cases} a + b = -1 \\ 4a + 2b = 2 \end{cases}$

$-2a - 2b = 2$
$4a - 2b = 2$
$\overline{2a = 4}$
$a = 2$
$b = -3$
$y = 2x^2 - 3x + 1$

47. x lbs peanuts, y lbs almonds,
    z lbs cashews

$$\begin{cases} x + y + z = 100 \\ x = 3y \\ 2.10x + 4.20y + 4.00z = 2.90(100) \end{cases}$$

$$\begin{cases} 4y + z = 100 \\ 105y + 40z = 2900 \end{cases}$$

$$\begin{array}{r} -160y - 40z = -4000 \\ 105y + 40z = 2900 \\ \hline -55y = -1100 \end{array}$$

$y = 20$, $x = 60$, $z = 20$
60 lbs peanuts, 20 lbs almonds,
20 lbs cashews

49. x one-dollar, y five-dollar,
    z ten-dollar bills

$$\begin{cases} x + y + z = 75 \\ x + 5y + 10z = 425 \\ x + 5 = z \end{cases}$$

$$\begin{array}{r} x + y + x + 5 = 75 \\ 2x + y = 70 \end{array}$$

$$\begin{array}{r} x + 5y + 10(x + 5) = 425 \\ 11x + 5y = 375 \\ -10x - 5y = -350 \\ \hline x = 25 \end{array}$$

$z = 30$, $y = 20$

25 one-dollar, 20 five-dollar, 30
ten-dollar bills

51. x workers at \$6, y workers at \$8,
    z workers at \$10

$$\begin{cases} x + y + z = 120 \\ y = 3z \\ 6x + 8y + 10z = 840 \end{cases}$$

$$\begin{array}{r} x + 4z = 120 \\ 6x + 34z = 840 \end{array}$$

$$\begin{array}{r} -6x - 24z = -720 \\ 6x + 34z = 840 \\ \hline 10z = 120 \\ z = 12 \end{array}$$

$y = 36$, $x = 72$

72 workers at \$6, 36 workers at
\$8, 12 workers at \$10

53. $\begin{vmatrix} 1 & y & 1 \\ x & y & z \\ x^2 & y^2 & z^2 \end{vmatrix}$

$= \begin{vmatrix} 1 & -1 & -1 \\ x & -y & -z \\ x^2 & -y^2 & -z^2 \end{vmatrix}$

$= \begin{vmatrix} 1 & 0 & 0 \\ x & x - y & x - z \\ x^2 & x^2 - y^2 & x^2 - z^2 \end{vmatrix}$

$= (1) \begin{vmatrix} x - y & x - z \\ x^2 - y^2 & x^2 - z^2 \end{vmatrix}$

$= (x - y)(x^2 - z^2) - (x - z)(x^2 - y^2)$
$= (x - y)(x - z)[(x + z) - (x + y)]$
$= (x - y)(x - z)(z - y)$
$= (x - y)(z - x)(y - z)$

55. $\begin{vmatrix} a & b + c & 1 \\ b & c + a & 1 \\ c & a + b & 1 \end{vmatrix}$ [add first to
    second column]

$= \begin{vmatrix} a & a + b + c & 1 \\ b & a + b + c & 1 \\ c & a + b + c & 1 \end{vmatrix} = 0$

[because second column is a
multiple of the third column]

57. $\begin{vmatrix} \cos\theta & -\sin\theta \\ \sin\theta & \cos\theta \end{vmatrix}$

$= (\cos\theta)(\cos\theta)$
$\quad - (\sin\theta)(\sin\theta)$
$= \cos^2\theta + \sin^2\theta$
$= 1$

59. $$\begin{vmatrix} 1 & x_1 & x_2 & x_3 \\ 1 & x & x_2 & x_3 \\ 1 & x_1 & x & x_3 \\ 1 & x_1 & x_2 & x \end{vmatrix} = 0$$

Multiplying the first row by -1 and adding to the second, then to the third, then to the fourth, obtain

$$\begin{vmatrix} 1 & x_1 & x_2 & x_3 \\ 0 & x - x_1 & 0 & 0 \\ 0 & 0 & x - x_2 & 0 \\ 0 & 0 & 0 & x - x_3 \end{vmatrix} = 0$$

hence

$(x - x_1)(x - x_2)(x - x_3) = 0.$
Thus the roots are x_1, x_2, x_3.

61. $\begin{cases} y = x^2 + 3 \\ x^2 + y^2 = 9 \end{cases}$

$$x^2 = y - 3$$
$$y - 3 + y^2 = 9$$
$$y^2 + y - 12 = 0$$
$$(y - 3)(y + 4) = 0$$

$y = 3$	$y = -4$
$x^2 = 0$	$x^2 = -7$
$x = 0$	No real
	solution

$x = 0, \ y = 3$

63. $\begin{cases} 3x^2 - 2y = 0 \\ 5x - 2y = -2 \end{cases}$

$$3x^2 - 2y = 0$$
$$-5x + 2y = 2$$
$$\overline{ 3x^2 - 5x = 2}$$
$$3x^2 - 5x - 2 = 0$$
$$(3x + 1)(x - 2) = 0$$

$x = -1/3$	$x = 2$
$5(-1/3) - 2y = -2$	$5(2) - 2y = -2$
$-5/3 - 2y = -2$	$10 - 2y = -2$
$-2y = -1/3$	$-2y = -12$
$y = 1/6$	$y = 6$

$x = -1/3, \ y = 1/6$ or $x = 2, \ y = 6$

The graph for exercises 65-71 are found in the answer section of your text.

73. $\begin{cases} x + y \le 4 \\ 2x - y \ge -3 \\ 0 \le x \le 3 \\ 0 \le y \le 3 \end{cases}$

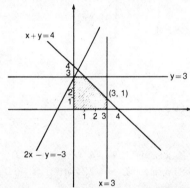

Vertex	L = 3x + 4y
(0, 0)	L = 0
(0, 3)	L = 12
(1, 3)	L = 15
(3, 1)	L = 13
(3, 0)	L = 9

Maximum L = 15 when x = 1, y = 3
Minimum L = 0 when x = 0, y = 0

75. $\begin{cases} x + y \geq -6 \\ x - 2y \geq 0 \\ -4 \leq x \leq 0 \\ y \leq 0 \end{cases}$

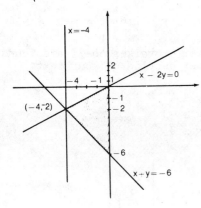

Vertex $L = x + 3y - 4$
(0, 0) $L = -4$
(0, -6) $L = -22$
(-4, -2) $L = -14$

Maximum $L = -4$ when $x = 0$, $y = 0$
Minimum $L = -22$ when $x = 0$, $y = -6$

77. $\begin{cases} x - y = 7 \\ xy = 450 \end{cases}$

$$x = y + 7$$
$$(y + 7)y = 450$$
$$y^2 + 7y - 450 = 0$$
$$(y + 25)(y - 18) = 0$$

$y = -25$	$y = 18$
$x = -18$	$x = 25$

The numbers are 18 and 25.

79. x number of Standard,
    y number of DeLuxe

$\begin{cases} 2x + 4y \leq 40 \\ 0 \leq x \leq 10 \\ 0 \leq y \leq 8 \end{cases}$
$P = 60x + 80y$

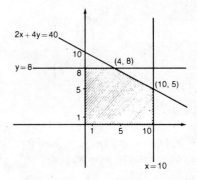

Vertex	P
(0, 0)	$P = 0$
(0, 8)	$P = 640$
(4, 8)	$P = 880$
(10, 5)	$P = 1000$
(10, 0)	$P = 600$

10 Standard, 5 DeLuxe, $1000

Chapter 10 Test

1. Solve the two by two linear system
$$\begin{cases} 4x - 2y = 8 \\ -5x + 8y = 1 \end{cases}$$

2. Find the solution of the system

$$\begin{cases} \dfrac{x}{3} - \dfrac{y}{4} = \dfrac{2}{3} \\ \\ 4x - \dfrac{y}{2} = 3 \end{cases}$$

3. Solve the system

$$\begin{cases} \dfrac{2}{x} + \dfrac{5}{y} = -1 \\ \\ -\dfrac{1}{3x} + \dfrac{2}{y} = 3 \end{cases}$$

4. With the aid of a tail wind, an airplane travels 1050 km in 2 hours and 20 minutes. Against the same wind, the airplane takes 3 hours to travel the same distance. What are the speed of the plane in still air and the speed of the wind?

5. How many liters of 3.5% butterfat milk must be mixed with 1% butterfat milk to get 600 liters of low-fat milk (2% butterfat)?

6. Solve by substitution the following system

$$\begin{cases} x \quad\quad + z = 1 \\ 2x + y + 3z = 1 \\ \quad\quad 3y - z = 5 \end{cases}$$

7. Solve by Gaussian elimination the three by three system

$$\begin{cases} 3x + 2y + z = 4 \\ x \quad\quad - 2z = -2 \\ \quad\quad 2y + z = -2 \end{cases}$$

8. Use row-admissible operations with the augmented matrix to solve the system

$$\begin{cases} x + y + z = 3 \\ -2x \quad\quad + 5z = 1 \\ 2x - y - 2z = 2 \end{cases}$$

9. Does the homogeneous system

$$\begin{cases} 3x - 2y + z = 0 \\ x + z = 0 \\ 4x - y + 3z = 0 \end{cases}$$

have non-trivial solution? If yes, find them.

10. Judy has $11.25 in nickels, dimes, and quarters. The number of quarters is the sum of the number of nickels and dimes, and it is seven times the difference between the number of nickels and dimes. How many nickels, dimes, and quarters does Judy have?

11. Let

$$A = \begin{bmatrix} 1 & -1 \\ 2 & -2 \\ 3 & -3 \end{bmatrix} \quad \text{and} \quad B = \begin{bmatrix} 1 & -2 \\ 0 & 0 \\ -2 & 1 \end{bmatrix}$$

Find $A - B$ and $2A + 3B$.

12. Let

$$A = \begin{bmatrix} 1 & 2 & 3 \\ 0 & 1 & 2 \\ 0 & 0 & 1 \end{bmatrix} \quad \text{and} \quad B = \begin{bmatrix} 1 & 1 & 1 \\ 0 & 1 & 1 \\ 0 & 0 & 1 \end{bmatrix}$$

Find AB and BA.

13. Find the inverse of the matrix $\begin{bmatrix} 1 & 0 \\ 2 & 1 \end{bmatrix}$.

14. Find the inverse of the matrix

$$A = \begin{bmatrix} 1 & 0 & 2 \\ 0 & 1 & 0 \\ 2 & 0 & 1 \end{bmatrix}$$

by transforming the augmented matrix $[A|I_3]$ to the form $[I_3|B]$.

15. Use the inverse matrix obtained in Problem 14 to solve the system

$$\begin{cases} x + 2z = 1 \\ y = 2 \\ 2x + z = -1 \end{cases}$$

16. Find the complementary matrix and minor of the element a_{32} in the matrix

$$\begin{bmatrix} 2 & -1 & 0 \\ 1 & 3 & 0 \\ 4 & 0 & 1 \end{bmatrix}$$

17. Compute by row or column expansion the determinant

$$\begin{vmatrix} 3 & 2 & 1 \\ -2 & 1 & 0 \\ 1 & 0 & 2 \end{vmatrix}$$

18. Show that

$$\begin{vmatrix} 1 & 0 & -1 \\ 2 & -2 & 2 \\ 5 & -2 & -1 \end{vmatrix} = 0$$

Use only properties of determinants.

19. Find all solutions of the equation

$$\begin{vmatrix} x-1 & 0 & 1 \\ 0 & x+3 & 0 \\ 3 & 0 & x+1 \end{vmatrix} = 0$$

20. Use Cramer's rule to solve the system

$$\begin{cases} x - y & = 1 \\ x & + z = 2 \\ y - 3z = 5 \end{cases}$$

In Problems 21-23, find all solutions of each of the given system

21. $\begin{cases} x^2 - y = 0 \\ 2x + y = 0 \end{cases}$

22. $\begin{cases} 2xy - 3 = 0 \\ 3x - 2y = 0 \end{cases}$

23. $\begin{cases} x^2 + y = 4 \\ x^2 + 2 = y \end{cases}$

24. Find two integers whose sum is 8 and whose product is -240.

25. The area of a rectangle is 48 m^2 and the sum of squares of the sides is 100 m^2. What are the dimensions of the rectangle?

26. Determine whether or not the points (-2, 3) and (4, -2) satisfy the inequality $3x - 4y < 5$.

27. Find the maximum and minimum values of the objective function $L = 3x - y$ on the following set of constraints

$$\begin{cases} x - y \leq 4 \\ 2x + y \leq 2 \\ -3 \leq x \leq 1 \\ 0 \leq y \leq 2 \end{cases}$$

28. Let $L = 2x + 3y - 1$ be an objective function and consider the set of constraints

$$\begin{cases} 3x + y \geq 6 \\ 3x + 4y \geq 15 \\ x \geq 0 \\ y \geq 0 \end{cases}$$

   Find the minimum value of L.

29. A manufacturer assembles two types of blenders: standard and heavy duty. One hour of labor is needed to assemble the standard model, while four hours are required to assemble the heavy duty. The maximum numbers of standard and heavy duty blenders assembled per day is 60 and 15, respectively. There are 80 hours of labor available each day. The manufacturer can make a profit of $12 on each standard blender and $24 on each heavy duty one. What is the maximum daily profit?

30. The table below gives the number of units of fat and protein contained in one kilogram of each of two ingredients A and B used in the preparation of a certain brand of dog food.

	A	B
Fat	3	2
Protein	3	4

   Each bag of dog food contains at least 12 units of fat and 18 units of protein. If each kilogram of ingredient A costs $.75 and each kilogram of ingredient B costs $.80, find the number of kilograms of each ingredient that should be used to minimize the cost.

Chapter 11
Zeros of Polynomials

Exercises 11.1

1. a) $P(1) = 2(1)^3 - (1)^2 + 2(1) - 1$
$= 2 - 1 + 2 - 1$
$= 2$

b) $P(2) = 2(2)^3 - (2)^2 + 2(2) - 1$
$= 2 \cdot 8 - 4 + 4 - 1$
$= 16 - 4 + 4 - 1$
$= 15$

c) $\dfrac{P(1)}{P(2)} = \dfrac{2}{15}$

d) $P(\dfrac{1}{2}) = 2(\dfrac{1}{2})^3 - (\dfrac{1}{2})$

$+ 2(\dfrac{1}{2}) - 1$

$= 2 \cdot \dfrac{1}{8} - \dfrac{1}{4} + 1 - 1$

$= \dfrac{1}{4} - \dfrac{1}{4} + 1 - 1$

$= 0$

e) $2P(1) + 3P(2) = 2(2) + 3(15)$
$= 4 + 45$
$= 49$

3. $A(1) = 1^3 - 2(1)^2 + a(1) - 1 = 4$
$1 - 2 + a - 1 = 4$
$-2 + a = 4$
$a = 6$

5. $4x^2 - 3x + 2$
$= (A + 1)x^2 + (B + 1)x + (C - 3)$
$A + 1 = 4$, so $A = 3$
$B + 1 = -3$, so $B = -4$
$C - 3 = 2$, so $C = 5$

7. $3x^2 - 4x + 6 = Ax^2 + (B - 2A)x$
$+ (C + 1)$

$A = 3$

$B - 2A = -4$
$B - 2(3) = -4$
$B - 6 = -4$
$B = 2$

$C + 1 = 6$
$C = 5$

9. $x^3 - 2x + 1 = (A - B)x^3 + (B + C)x^2$
$+ (C - 2D)x + D$

$D = 1$

$C - 2D = -2$
$C - 2(1) = -2$
$C - 2 = -2$
$C = 0$

$B + C = 0$
$B + 0 = 0$
$B = 0$

$A - B = 1$
$A - 0 = 1$
$A = 1$

11.
$$
\begin{array}{r}
3x - 2 \\
2x + 5 \overline{)6x^2 + 11x + 4} \\
6x^2 + 15x \\
\hline
-4x + 4 \\
-4x - 10 \\
\hline
14
\end{array}
$$

$Q(x) = 3x - 2, \ R = 14$

13.
$$
\begin{array}{r}
\frac{1}{3}x^2 - \frac{2}{9}x + \frac{28}{27} \\
3x - 4 \overline{)x^3 - 2x^2 + 4x - 1} \\[4pt]
x^3 - \frac{4}{3}x^2 \\
\hline
-\frac{2}{3}x^2 + 4x \\[4pt]
-\frac{2}{3}x^2 + \frac{8}{9}x \\
\hline
\frac{28}{9}x - 1 \\[4pt]
\frac{28}{9}x - \frac{112}{27} \\
\hline
\frac{85}{27}
\end{array}
$$

$Q(x) = \frac{1}{3}x^2 - \frac{2}{9}x + \frac{28}{27},$

$R = 85/27$

15.
$$
\begin{array}{r}
x^2 + 3x - 1 \\
x^2 - x + 2 \overline{)x^4 + 2x^3 - 2x^2 + 7x + 2} \\
x^4 - x^3 + 2x^2 \\
\hline
3x^3 - 4x^2 + 7x \\
3x^3 - 3x^2 + 6x \\
\hline
-x^2 + x + 2 \\
-x^2 + x - 2 \\
\hline
4
\end{array}
$$

$Q(x) = x^2 + 3x - 1, \ R = 4$

17.
$$
\begin{array}{r}
2x^3 + \frac{1}{2}x + 1 \\
4x^3 - x - 2 \overline{)8x^6 + 0x^5 + 0x^4 + 0x^3 - 3x^2 + 0x + 1} \\
8x^6 - 2x^4 - 4x^3 \\
\hline
2x^4 + 4x^3 - 3x^2 + 0x \\
2x^4 - \frac{1}{2}x^2 - x \\
\hline
4x^3 - \frac{5}{2}x^2 + x + 1 \\
4x^3 - x - 2 \\
\hline
-\frac{5}{2}x^2 + 2x + 3
\end{array}
$$

$Q(x) = 2x^3 + \frac{1}{2}x + 1,$

$R(x) = -\frac{5}{2}x^2 + 2x + 3$

19.
$$
\begin{array}{r}
\frac{1}{2}x^3 + 1 \\
2x^3 - 4 \overline{)x^6 + 0x^5 + 0x^4 + 0x^3 - x^2 + 0x - 1} \\
x^6 - 2x^3 \\
\hline
2x^3 - x^2 + 0x - 1 \\
2x^3 - 4 \\
\hline
-x^2 + 3
\end{array}
$$

$Q(x) = \frac{1}{2}x^3 + 1, \ R(x) = -x^2 + 3$

21.
$$
\begin{array}{r}
x - 3 \\
x - 1 \overline{)x^2 - 4x + 5} \\
x^2 - x \\
\hline
-3x + 5 \\
-3x + 3 \\
\hline
2
\end{array}
$$

Thus

$$\frac{x^2 - 4x + 5}{x - 1} = x - 3 + \frac{2}{x - 1}$$

23.

$$
\begin{array}{r}
x^2 - 2x + 2 \\
2x + 4 \overline{)2x^3 \quad\;\; - 4x - 2} \\
\underline{2x^3 + 4x^2} \\
-4x^2 - 4x \\
\underline{-4x^2 - 8x} \\
4x - 2 \\
\underline{4x + 8} \\
-10
\end{array}
$$

$$\frac{2x^3 - 4x - 2}{2x + 4}$$

$$= x^2 - 2x + 2 - \frac{10}{2x + 4}$$

25.

$$
\begin{array}{r}
\frac{1}{2}x^3 + \frac{1}{4}x^2 - \frac{19}{8}x - \frac{19}{16} \\
2x - 1 \overline{)\; x^4 + 0x^3 - 5x^2 + 0x + 3} \\
\underline{x^4 - \frac{1}{2}x^3} \\
\frac{1}{2}x^3 - 5x^2 \\
\underline{\frac{1}{2}x^3 - \frac{1}{4}x^2} \\
-\frac{19}{4}x^2 + 0x \\
\underline{-\frac{19}{4}x^2 + \frac{19}{8}x} \\
-\frac{19}{8}x + 3 \\
\underline{-\frac{19}{8}x + \frac{19}{16}} \\
\frac{29}{16}
\end{array}
$$

$$\frac{x^4 - 5x^2 + 3}{2x - 1} = \left(\frac{1}{2}\right)x^3 + \left(\frac{1}{4}\right)x^2$$

$$- \left(\frac{19}{8}\right)x - \left(\frac{19}{16}\right) + \frac{29}{16(2x - 1)}$$

27.

$$
\begin{array}{r}
5x^2 + \dfrac{5}{2} \\
2x^4 - x^2 - 1 \overline{)10x^6 + 0x^5 + 0x^4 + 5x^3 + 0x^2 - x + 0} \\
\underline{10x^6 \qquad -5x^4 \qquad -5x^2} \\
5x^4 + 5x^3 + 5x^2 - x + 0 \\
\underline{5x^4 \qquad -\frac{5}{2}x^2 \qquad -\frac{5}{2}} \\
5x^3 + \frac{15}{2}x^2 - x + \frac{5}{2}
\end{array}
$$

$$\frac{10x^6 + 5x^3 - x}{2x^4 - x^2 - 1} = 5x^2 + \frac{5}{2}$$

$$+ \frac{5x^3 + \dfrac{15}{2}x^2 - x + \dfrac{5}{2}}{2x^4 - x^2 - 1}$$

29.

$$
\begin{array}{r}
3x^2 + x + 5 \\
x^2 - x - 1 \overline{)3x^4 - 2x^3 + x^2 + 2x + 3} \\
\underline{3x^4 - 3x^3 - 3x^2} \\
x^3 + 4x^2 + 2x \\
\underline{x^3 - x^2 - x} \\
5x^2 + 3x + 3 \\
\underline{5x^2 - 5x - 5} \\
8x + 8
\end{array}
$$

$$\frac{3x^4 - 2x^3 + x^2 + 2x + 3}{x^2 - x + 1} = 3x^2 + x$$

$$+ 5 + \frac{8x + 8}{x^2 - x - 1}$$

Exercises 11.2

1. $A(-2) = 2(-2)^2 - 6(-2) - 4$
   $= 2 \cdot 4 + 12 - 4$
   $= 8 + 12 - 4$
   $= 16$

3. $A(-\frac{1}{2}) = -(-\frac{1}{2})^3 - 2(-\frac{1}{2})^2$

$\qquad\qquad - \frac{1}{2} + 1$

$\qquad = -(-\frac{1}{8}) - 2(\frac{1}{4}) - \frac{1}{2} + 1$

$\qquad = \frac{1}{8} - \frac{1}{2} - \frac{1}{2} + 1$

$\qquad = \frac{1}{8} - 1 + 1$

$\qquad = \frac{1}{8}$

5. $A(2) = 2^5 - 2^3 - 2(2) + 2$
$\qquad = 32 - 8 - 4 + 2$
$\qquad = 22$

7. $A(0) = -4$

9. $P(x) = 0.2x^3 - 0.03x^2$
$\qquad\qquad + 0.21x - 0.18$

$\qquad = [0.2x^2 - 0.03x + 0.21]x$
$\qquad\qquad - 0.18$

$\qquad = [(0.2x - 0.03)x + 0.21]x$
$\qquad\qquad - 0.18$

$P(-0.5) = [((0.2)(-0.5)-0.03)(-0.5)$
$\qquad\qquad + 0.21](-0.5) - 0.18$

$\qquad = -0.3175$

11.

2	2	-6	-4
		4	-4
	2	-2	-8

$Q(x) = 2x - 2, \quad R = -8$

13.

$-\frac{1}{2}$	-1	-2	1	1
		$\frac{1}{2}$	$\frac{3}{4}$	$-\frac{7}{8}$
	-1	$-\frac{3}{2}$	$\frac{7}{4}$	$\frac{1}{8}$

$Q(x) = -x^2 - \frac{3}{2}x + \frac{7}{4},$

$R = \frac{1}{8}$

15.

4	4	0	-3	4
		16	64	244
	4	16	61	248

$Q(x) = 4x^2 + 16x + 61, \quad R = 248$

17.

-2	1	0	-1	0	-2	2
		-2	4	-6	12	-20
	1	-2	3	-6	10	-18

$Q(x) = x^4 - 2x^3 + 3x^2 - 6x^2 + 10,$
$R = -18$

19.

-1	1	0	0	0	-1
		-1	1	-1	1
	1	-1	1	-1	0

$Q(x) = x^3 - x^2 + x - 1, \quad R = 0$

21.

-0.5	0.2	-0.03	0.21	-0.18
		-0.1	0.065	-0.1375
	0.2	-0.13	0.275	-0.3175

$Q(x) = 0.2x^2 - 0.13x + 0.275,$
$R = -0.3175$

23.

3	1	-6	11	-6
		3	-9	6
	1	-3	2	0

$R = 0$
$x^3 - 6x^2 + 11x - 6$
$\qquad = (x - 3)(x^2 - 3x + 2)$
$\qquad = (x - 3)(x - 1)(x - 2)$

25.

-1	2	-1	-5	-2
		-2	3	2
	2	-3	-2	0

$2x^3 - x^2 - 5x - 2$
$\qquad = (x + 1)(2x^2 - 3x - 2)$
$\qquad = (x + 1)(x - 2)(2x + 1)$

27. $P(x) = 4x^3 - 3x^2 + 5x - 1$
$\qquad = [4x^2 - 3x + 5]x - 1$
$\qquad = [(4x - 3)x + 5]x - 1$

$\quad P(2) = [(4 \cdot 2 - 3)2 + 5]2 - 1$
$\qquad = 29$

29. $A(x) = x^3 - 2x^2 + 5x - 2$
$\qquad = [x^2 - 2x + 5]x - 2$
$\qquad = [(x - 2)x + 5]x - 2$
$\quad A(0.18)$
$\quad = [(0.18 - 2)(0.18) + 5](0.18) - 2$
$\quad = -1.158968$

31. $A(t) = 0.1t^4 - 0.4t^3 + 0.3t^2 - 0.5$
$\qquad = (0.1t^3 - 0.4t^2 + 0.3t + 0)t - 0.5$
$\qquad = [(0.1t^2 - 0.4)t + 0.3)t + 0]t - 0.5$
$\qquad = \{[(0.1t - 0.4)t + 0.3]t + 0\}t - 0.5$
$\quad A(1.2) = -0.55184$

33.

-4	1	1	-17	a
		-4	12	20
	1	-3	-5	0

$a + 20 = 0$
$\qquad a = -20$

35.

3	1	-7	$3a$	-18
		3	-12	$3(-12 + 3a)$
	1	-4	$-12 + 3a$	0

$-18 + 3(-12 + 3a) = 0$
$\quad -18 - 36 + 9a = 0$
$\qquad\qquad\quad 9a = 54$
$\qquad\qquad\quad\ a = 6$

37.

2	1	$-a$	0	5
		2	$2(2 - a)$	$4(2 - a)$
	1	$2 - a$	$2(2 - a)$	-3

$5 + 4(2 - a) = -3$
$\quad 5 + 8 - 4a = -3$
$\quad\ 13 - 4a = -3$
$\qquad\quad -4a = -16$
$\qquad\qquad a = 4$

39.

a	3	-2	1
		$3a$	$a(3a - 2)$
	3	$3a - 2$	$1 + a(3a - 2)$
			$1 + 3a^2 - 2a$

Quotient: $3x + 3a - 2$
Remainder: $3a^2 - 2a + 1$

Exercises 11.3

1. $P(1) = 1^2 - 3(1) + 3$
$\qquad = 1 - 3 + 3$
$\qquad = 1$
no

3. $P(-2) = 3(-2)^3 - 7(-2) + 10$
$\qquad = 3(-8) + 14 + 10$
$\qquad = -24 + 14 + 10$
$\qquad = 0$
yes

5. $P(-\frac{1}{2})$

$\quad = 2(-\frac{1}{2})^2 - 5(-\frac{1}{2}) - 3$

$\quad = 2(\frac{1}{4}) + \frac{5}{2} - 3$

$\quad = \frac{1}{2} + \frac{5}{2} - 3$

$\quad = \frac{6}{2} - 3 = 0$

yes

7. $P(-2)$

$\quad = (-2)^4 - 2(-2)^3$
$\qquad + 3(-2)^2 - 3$
$\quad = 16 - 2(-8) + 3(4) - 3$
$\quad = 16 + 16 + 12 - 3$
$\quad = 41$
no

9. $P(1 - 2i)$
    $= (1 - 2i)^2 - 2(1 - 2i) + 5$
    $= 1 - 4i + 4i^2 - 2 + 4i + 5$
    $= 1 - 4i - 4 - 2 + 4i + 5$
    $= 1 - 6 + 5$
    $= 0$
    yes

11.
$$
\begin{array}{r|rrrr}
2 & 2 & -4 & 3 & -6 \\
  &   & 4 & 0 & 6 \\
\hline
  & 2 & 0 & 3 & 0
\end{array}
$$

$2x^3 - 4x^2 + 3x - 6$
$\quad = (x - 2)(2x^2 + 3)$

13.
$$
\begin{array}{r|rrrrr}
-3 & 2 & 6 & 5 & 14 & -3 \\
   &   & -6 & 0 & -15 & 3 \\
\hline
   & 2 & 0 & 5 & -1 & 0
\end{array}
$$

$2x^4 + 6x^3 + 5x^2 + 14x - 3$
$\quad = (x + 3)(2x^3 + 5x - 1)$

15. $P(x) = x^2 - 4x + 5$ has real
     coefficients and $2 - i$ is a
     complex zero. By the theorem
     about complex zeros of polynomials
     with real coefficients, $2 + i$
     (complex conjugate of $2 - i$) is
     the other zero.

17. $2 + 3i$ is a zero of the polynomial
     $P(x) = 2x^3 - 7x^2 + 22x + 13$
     whose coefficients are real. Thus
     $2 - 3i$ is another zero.

19.
$$
\begin{array}{r|rrrr}
4 & 1 & -1 & -9 & -12 \\
  &   & 4 & 12 & 12 \\
\hline
  & 1 & 3 & 3 & 0
\end{array}
$$

$x^3 - x^2 - 9x - 12$
$\quad = (x - 4)(x^2 + 3x + 3)$
$x^2 + 3x + 3 = 0$

$$x = \frac{-3\pm\sqrt{9 - 4(3)}}{2} = \frac{-3\pm\sqrt{-3}}{2}$$
$$ = \frac{-3\pm i\sqrt{3}}{2}$$

The other roots are

$$\frac{-3 + i\sqrt{3}}{2} \text{ and } \frac{-3 - i\sqrt{3}}{2}$$

21.
$$
\begin{array}{r|rrrrr}
2i & 1 & 0 & 3 & 0 & -4 \\
   &   & 2i & -4 & -2i & 4 \\
\hline
2i & 1 & 2i & -1 & -2i & 0 \\
   &   & -2i & 0 & 2i & 0 \\
\hline
   & 1 & 0 & -1 & 0 & 0
\end{array}
$$

$x^4 + 3x^2 - 4$
$\quad = (x - 2i)(x + 2i)(x^2 - 1)$
$\quad = (x - 2i)(x + 2i)(x + 1)(x - 1)$

The other roots are $-2i$, 1,
and -1.

23.
$$
\begin{array}{r|rrrrr}
3 & 1 & -5 & 4 & 3 & 9 \\
  &   & 3 & -6 & -6 & -9 \\
\hline
3 & 1 & -2 & -2 & -3 & 0 \\
  &   & 3 & 3 & 3 & \\
\hline
  & 1 & 1 & 1 & 0 &
\end{array}
$$

$x^4 - 5x^3 + 4x^2 + 3x + 9$
$\quad = (x - 3)^2(x^2 + x + 1)$
$x^2 + x + 1 = 0$

$$x = \frac{-1\pm\sqrt{1 - 4}}{2} = \frac{-1\pm\sqrt{-3}}{2}$$
$$ = \frac{-1\pm i\sqrt{3}}{2}$$

The other roots are

$$\frac{-1 + i\sqrt{3}}{2} \text{ and } \frac{-1 - i\sqrt{3}}{2}$$

25. The zeros are 5 with multiplicity
     one and -2/3 with multiplicity
     one.

27. The zeros are -3/2 with
     multiplicity two, 1 with
     multiplicity three, and 4/3 with
     multiplicity one.

29. $x^4 + 2x^3 + 2x^2 = 0$
$x^2(x^2 + 2x + 2) = 0$
$x^2 = 0 \quad x^2 + 2x + 2 = 0$

$$x = \frac{-2 \pm \sqrt{-4}}{2}$$

$$= \frac{-2 \pm 2i}{2}$$

$$= -1 \pm i$$

The zeros are $-1 + i$, $-1 - i$ with multiplicity one, and 0 with multiplicity two.

31. $(x^2 - 2x + 4)^2 = 0$
$x^2 - 2x + 4 = 0$

$$x = \frac{2 \pm \sqrt{4 - 4(4)}}{2}$$

$$= \frac{2 \pm 2i\sqrt{3}}{2}$$

$$= 1 \pm i\sqrt{3}$$

The zeros are $1 + i\sqrt{3}$ and $1 - i\sqrt{3}$ both with multiplicity two.

33. $P(x) = a(x + 2)(x - 2)(x - 3)$
$P(1) = a(1 + 2)(1 - 2)(1 - 3) = 6$
$a(3)(-1)(-2) = 6$
$6a = 6$
$a = 1$
$P(x) = (x + 2)(x - 2)(x - 3)$
$= (x^2 - 4)(x - 3)$
$= x^3 - 3x^2 - 4x + 12$

35. $P(x) = a(x + 1)(x - 2)(x - 4)$
$P(0) = a(0 + 1)(0 - 2)(0 - 4) = 8$
$a(1)(-2)(-4) = 8$
$8a = 8$
$a = 1$
$P(x) = (x + 1)(x - 2)(x - 4)$
$= (x^2 - x - 2)(x - 4)$
$= x^3 - 4x^2 - x^2 + 4x - 2x + 8$
$= x^3 - 5x^2 + 2x + 8$

37. $P(x) = (x + 1)^2(x - 3)^2$
$= (x^2 + 2x + 1)(x^2 - 6x + 9)$
$= x^4 - 6x^3 + 9x^2 + 2x^3 - 12x^2$
$\quad + 18x + x^2 - 6x + 9$
$= x^4 - 4x^3 - 2x^2 + 12x + 9$

39. $P(x) = a(x-(1 + 2i))(x -(1 - 2i)) \times$
$\quad (x - 3)$
$P(0) = a[(0 - 1) - 2i] \times$
$\quad [(0 - 1) + 2i](0 - 3)$
$= -15$
$a(-1 - 2i)(-1 + 2i)(-3) = 15$
$a(1 - 4i^2)(-3) = 15$
$a(5)(-3) = 15$
$-15a = 15$
$a = -1$

$P(x) = -1[(x - 1) - 2i][(x - 1)$
$\quad + 2i](x - 3)$
$= -1[(x - 1)^2 - 4i^2](x - 3)$
$= -1(x^2 - 2x + 1 + 4)(x - 3)$
$= -1(x^2 - 2x + 5)(x - 3)$
$= -1(x^3 - 3x^2 - 2x^2 + 6x$
$\quad + 5x - 15)$
$= -1(x^3 - 5x^2 + 11x - 15)$
$= -x^3 + 5x^2 - 11x + 15$

Exercises 11.4

1. $x^3 + x^2 - 4x - 4$ possible rational roots ± 1, ± 2, ± 4
$x^2(x + 1) - 4(x + 1)$
$= (x + 1)(x^2 - 4)$
$= (x + 1)(x + 2)(x - 2)$

Roots are 2, -2, and -1.

3. $x^3 - 3x^2 - 4x + 12$ possible rational roots ± 1, ± 2, ± 3, ± 4, ± 6, ± 12
$x^2(x - 3) - 4(x - 3)$
$= (x - 3)(x^2 - 4)$
$= (x - 3)(x + 2)(x - 2)$

Roots are 2, 3, and -2.

5. $x^3 - x = x(x^2 - 1)$
    $$= x(x + 1)(x - 1)$$

    Roots are 0, 1, and -1.

7. $x^3 + x^2 + x + 2$ possible
        rational roots $\pm 1, \pm 2$
    $P(1) = 1 + 1 + 1 + 2 = 5$
    $P(-1) = -1 + 1 - 1 + 2 = 1$
    $P(2) = 8 + 4 + 2 + 2 = 16$
    $P(-2) = -8 + 4 - 2 + 2 = -4$

    The polynomial has no rational roots.

9. $2x^3 + x^2 - 2x - 1$ possible
        rational roots $\pm 1, \pm 1/2$
    $x^2(2x + 1) - 1(2x + 1)$
    $(2x + 1)(x^2 - 1)$
    $(2x + 1)(x + 1)(x - 1)$

    Roots are 1, -1, and -1/2.

11. $4x^3 - x = x(4x^2 - 1)$
    $$= x(2x + 1)(2x - 1)$$

    Roots are 0, 1/2, -1/2.

13. $x^3 - 4x^2 + 4x - 4$ possible roots
        $\pm 1, \pm 2, \pm 4$
    $P(1) = 1 - 4 + 4 - 4 = -3$
    $P(-1) = -1 - 4 - 4 - 4 = -13$
    $P(2) = 8 - 16 + 8 - 4 = -4$
    $P(-2) = -8 - 16 - 8 - 4 = -36$
    $P(4) = 64 - 64 + 16 - 4 = 12$
    $P(-4) = -64 - 64 - 16 - 4 = -148$

    The polynomial has no rational roots.

15. $x^3 - x = x(x^2 - 1)$
    $$= x(x + 1)(x - 1)$$

17. $\dfrac{1}{2}x^3 - 2x = \dfrac{1}{2}x(x^2 - 4)$

    $$= \dfrac{1}{2}x(x - 2)(x + 2)$$

19. $x^3 + x^2 - 4x - 4 = x^2(x + 1)$
    $$- 4(x + 1)$$
    $$= (x + 1)(x^2 - 4)$$
    $$= (x + 1)(x + 2)(x - 2)$$

21. $x^3 + x^2 - 6x = x(x^2 + x - 6)$
    $$= x(x + 3)(x - 2)$$

23. $x^4 - 16 = (x^2 + 4)(x^2 - 4)$
    $$= (x^2 + 4)(x + 2)(x - 2)$$
    $$= (x + 2i)(x - 2i)(x + 2)(x - 2)$$

25. $x^3 - 4x^2 + x - 4 = x^2(x - 4)$
    $$+ 1(x - 4)$$
    $$= (x - 4)(x^2 + 1)$$
    $$= (x - 4)(x + i)(x - i)$$

27. $2x^3 - 3x^2 + 4x - 6$
    $$= x^2(2x - 3) + 2(2x - 3)$$
    $$= (2x - 3)(x^2 + 2)$$
    $$= (2x - 3)(x + i\sqrt{2})(x - i\sqrt{2})$$

29. $\frac{3}{4}x^4 - \frac{3}{2}x^3 - \frac{3}{2}x^2 + 6x - 6$

$$= \frac{3}{4}(x^4 - 2x^3 - 2x^2 + 8x - 8)$$

```
 2 | 1  -2  -2   8  -8
   |     2   0  -4   8
-2 | 1   0  -2   4   0
   |    -2   4  -4
     1  -2   2   0
```

$\frac{3}{4}(x^4 - 2x^3 - 2x^2 + 8x - 8)$

$$= \frac{3}{4}(x + 2)(x - 2)(x^2 - 2x + 2)$$

$x^2 - 2x + 2 = 0$

$$x = \frac{2 \pm\sqrt{4 - 4(2)}}{2}$$

$$= \frac{2 \pm\sqrt{-4}}{2}$$

$$= \frac{2 \pm 2i}{2} = 1 \pm i$$

$\frac{3}{4}x^4 - \frac{3}{2}x^3 - \frac{3}{2}x^2 + 6x - 6$

$$= (\frac{3}{4})(x + 2)(x - 2)(x - 1 - i)\times$$

$$(x - 1 + i)$$

31. $\frac{1}{4}x^3 - \frac{5}{4}x^2 + \frac{3}{2}x - \frac{1}{2}$

$$= \frac{1}{4}(x^3 - 5x^2 + 6x - 2)$$

```
 1 | 1  -5   6  -2
   |     1  -4   2
     1  -4   2   0
```

$\frac{1}{4}(x^3 - 5x^2 + 6x - 2)$

$$= \frac{1}{4}(x - 1)(x^2 - 4x + 2)$$

$x^2 - 4x + 2 = 0$

$$x = \frac{4 \pm\sqrt{16 - 4(2)}}{2}$$

$$= \frac{4 \pm\sqrt{8}}{2}$$

$$= \frac{4 \pm 2\sqrt{2}}{2} = 2\pm\sqrt{2}$$

$\frac{1}{4}(x^3 - 5x^2 + 6x - 2)$

$$= (\frac{1}{4})(x - 1)(x - 2 - \sqrt{2})(x - 2 + \sqrt{2})$$

33. The possible rational roots of $x^2 - 2 = 0$ are ± 1, ± 2. However

$$x^2 - 2 = 0$$
$$x^2 = 2$$
$$x = \pm\sqrt{2},$$

That is the given polynomial has two irrational roots.

35. The possible rational roots of $2x^2 - x + 3$ are ± 1, ± 3, $\pm 1/2$, $\pm 3/2$. Since the discriminant $\Delta = (-1)^2 - 4(2)(3) = -23 < 0$, it follows that the given polynomial has <u>no</u> real roots. The complex roots are

$$\frac{1 \pm i\sqrt{23}}{4}.$$

37. Possible rational roots of
$x^3 - 3x^2 + 4x - 6$ are: $\pm 1, \pm 2, \pm 3,$
$\pm 6.$ Since

$P(1) = 1 - 3 + 4 - 6 = -4$
$P(-1) = -1 - 3 - 4 - 6 = -14$
$P(2) = 8 - 12 + 8 - 6 = -2$
$P(-2) = -8 - 12 - 8 - 6 = -34$
$P(3) = 27 - 27 + 12 - 6 = 6$
$P(-3) = -27 - 27 - 12 - 6 = -72$
$P(6) = 216 - 108 + 24 - 6 = 126$
$P(-6) = -216 - 108 - 24 - 6 = -354$

are all different from zero, it
follows that the given polynomial
has no rational root.

39. Possible rational roots of
$x^4 + x^3 - x^2 - 2x - 2$ are: $\pm 1,$
$\pm 2.$

$P(1) = 1 + 1 - 1 - 2 - 2 = -3$
$P(-1) = 1 - 1 - 1 + 2 - 2 = -1$
$P(2) = 16 + 8 - 4 - 4 - 2 = 14$
$P(-2) = 16 - 8 - 4 + 4 - 2 = 6$

Thus the polynomial has no
rational roots.

Exercises 11.5

1. $\dfrac{4x - 14}{(x - 2)(x - 4)} = \dfrac{A}{x - 2}$
   $\qquad + \dfrac{B}{x - 4}$

$4x - 14 = A(x - 4) + B(x - 2)$
$\qquad\quad = Ax - 4A + Bx - 2B$
$\qquad\quad = (A + B)x - (4A + 2B)$
$\quad A + B = 4$
$\quad 4A + 2B = 14$
$\quad\ \ 4A + 4B = 16$
$\quad -4A - 2B = -14$
$\overline{\qquad\qquad 2B = 2}$
$\qquad\qquad\ B = 1$
$\ A + 1 = 4$
$\ A = 3$

3. $\dfrac{x + 3}{(x - 4)(3x + 2)} = \dfrac{A}{x - 4}$
   $\qquad + \dfrac{B}{3x + 2}$

$x + 3 = A(3x + 2) + B(x - 4)$
$\qquad = 3Ax + 2A + Bx - 4B$
$\qquad = (3A + B)x + 2A - 4B$

$3A + B = 1$
$2A - 4B = 3$

$\quad 2A - 4B = 3$
$\quad 12A + 4B = 4$
$\overline{\quad 14A \qquad = 7}$

$A = \dfrac{1}{2}$

$B = -\dfrac{1}{2}$

5. $\dfrac{5x^2 + 9x + 3}{x(x + 1)^2} = \dfrac{A}{x} + \dfrac{B}{x + 1}$
   $\qquad\qquad + \dfrac{C}{(x + 1)^2}$

$5x^2 + 9x + 3 = A(x + 1)^2$
$\qquad\qquad\qquad + Bx(x + 1) + Cx$
$\quad = Ax^2 + 2Ax + A + Bx^2 + Bx + Cx$
$\quad = (A + B)x^2 + (2A + B + C)x + A$

$A = 3$
$A + B = 5$
$3 + B = 5$
$\quad\ B = 2$

$2A + B + C = 9$
$6 + 2 + C = 9$
$\qquad\quad C = 1$

7. $\dfrac{2x + 3}{(x + 1)(x^2 + x + 1)} = \dfrac{A}{x + 1}$

$\qquad\qquad\qquad\quad + \dfrac{Bx + C}{x^2 + x + 1}$

$2x + 3 = A(x^2 + x + 1)$
$\qquad\quad + (Bx + C)(x + 1)$
$\qquad = Ax^2 + Ax + A + Bx^2 + Bx$
$\qquad\quad + Cx + C$
$\qquad = (A + B)x^2 + (A + B + C)x$
$\qquad\quad + (A + C)$

$A + B = 0 \qquad\quad B = -A$
$A + B + C = 2 \qquad C = 2$
$A + C = 3 \qquad\quad A = 1$

Thus, $A = 1$, $B = -1$, $C = 2$.

9. $\dfrac{2x^2 - x + 6}{(x^2 - x + 2)^2} = \dfrac{Ax + B}{x^2 - x + 2}$

$\qquad\qquad\qquad\quad + \dfrac{Cx + D}{(x^2 - x + 2)^2}$

$2x^2 - x + 6 = (Ax + B)(x^2 - x + 2)$
$\qquad\qquad\quad + Cx + D$
$\qquad\qquad = Ax^3 + (B - A)x^2 +$
$\qquad\qquad\quad (2A - B + C)x +$
$\qquad\qquad\quad (2B + D)$

$A = 0$
$B - A = 2$
$2A - B + C = -1$
$2B + D = 6$

$A = 0$, $B = 2$, $C = 1$, $D = 2$

11. $\dfrac{x - 8}{x^2 - x - 12} = \dfrac{x - 8}{(x + 3)(x - 4)}$

$\qquad\qquad\quad = \dfrac{A}{x + 3} + \dfrac{B}{x - 4}$

$x - 8 = A(x - 4) + B(x + 3)$
$\qquad = (A + B)x - 4A + 3B$

$A + B = 1$
$-4A + 3B = -8$

$\quad 4A + 4B = 4$
$-4A + 3B = -8$
$\overline{\qquad 7B = -4}$

$B = -4/7$
$A = 11/7$

$\dfrac{x - 8}{x^2 - x - 12} = \dfrac{11}{7(x + 3)}$

$\qquad\qquad\qquad - \dfrac{4}{7(x - 4)}$

13. $\dfrac{-7x + 5}{2x^2 + 5x - 12} = \dfrac{-7x + 5}{(2x - 3)(x + 4)}$

$\qquad\quad = \dfrac{A}{2x - 3} + \dfrac{B}{x + 4}$

$-7x + 5 = A(x + 4) + B(2x - 3)$
$\qquad\qquad = Ax + 4A + 2Bx - 3B$
$\qquad\qquad = (A + 2B)x + 4A - 3B$

$\quad A + 2B = -7$
$\quad 4A - 3B = 5$
$-4A - 8B = 28$
$\quad 4A - 3B = 5$
$\overline{\quad -11B = 33}$
$\qquad B = -3$
$\quad A - 6 = -7$
$\qquad A = -1$

$\dfrac{-7x + 5}{2x^2 + 5x - 12} = \dfrac{-1}{2x - 3} + \dfrac{-3}{x + 4}$

15. $\dfrac{9x^2 + 9x - 12}{2x^3 - x^2 - 6x} = \dfrac{9x^2 + 9x - 12}{x(2x + 3)(x - 2)}$

$\qquad = \dfrac{A}{x} + \dfrac{B}{2x + 3} + \dfrac{C}{x - 2}$

$9x^2 + 9x - 12 = A(2x + 3)(x - 2)$
$\qquad\qquad\quad + Bx(x - 2) + Cx(2x + 3)$
$\qquad\quad = (2C + 2A + B)x^2$
$\qquad\qquad + (-A - 2B + 3C)x - 6A$
$-6A = -12$
$-A - 2B + 3C = 9$
$2C + 2A + B = 9$

$A = 2$, $B = -1$, $C = 3$

$\dfrac{9x^2 + 9x - 12}{2x^3 - x^2 - 6x} = \dfrac{2}{x} - \dfrac{1}{2x + 3}$

$\qquad\qquad\qquad + \dfrac{3}{x - 2}$

17. $\dfrac{4x^2 + x - 2}{x^3 - x^2} = \dfrac{4x^2 + x - 2}{x^2(x - 1)} = \dfrac{A}{x}$

$$+ \dfrac{B}{x^2} + \dfrac{C}{x - 1}$$

$4x^2 + x - 2$
$\quad = Ax(x - 1) + B(x - 1) + Cx^2$
$\quad = (A + C)x^2 + (B - A)x - B$

$\quad -B = -2$
$B - A = 1$
$A + C = 4$

$A = 1,\ B = 2,\ C = 3$

$$\dfrac{4x^2 + x - 2}{x^3 - x^2} = \dfrac{1}{x} + \dfrac{2}{x^2} + \dfrac{3}{x - 1}$$

19. $\dfrac{2x^3 + 5x^2 + 8x + 1}{(x^2 + 2x + 2)^2} = \dfrac{Ax + B}{x^2 + 2x + 2}$

$$+ \dfrac{Cx + D}{(x^2 + 2x + 2)^2}$$

$2x^3 + 5x^2 + 8x + 1$
$= (Ax + B)(x^2 + 2x + 2) + Cx + D$
$= Ax^3 + (2A + B)x^2 + (2A + 2B + C)x$
$\quad + (2B + D)$

$\qquad\qquad A = 2$
$\qquad 2A + B = 5$
$\quad 2A + 2B + C = 8$
$\qquad\quad 2B + D = 1$

$A = 2,\ B = 1,\ C = 2,\ D = -1$

$$\dfrac{2x^3 + 5x^2 + 8x + 1}{(x^2 + 2x + 2)^2} = \dfrac{2x + 1}{x^2 + 2x + 2}$$

$$+ \dfrac{2x - 1}{(x^2 + 2x + 2)^2}$$

Review Exercises - Chapter 11

1.
$$
\begin{array}{r}
3x - 2 \\
2x^2 - 1\overline{)6x^3 - 4x^2 + 5x - 2} \\
6x^3 \qquad\quad - 3x \\
\hline
-4x^2 + 8x - 2 \\
-4x^2 \qquad + 2 \\
\hline
8x - 4
\end{array}
$$

Quotient: $3x - 2$, remainder: $8x - 4$

3.
$$
\begin{array}{r}
2x^4 + 3x^3 + 8x^2 + 16x + 36 \\
x^2 - 3x + 1\overline{)2x^6 - 3x^5 + x^4 - 5x^3 - 4x^2 + 2x - 3} \\
2x^6 - 6x^5 + 2x^4 \\
\hline
3x^5 - x^4 - 5x^3 \\
3x^5 - 9x^4 + 3x^3 \\
\hline
8x^4 - 8x^3 - 4x^2 \\
8x^4 - 24x^3 + 8x^2 \\
\hline
16x^3 - 12x^2 + 2x \\
16x^3 - 48x^2 + 16x \\
\hline
36x^2 - 14x - 3 \\
36x^2 - 108x + 36 \\
\hline
94x - 39
\end{array}
$$

Quotient: $2x^4 + 3x^3 + 8x^2 + 16x + 36$,
remainder: $94x - 39$.

5.
$$
\begin{array}{r|rrrr}
4 & 2 & 0 & 0 & -3 \\
  &   & 8 & 32 & 128 \\
\hline
  & 2 & 8 & 32 & 125
\end{array}
$$

$Q(x) = 2x^2 + 8x + 32,\ R = 125$

7.
$$
\begin{array}{r|rrrrr}
-3 & 2 & 0 & -3 & 0 & 1 \\
   &   & -6 & 18 & -45 & 135 \\
\hline
   & 2 & -6 & 15 & -45 & 136
\end{array}
$$

$Q(x) = 2x^3 - 6x^2 + 15x - 45,$
$R = 136$

9.
$$
\begin{array}{r|rrrr}
-0.4 & 0.3 & -0.2 & 0.1 & 1 \\
     &   & -0.12 & 0.128 & -0.0912 \\
\hline
     & 0.3 & -0.32 & 0.228 & 0.9088
\end{array}
$$

$Q(x) = 0.3x^2 - 0.32x + 0.228,$
$R = 0.9088$

11.
$$
\begin{array}{r|rrrr}
5 & 2 & -13 & 16 & -5 \\
  &   & 10 & -15 & 5 \\
\hline
  & 2 & -3 & 1 & 0
\end{array}
$$

$2x^3 - 13x^2 + 16x - 5$
$\quad = (x - 5)(2x^2 - 3x + 1)$
$\quad = (x - 5)(2x - 1)(x - 1)$

13.
$$\begin{array}{r|rrrr} 1/2 & 2 & 3 & -8 & 3 \\ & & 1 & 2 & -3 \\ \hline & 2 & 4 & -6 & 0 \end{array}$$

$2x^3 + 3x^2 - 8x + 3$
$\quad = (x - 1/2)(2x^2 + 4x - 6)$
$\quad = (2x - 1)(x^2 + 2x - 3)$
$\quad = (2x - 1)(x + 3)(x - 1)$

15.
$$\begin{array}{r|rrrr} -2 & 0.5 & 0.8 & -0.7 & -0.6 \\ & & -1 & 0.4 & 0.6 \\ \hline & 0.5 & -0.2 & -0.3 & 0 \end{array}$$

$0.5x^3 + 0.8x^2 - 0.7x - 0.6$
$\quad = (x + 2)(0.5x^2 - 0.2x - 0.3)$
$\quad = 0.1(x + 2)(5x^2 - 2x - 3)$
$\quad = 0.1(x + 2)(5x + 3)(x - 1)$

17.
$$\begin{array}{r|rrr} 2/5 & 5 & 13 & -6 \\ & & 2 & 6 \\ \hline & 5 & 15 & 0 \end{array}$$

$5x^2 + 13x - 6 = (x - 2/5)(5x + 15)$
$\qquad\qquad\qquad = 5(x - (2/5))(x + 3)$
$\qquad\qquad\qquad = (5x - 2)(x + 3)$

19.
$$\begin{array}{r|rrrrr} 1 - 2i & 1 & -2 & 4 & 2 & -5 \\ & & 1 - 2i & -5 & -1 + 2i & 5 \\ \hline & 1 & -1 - 2i & -1 & 1 + 2i & 0 \end{array}$$

$x^4 - 2x^3 + 4x^2 + 2x - 5$
$\quad = (x - 1 + 2i)(x - 1 - 2i) \times$
$\qquad (x - 1)(x + 1)$

21. $P(x) = (x + 1)(x - [3 - 2i]) \times$
$\qquad\qquad (x - [3 + 2i])$
$\quad = (x + 1)[(x - 3) + 2i] \times$
$\qquad [(x - 3) - 2i]$
$\quad = (x + 1)[(x - 3)^2 - 4i^2]$
$\quad = (x + 1)(x^2 - 6x + 9 + 4)$
$\quad = (x + 1)(x^2 - 6x + 13)$
$\quad = x^3 - 6x^2 + 13x + x^2 - 6x$
$\qquad + 13$
$P(x) = x^3 - 5x^2 + 7x + 13$

23.
$$\begin{array}{r|rrrrr} 1 & 1 & -6 & 12 & -10 & 3 \\ & & 1 & -5 & 7 & -3 \\ \hline 1 & 1 & -5 & 7 & -3 & 0 \\ & & 1 & -4 & 3 & \\ \hline & 1 & -4 & 3 & 0 & \end{array}$$

$x^4 - 6x^3 + 12x^2 - 10x + 3$
$\quad = (x - 1)^2(x^2 - 4x + 3)$
$\quad = (x - 1)^2(x - 3)(x - 1)$
$\quad = (x - 1)^3(x - 3)$

25. $P(x) = a(x + 1)(x + 2)(x + 3)$
$P(1) = a(2)(3)(4) = 12$
$24a = 12$
$\quad a = 1/2$
$P(x) = 1/2(x + 1)(x + 2)(x + 3)$
$\quad = 1/2(x + 1)(x^2 + 5x + 6)$
$\quad = 1/2(x^3 + 5x^2 + 6x + x^2$
$\qquad + 5x + 6)$
$\quad = 1/2(x^3 + 6x^2 + 11x + 6)$

$P(x) = \dfrac{1}{2}x^3 + 3x^2 + \dfrac{11}{2}x + 3$

27. $P(x) = (x - 3)^2(x - [2 + i]) \times$
$\qquad\qquad (x - [2 - i])$
$\quad = (x^2 - 6x + 9)((x - 2)^2 - i^2)$
$\quad = (x^2 - 6x + 9) \times$
$\qquad (x^2 - 4x + 4 + 1)$
$\quad = (x^2 - 6x + 9)(x^2 - 4x + 5)$
$\quad = x^4 - 4x^3 + 5x^2 - 6x^3 + 24x^2$
$\qquad - 30x + 9x^2 - 36x + 45$
$P(x) = x^4 - 10x^3 + 38x^2 - 66x + 45$

29. Possible rational roots:
$\pm 1, \pm 3, \pm 9$

$$\begin{array}{r|rrrr} -1 & 1 & -5 & 3 & 9 \\ & & -1 & 6 & -9 \\ \hline & 1 & -6 & 9 & 0 \end{array}$$

$x^3 - 5x^2 + 3x + 9$
$\quad = (x + 1)(x^2 - 6x + 9)$
$\quad = (x + 1)(x - 3)(x - 3)$

Roots are -1 and 3 (multiplicity 2)

31. Possible rational roots: ± 1, ± 2, ± 4, $\pm 1/2$

$$
\begin{array}{r|rrrr}
2 & 2 & -7 & 4 & 4 \\
  &   & 4 & -6 & -4 \\
\hline
  & 2 & -3 & -2 & 0
\end{array}
$$

$2x^3 - 7x^2 + 4x + 4$
$\quad = (x - 2)(2x^2 - 3x - 2)$
$\quad = (x - 2)(2x + 1)(x - 2)$

Roots are 2 (multiplicity 2) and $-1/2$.

33. The possible rational roots of $2x^2 - 3$ are ± 1, ± 3, $\pm 1/2$, $\pm 3/2$. Since

$2x^2 - 3 = 0$
$\quad 2x^2 = 3$
$\quad\quad x = \pm\sqrt{6}/2$

it follows that the polynomial does not have rational roots.

35. The possible rational roots of $x^2 - x - 3$ are ± 1 and ± 3. Now, the discriminant $\Delta = 1 - 4(-3) = 13$ is not a perfect square. You may check that the given polynomial has two irrational roots: $(1 \pm \sqrt{13})/2$.

37. $\dfrac{2x^2 - x}{(x - 1)(x^2 - x + 1)} = \dfrac{A}{x - 1}$
$\quad\quad + \dfrac{Bx + C}{x^2 - x + 1}$

$2x^2 - x = A(x^2 - x + 1)$
$\quad\quad\quad\quad + (Bx + C)(x - 1)$
$\quad = (A + B)x^2 + (-A - B + C)x$
$\quad\quad\quad\quad + (A - C)$
$\quad\quad A - C = 0$
$-A - B + C = -1$
$\quad\quad A + B = 2$

$A = 1$, $B = 1$, $C = 1$

$\dfrac{2x^2 - x}{(x - 1)(x^2 - x + 1)} = \dfrac{1}{x - 1}$
$\quad\quad + \dfrac{x + 1}{x^2 - x + 1}$

39. $\dfrac{3x^2 + 7x + 1}{x(x + 1)^2} = \dfrac{A}{x} + \dfrac{B}{x + 1}$
$\quad\quad + \dfrac{C}{(x + 1)^2}$

$3x^2 + 7x + 1 = A(x + 1)^2$
$\quad\quad\quad\quad + Bx(x + 1) + Cx$
$\quad = Ax^2 + 2Ax + A + Bx^2 + Bx + Cx$
$\quad = (A + B)x^2 + (2A + B + C)x + A$
$A = 1$
$A + B = 3$
$1 + B = 3$
$B = 2$
$2A + B + C = 7$
$2 + 2 + C = 7$
$C = 3$

$\dfrac{3x^2 + 7x + 1}{x(x + 1)^2} = \dfrac{1}{x}$
$\quad\quad + \dfrac{2}{x + 1} + \dfrac{3}{(x + 1)^2}$

41. The polar form of a complex number and DeMoivre's formula is needed here. Let

$$\omega_k = r^{1/n}[\cos(\frac{\theta + 2k\pi}{n})$$

$$+ i \sin(\frac{\theta + 2k\pi}{n})]$$

where $n = 5$, $r = 2$, $\theta = 0$, and $k = 0, 1, 2, 3, 4$.

$$\omega_0 = \sqrt[5]{2}$$

$$\omega_1 = \sqrt[5]{2}(\cos\frac{2\pi}{5} + i \sin\frac{2\pi}{5})$$

$$\omega_2 = \sqrt[5]{2}(\cos\frac{4\pi}{5} + i \sin\frac{4\pi}{5})$$

$$\omega_3 = \sqrt[5]{2}(\cos\frac{6\pi}{5} + i \sin\frac{6\pi}{5})$$

$$\omega_4 = \sqrt[5]{2}(\cos\frac{8\pi}{5} + i \sin\frac{8\pi}{5})$$

43. $P(i) = i^2 - 3i(i) - 2$
    $= i^2 - 3i^2 - 2$
    $= -2i^2 - 2$
    $= 2 - 2 = 0$

$P(-i) = (-i)^2 - 3i(-i) - 2$
    $= i^2 + 3i^2 - 2$
    $= 4i^2 - 2$

    $= -4 - 2 = -6$

The theorem is true only for polynomials with real coefficients. Next,

$x^2 - 3ix - 2$
  $= (x - i)(x - 2i)$.
So the other root is $2i$.

45. Every polynomial $P(x)$ of degree n has exactly n roots (counting multiplicities). If the coefficients of $P(x)$ are real, then for every complex root, its complex conjugate is also a root. Thus if all the roots of a polynomial with real coefficients are complex, then the polynomial must have even degree. Therefore, every polynomial of odd degree with real coefficients must have at least one real root.

47. Let $P(x) = x^n + a^n$ where n is an odd number. Let us show that $-a$ is a root of $P(x)$. We have

$P(-a) = (-a)^n + a^n$
    $= -a^n + a^n$ [because n
    $= 0$ is odd]

Since $-a$ is a root of $P(x)$, it follows that $x - (-a) = x + a$ is a factor of $P(x)$.

49. If $r_1, r_2, \ldots, r_n$ are the zeros of the polynomial $P(x) = x^n + a_{n-1}x^{n-1} + \ldots + a_1x + a_0$, then we have
$x^n + a_{n-1}x^{n-1} + \ldots + a_1x + a_0$
$= (x - r_1)(x - r_2) \ldots (x - r_n)$.
Now the coefficient of x^{n-1} on the left-hand side is a_{n-1},
while the coefficient of x^{n-1} after computing the product on the right-hand side is
$(-r_1 - r_2 - \ldots - r_n)$
$= -(r_1 + r_2 + \ldots + r_n)$.
Since both coefficients must be equal, we obtain

$-(r_1 + r_2 + \ldots + r_n) = a_{n-1}$
so
$r_1 + r_2 + \ldots + r_n = -a_{n-1}$.

Chapter 11 Test

1. If $P(x) = x^3 + 2x^2 - 4x + 1$, find a) $P(1)$, b) $P(-2)$, c) $P(1)/P(-2)$,
    d) $P(-1/2)$.

2. Let $P(x) = 5x^2 - 2x + 3$ and $Q(x) = (A + B)x^2 + (B - C)x + C - 3$. Find A, B
    and C so that $P(x) = Q(x)$.

3. Use long division to find the quotient and remainder in the division of
    $3x^3 - x + 6$ by $x - 4$.

4. Find a if $P(x) = 4x^3 + ax^2 - 2x + 3$ and $P(-1) = 7$.

5. Perform the division $(6x^4 + 2x^3 - 5) \div (3x^2 + 2)$.

6. Write the rational expression

    $$\frac{4x^3 - 2x + 3}{x^2 + 1}$$

    as a sum of a polynomial and a proper rational expression.

7. Without performing the division, find the remainder in the following division

    $(5x^3 - 2x + 6) \div (x + 2)$.

In Problems 8 and 9, find the quotient and remainder by synthetic division.

8. $(3x^3 - 4x^2 + 5) \div (x + 3)$ 9. $(x^3 - \frac{10}{3}x^2 + 5x - \frac{4}{3}) \div (x - \frac{1}{3})$

10. Let $P(x) = 3x^4 + x^3 - 2x^2 + 1$. Apply Horner's method to find $P(2)$.

11. Find a such that $x - 3$ is a factor of $x^3 - 4x^2 - 2x + a$.

12. Find a such that the remainder of $3x^3 + 8x^2 + ax + 2$ divided by $x + 3$ is -1.

13. Verify that $4 - i$ is a zero of $x^2 - 8x + 17$. What is the other zero?

14. Verify whether or not the linear polynomial $x + 2$ is a factor of the
    polynomial $x^4 - x^3 - 7x^2 + x + 6$. If possible, find the other factors.

15. Show that $2 + i$ is a zero of the polynomial $x^3 - 5x^2 + 9x - 5$ and find the
    other zeros.

16. Show that 2 is a double root of $x^4 - 6x^3 + 9x^2 + 4x - 12$ and find the other
    roots.

In Problems 17-18, find a polynomial $P(x)$ of lowest possible degree with indicated
zeros and satisfying the given condition.

17. -1, 0, 3; $P(1) = -8$

18. $1 + i$, $1 - i$, -2 (multiplicity 2); leading coefficient 1.

In Problems 19 and 20, use the test for rational zeros to find all roots of the given polynomials.

19. $2x^3 + 13x^2 + 17x - 12$ 20. $x^4 + x^3 - x^2 + x - 2$

In Problems 21 and 22, show that the given equations have no rational roots.

21. $x^2 - 2x + 17 = 0$ 22. $2x^3 + x^2 - 6x + 1 = 0$

In Problems 23 and 24, find all rational roots and factor completely each polynomial.

23. $x^3 - 4x^2 + x - 4$ 24. $x^3 + 6x + 20$

In Problems 25-26, find the constants A, B, C for which the left-hand side is equal to the right-hand side.

25. $\dfrac{3x - 5}{(x + 3)(2x - 1)} = \dfrac{A}{x + 3} + \dfrac{B}{2x - 1}$

26. $\dfrac{-2x^2 + 8x - 3}{(x + 2)(x - 1)^2} = \dfrac{A}{x - 1} + \dfrac{B}{x + 2} + \dfrac{C}{(x - 1)^2}$

In Problems 27-30, decompose each rational expression into partial fractions.

27. $\dfrac{2x + 3}{x^2 + 3x - 10}$ 28. $\dfrac{5x^2 + x - 1}{x^3 + x^2}$

29. $\dfrac{3x^2 + 5}{x^3 + x}$ 30. $\dfrac{6x^3 + 3x - 1}{(2x^2 + 1)^2}$

Chapter 12
Mathematical Induction, Sequences, Series, and Probability

Exercises 12.1

1. $P(n)$ is the statement "2 + 4 + 6 + ... + 2n = n(n + 1)". If $n = 1$, then 2 = 1(1 + 1), so P(1) is true. Next, we have to show that whenever we assume that $P(k)$ is true for a certain k — that is, whenever we assume that 2 + 4 + 6 + ... + 2k = k(k + 1) — we can deduce that P(k + 1) is also true. To do this, we write

   $$2 + 4 + 6 + ... + 2k + 2(k + 1)$$
   $$= k(k + 1) + 2(k + 1)$$
   [Using the assumption that $P(k)$ is true]
   $$= (k + 1)(k + 2)$$
   [Factoring $k + 1$]
   $$= (k + 1)[(k + 1) + 1].$$

   This last equation is the statement P_{k+1}. Thus, assuming that $P(k)$ is true, we deduced that P(k + 1) is also true, and the induction proof is complete.

3. $P(n)$ is the statement
   "$1 \cdot 2 + 2 \cdot 3 + ... + n(n + 1) =$

   $$\frac{n(n + 1)(n + 2)}{3} "$$. If $n = 1$,

   then $1 \cdot 2 = \dfrac{1(2)(\cancel{3})}{\cancel{3}}$, so P(1)

   is true. Next, we have to show that whenever

   $$1 \cdot 2 + 2 \cdot 3 + ... + k(k + 1)$$

   $$= \frac{k(k + 1)(k + 2)}{3}$$

   is true for a certain k - that is, whenever $P(k)$ is true - we can deduce that P(k + 1) is also true. To do this, we write

$1 \cdot 2 + 2 \cdot 3 + \ldots + k(k + 1)$
$\quad + (k + 1)(k + 2)$

$= \dfrac{k(k + 1)(k + 2)}{3}$

$\quad + (k + 1)(k + 2)$

$= \dfrac{k(k + 1)(k + 2) + 3(k + 1)(k + 2)}{3}$

$= \dfrac{(k + 1)(k + 2)(k + 3)}{3}$

$= \dfrac{(k + 1)[(k + 1) + 1][(k + 1) + 2]}{3}$

which is the statement P(k + 1). Thus, we have proved that P(k + 1) follows from P(k), which completes the induction proof.

In the induction proofs that follow, we leave to you the proof of the first step and show only how the statement P(k + 1) can be deduced from the statement P(k).

5. $1^2 + 2^2 + \ldots + j^2 + (j + 1)^2$

$= \dfrac{j(j + 1)(2j + 1)}{6} + (j + 1)^2$

$= \dfrac{j(j + 1)(2j + 1) + 6(j + 1)^2}{6}$

$= \dfrac{(j + 1)[j(2j + 1) + 6(j + 1)]}{6}$

$= \dfrac{(j + 1)(2j^2 + 7j + 6)}{6}$

$= \dfrac{(j + 1)(j + 2)(2j + 3)}{6}$

$= \dfrac{(j + 1)[(j + 1) + 1][2(j + 1) + 1]}{6}$

7. $1^3 + 3^3 + \ldots + (2j - 1)^3 + [2(j + 1) - 1]^3$
$\quad = j^2(2j^2 - 1) + (2j + 1)^3$
$\quad = 2j^4 + 8j^3 + 11j^2 + 6j + 1$
$\quad = (j + 1)^2(2j^2 + 4j + 1)$

[Use synthetic division to obtain the last factorization]

$\quad = (j + 1)^2[2(j + 1)^2 - 1]$

9. $\dfrac{1}{2} + 1 + \dfrac{3}{2} + 2 + \ldots + \dfrac{k}{2} + \dfrac{k+1}{2}$

$= \dfrac{k(k + 1)}{4} + \dfrac{k + 1}{2}$

$= \dfrac{k(k + 1) + 2(k + 1)}{4}$

$= \dfrac{(k + 1)(k + 2)}{4}$

11. $7^{k + 1} - 3^{k + 1} = 7^k \cdot 7 - 3^k \cdot 3$
$\quad = 7^k(4 + 3) - 3^k \cdot 3$
$\quad = 7^k \cdot 4 + 7^k \cdot 3 - 3^k \cdot 3$
$\quad = 7^k \cdot 4 + (7^k - 3^k) \cdot 3$

The first term, $7^k \cdot 4$, is divisible by 4. The second one $(7^k - 3^k) \cdot 3$ is also divisible by 4, because we are assuming that $7^k - 3^k$ is divisible by 4.

13. $4^{2(n+1)} - 1 = 4^{2n} \cdot 4^2 - 1$
$= 4^{2n} \cdot 16 - 1$
$= 4^{2n}(15 + 1) - 1$
$= 4^{2n} \cdot 15 + (4^{2n} - 1)$

where the two terms are divisible by 5.

15. $(m + 1)^3 - (m + 1) + 3$
$= m^3 + 3m^2 + 3m + 1 - m - 1 + 3$
$= (m^3 - m + 3) + 3m(m + 1)$

The term $3m(m + 1)$ is clearly a multiple of 3, while the term $m^3 - m + 3$ is also a multiple of 3 by the induction assumption.

17. $3^{2(j+1)} - 1 = 3^{2j} \cdot 3^2 - 1$
$= 3^{2j}(8 + 1) - 1$
$= 3^{2j} \cdot 8 + (3^{2j} - 1)$

where both terms are divisible by 8.

19. $x^{n+1} - a^{n+1}$
$= (x^n - a^n)x + a^n(x - a)$

The term $a^n(x - a)$ is divisible by $x - a$. Since we are assuming than $P(n)$ is true, it follows that $(x^n - a^n)x$ is also divisible by $x - a$.

21. $\sum_{j=1}^{6} (j + 1) = 2 + 3 + 4 + 5 + 6 + 7$

$= 27$

23. $\sum_{j=1}^{5} (2j - 1) = 1 + 3 + 5 + 7 + 9$

$= 25$

25. $\sum_{j=1}^{4} (2j+1)/2 = \dfrac{3}{2} + \dfrac{5}{2} + \dfrac{7}{2} + \dfrac{9}{2}$

$= \dfrac{24}{2} = 12$

27. $\sum_{j=1}^{6} j^2 = 1 + 4 + 9 + 16 + 25 + 36$

$= 91$

29. $\sum_{j=3}^{5} (j^2 - 2j) = \sum_{j=3}^{5} j(j - 2)$

$= 3(1) + 4(2) + 5(3) = 26$

31. Assuming that $2^k > k$ for a certain k, we obtain

$2^{k+1} = 2 \cdot 2^k > 2k.$

Now $2k = k + k \geq k + 1$, thus $2^{k+1} > k + 1.$

33. $2^{(k+1)+3} = 2^{1+(k+3)}$
$= 2 \cdot 2^{k+3}$
$< 2(k + 3)!$

using the assumption that the statement $P(k)$ is true. Now $2 < k + 4$ for all natural number k. Thus

$2^{(k+1)+3} < (k + 4)(k + 3)!$
$= (k + 4)! = [(k + 1) + 3]!$

which is the statement $P(k + 1)$.

35. $(k + 1)^2 = k^2 + 2k + 1$
$> 2k + (2k + 1),$

assuming that $k^2 > 2k$. Now $2k + 1 > 2$, thus

$(k + 1)^2 > 2k + 2 = 2(k + 1)$

which is the statement $P(k + 1)$.

37. $2^{(k+1)-1} = 2^k = 2 \cdot 2^{k-1}$

$< 2k!$
$< (k + 1)k! = (k + 1)!$

39. Let $a > 1$ and let $P(n)$ be the statement "$a^n > 1$ for all natural numbers n". If $n = 1$, then $a^1 = a > 1$, so $P(1)$ is true. Assume that $P(k)$ is true for a certain k. We have

$$a^{k+1} = a \cdot a^k > 1 \cdot 1 = 1,$$

thus $P(k + 1)$ is also true.

41. We want to prove the formula

$$\sum_{j=1}^{n} (a_j + b_j)$$

$$= \sum_{j=1}^{n} a_j + \sum_{j=1}^{n} b_j$$

by mathematical induction. For $n = 1$, we have

$$\sum_{j=1}^{1} (a_j + b_j) = a_1 + b_1$$

$$= \sum_{j=1}^{1} a_j + \sum_{j=1}^{1} b_j.$$

Thus the formula is true for $n = 1$. Assuming it to be true for k, let us prove it for $k + 1$.

$$\sum_{j=1}^{k+1} (a_j + b_j)$$

$$= \sum_{j=1}^{k} (a_j + b_j) + (a_{k+1} + b_{k+1})$$

$$= \sum_{j=1}^{k} a_j + \sum_{j=1}^{k} b_j + (a_{k+1} + b_{k+1})$$

$$= \sum_{j=1}^{k} a_j + a_{k+1} + \sum_{j=1}^{k} b_j + b_{k+1}$$

$$= \sum_{j=1}^{k+1} a_j + \sum_{j=1}^{k+1} b_j.$$

Thus the formula is true for $k+1$.

43. $(ab)^{k+1} = (ab)(ab)^k$

$$= (ab)a^k b^k$$
$$= (a \cdot a^k)(b \cdot b^k)$$
$$= a^{k+1} b^{k+1}$$

45. Let a be a real number. We want to prove that $(a^n)^m = a^{nm}$ for every pair of natural numbers n and m. We fix n and do induction on m. For $m = 1$,

$$(a^n)^1 = a^n = a^{n \cdot 1}$$

and the formula is true for $m = 1$. Assume it to be true for a certain k and let us prove it for $k + 1$:

$$(a^n)^{k+1} = (a^n)^k (a^n)$$
$$= a^{nk} a^n = a^{nk+n}$$
$$= a^{n(k+1)}.$$

47. $\overline{z_1 \cdot z_2 \cdots z_k \cdot z_{k+1}}$

$$= \overline{z_1 \cdot z_2 \cdots z_k} \cdot \overline{z_{k+1}}$$

$$= \overline{z_1} \cdot \overline{z_2} \cdots \overline{z_k} \cdot \overline{z_{k+1}}$$

49. $\sin[\alpha + (k+1)\pi] = \sin[(\alpha+k\pi) + \pi]$
$$= \sin(\alpha + k\pi)\cos \pi$$
$$\quad + \cos(\alpha + k\pi)\sin \pi$$

$$= (-1)^k \sin \alpha(-1)$$
$$\quad + \cos(\alpha + k\pi)(0)$$

$$= (-1)^{k+1} \sin \alpha.$$

Exercises 12.2

1. $\binom{9}{6} = \dfrac{9!}{6!3!} = \dfrac{9 \cdot 8 \cdot 7}{3 \cdot 2 \cdot 1} = 84$

3. $\binom{12}{9} = \dfrac{12!}{9!3!} = \dfrac{12 \cdot 11 \cdot 10}{3 \cdot 2 \cdot 1} = 220$

5. $(a + b)^4 = a^4 + 4a^3 b + 6a^2 b^2 + 4ab^3 + b^4$

7. $(x + y)^6 = x^6 + 6x^5 y + 15x^4 y^2 + 20x^3 y^3 + 15x^2 y^4 + 6xy^5 + y^6$

9. $(2a - b)^5 = \binom{5}{0}(2a)^5 + \binom{5}{1}(2a^4)(-b)$

$+ \binom{5}{2}(2a)^3(-b)^2 + \binom{5}{3}(2a)^2(-b)^3$

$+ \binom{5}{4}(2a)(-b)^4 + \binom{5}{5}(-b)^5$

$= 32a^5 - 80a^4b + 80a^3b^2 - 40a^2b^3$
$+ 10ab^4 - b^5$

11. $(x - 2)^5 = \binom{5}{0}x^5 + \binom{5}{1}x^4(-2)$

$+ \binom{5}{2}x^3(-2)^2 + \binom{5}{3}x^2(-2)^3$

$+ \binom{5}{4}x(-2)^4 + \binom{5}{5}(-2)^5$

$= x^5 - 10x^4 + 40x^3 - 80x^2$
$+ 80x - 32$

13. $(a - \dfrac{1}{a})^6 = \binom{6}{0}a^6 + \binom{6}{1}a^5\left(-\dfrac{1}{a}\right)$

$+ \binom{6}{2}a^4\left(-\dfrac{1}{a}\right)^2 + \binom{6}{3}a^3\left(-\dfrac{1}{a}\right)^3$

$+ \binom{6}{4}a^2\left(-\dfrac{1}{a}\right)^4 + \binom{6}{5}a\left(-\dfrac{1}{a}\right)^5$

$+ \binom{6}{6}\left(-\dfrac{1}{a}\right)^6$

$= a^6 - 6a^4 + 15a^2 - 20$

$+ \dfrac{15}{a^2} - \dfrac{6}{a^4} + \dfrac{1}{a^6}$

15. $\binom{7}{5-1}(2x)^{7-5+1}(-a)^{5-1}$

$= \binom{7}{4}(2x)^3(-a)^4 = 280a^4x^3$

17. $\binom{10}{7}\left(\dfrac{u}{2}\right)^3(-2)^7 = 120\,\dfrac{u^3}{8}(-128)$

$= -1920u^3$

19. The term independent of y is the middle (5th term)

$\binom{8}{4}(2y)^4\left(-\dfrac{1}{2y}\right)^4 = 70$

21. The middle (5th) term is

$\binom{8}{4}(a^{1/4})^4(b^{1/4})^4 = 70ab$

23. The two middle terms are

$\binom{7}{3}(xy)^4(-a)^3 + \binom{7}{4}(xy)^3(-a)^4$

$= -35x^4y^4a^3 + 35x^3y^3a^4$

25. The term containing b^6 is

$\binom{5}{3}a^2(-3b^2)^3 = 10a^2(-27b^6)$

$= -270a^2b^6$

27. The first five terms in the binomial expansion of $(1 + 0.01)^8$ are

$\binom{8}{0}(1)^8 + \binom{8}{1}(1)^7(0.01)$

$+ \binom{8}{2}(1)^6(0.01)^2$

$+ \binom{8}{3}(1)^5(0.01)^3$

$+ \binom{8}{4}(1)^4(0.01)^4$

$= 1 + 0.08 + 0.0028 + 0.000056$
$+ 0.00000070$

$= 1.08285670$

$(1.01)^8 \simeq 1.0828567$

29. $(1.99)^6 = (2 - 0.01)^6$

$\simeq 2^6 + \binom{6}{1}2^5(-0.01)$

$+ \binom{6}{2}2^4(-0.01)^2 + \binom{6}{3}2^3(-0.01)^3$

$= 64 - 1.92 + 0.024 - 0.00016$

$= 62.10384$

$\simeq 62.104$

Note. The better approximation 62.10384 is indeed <u>exact</u> to five decimal places, because it can be shown that the error we are making in using this approximation is <u>less than</u> the next term of the expansion

$\binom{6}{4}2^2(-0.01)^4 = 6 \times 10^{-7} < 10^{-6}.$

31. $(1.02)^{10} = (1 + 0.02)^{10}$

$\simeq 1 + \binom{10}{1}(0.02) + \binom{10}{2}(0.02)^2$

$+ \binom{10}{3}(0.02)^3$

$= 1 + 0.2 + 0.018 + 0.00096$

$= 1.21896$

$\simeq 1.219$

33. $12n! = (n + 2)!$
$12 = (n + 2)(n + 1)$
$12 = n^2 + 3n + 2$
$n^2 + 3n - 10 = 0$
$(n + 5)(n - 2) = 0$
$n = 2, \ n = -5$ (extraneous)

35. $\dfrac{(n + 4)!}{(n + 2)!} = 30$

$(n + 4)(n + 3) = 30$
$n^2 + 7n + 12 = 30$
$n^2 + 7n - 18 = 0$

$(n + 9)(n - 2) = 0$
$n = 2, \ n = -9$ (extraneous)

37. $(n + 1)! = 1 \cdot 2 \cdot \ldots \cdot n(n + 1)$
$\qquad\quad = n!(n + 1)$

39. $\binom{n}{0} = \dfrac{n!}{n!0!} = 1$ and

$\binom{n + 1}{0} = \dfrac{(n + 1)!}{(n + 1)!0!} = 1$

Exercises 12.3

1. $a_n = 5 - 2n$
$a_1 = 5 - 2 = 3$
$a_2 = 5 - 4 = 1$
$a_3 = 5 - 6 = -1$
$a_4 = 5 - 8 = -3$
$a_5 = 5 - 10 = -5$

3. $a_n = (-1)^n n^2$
$a_1 = (-1)(1) = -1$
$a_2 = (1)(4) = 4$
$a_3 = (-1)(9) = -9$
$a_4 = (1)(16) = 16$
$a_5 = (-1)(25) = -25$

5. $a_n = (-3)^n$
$a_1 = -3, \ a_2 = 9, \ a_3 = -27,$
$a_4 = 81, \ a_5 = -243$

7. $a_n = \left(1 + \dfrac{1}{n}\right)^n$

$a_1 = 2, \ a_2 = 9/4,$
$a_3 = 64/27, \ a_4 = 625/256$
$a_5 = 7776/3125$

9. $a_n = \dfrac{(-1)^n 2^n}{2^n - 1}$

$a_1 = -2, \ a_2 = 4/3,$
$a_3 = -8/7, \ a_4 = 16/15,$
$a_5 = -32/31$

11. $a_n = 2n, \ n \geq 1$

$2, \ 4, \ 6, \ 8, \ 10$

13. $a_n = -4n$, $n \geq 1$

    -4, -8, -12, -16, -20

15. $a_n = 3n - 1$, $n \geq 1$

    2, 5, 8, 11, 14

17. $a_1 = -5$, $a_k = 2a_{k-1} + 7$
    $a_2 = 2(-5) + 7 = -3$
    $a_3 = 2(-3) + 7 = 1$
    $a_4 = 2(1) + 7 = 9$
    $a_5 = 2(9) + 7 = 25$

19. $a_1 = 0$, $a_k = 4 - 3a_{k-1}$
    $a_2 = 4$, $a_3 = -8$, $a_4 = 28$,
    $a_5 = -80$

21. $a_1 = 3$, $a_k = 1/a^2_{k-1}$
    $a_2 = a/(3)^2 = 1/9$
    $a_3 = 1/(1/9)^2 = 81$
    $a_4 = 1/(81)^2 = 1/6561$
    $a_5 = 1/(1/6561)^2 = 43046721$

23. $a_1 = 3$, $a_k = (k-1)a_{k-1}$
    $a_2 = 3$, $a_3 = 6$, $a_4 = 18$,
    $a_5 = 72$

25. $a_1 = 2$, $a_k = \dfrac{3}{4}a_{k-1}$

    $a_2 = 3/2$, $a_3 = 9/8$,
    $a_4 = 27/32$, $a_5 = 81/128$

27. $a_1 = 1$, $a_k = a_{k-1} + 4$
    1, 5, 7, 9, 13, 17

29. $s_n = 2^{n+1} - 2$
    The general term a_n of the
    sequence whose nth partial sum is
    s_n is obtained as follows

    $\begin{aligned} a_n &= s_n - s_{n-1}, \ n \geq 2 \\ &= (2^{n+1}) - 2) - (2^n - 2) \\ &= 2^{n+1} - 2^n \\ &= 2^n(2 - 1) \\ &= 2^n \end{aligned}$

31. $s_n = n(n+1)$
    $a_n = s_n - s_{n-1}$, $n \geq 2$

$\begin{aligned} &= n(n+1) - (n-1)n \\ &= n[\not{n} + 1 - \not{n} + 1] \\ &= 2n \end{aligned}$

33. $s_n = n(n+1)/4$

    $a_n = \dfrac{n(n+1)}{4} - \dfrac{(n-1)n}{4}$, $n \geq 2$

    $= \dfrac{n[n+1-n+1]}{4}$

    $= \dfrac{n}{2}$

35. $s_n = 3^n - 1$
    $\begin{aligned} a_n &= s_n - s_{n-1}, \ n \geq 2 \\ &= (3^n - 1) - (3^{n-1} - 1) \\ &= 3^{n-1}(3 - 1) \\ &= 2(3^{n-1}) \end{aligned}$

37. $x_1 = 2$

    $x_2 = \dfrac{1}{2}(2 + \dfrac{5}{2}) = 2.25$

    $x_3 = \dfrac{1}{2}(2.25 + \dfrac{5}{2.25}) = 2.236\overline{1}$

    $x_4 = \dfrac{1}{2}(2.236\overline{1} + \dfrac{5}{2.236\overline{1}})$

    $= 2.236068$

39. $x_1 = 3.2$

    $x_2 = \dfrac{1}{2}(3.2 + \dfrac{10}{3.2}) = 3.1625$

    $x_3 = 3.16227767$

    $x_4 = 3.16227766$

Exercises 12.4

1. $d = 2$; 8, 10

3. $d = -1.5$; -2.5, -4

5. $\log 2$, $\log 4$, $\log 8$, ...

$d = \log 4 - \log 2$
$\quad = 2 \log 2 - \log 2$
$\quad = \log 2$

The next two terms are log 16 and log 32.

7. $15, 19, 23, \ldots$
$d = 4$
$a_n = 15 + 4(n - 1)$

9. $-13, -2, 9, \ldots$
$d = 11$
$a_n = -13 + 11(n - 1)$

11. $a, a + 2r, a + 4r, \ldots$
$d = 2r$
$a_n = a + 2r(n - 1)$

13. $a_1 = 4, a_2 = 7$
$d = a_2 - a_1 = 3$
$a_{15} = 4 + 14(3) = 46$

15. $a_2 = 15, a_5 = 6$
$15 = a_1 + d$
$6 = a_1 + 4d$
$d = -3, a_1 = 18$
$a_9 = 18 + 8(-3) = -6$

17. $a_1 = -10, d = 4, n = 12$

$s_{12} = \dfrac{(a_1 + a_{12}) \cdot 12}{2}$

$a_{12} = -10 + 11(4) = 34$

$s_{12} = \dfrac{(-10 + 34) \cdot 12}{2} = 144$

19. $a_1 = 1/2, d = 1/3, n = 16$

$a_{16} = \dfrac{1}{2} + 15(\dfrac{1}{3}) = 11/2$

$s_{16} = \dfrac{(\dfrac{1}{2} + \dfrac{11}{2}) \cdot 16}{2} = 48$

21. The terms of the sum

$\sum_{n=1}^{25} (2n + 3)$

form an arithmetic sequence whose first term is $a_1 = 2(1) + 3 = 5$ and whose last term is $a_{25} = 2(25) + 3 = 53$. Thus

$\sum_{n=1}^{25} (2n + 3) = \dfrac{(a_1 + a_{25}) \cdot 25}{2}$

$= \dfrac{(5 + 53) \cdot 25}{2} = 725$

23. $\sum_{n=1}^{30} (4 - 3n)$

$= \dfrac{[(4 - 3(1)) + (4 - 3(30))] \cdot 30}{2}$

$= \dfrac{(1 - 86) \cdot 30}{2} = -1275$

25. $\sum_{n=1}^{15} (\dfrac{n}{2} + 3)$

$= \dfrac{[(\dfrac{1}{2} + 3) + (\dfrac{15}{2} + 3)] \cdot 15}{2}$

$= \dfrac{(\dfrac{7}{2} + \dfrac{21}{2}) \cdot 15}{2}$

$= 105$

27. $a_1 = 22, a_n = 94, d = 2$
$94 = 22 + (n - 1)(2)$
$72 = 2(n - 1)$
$n - 1 = 36$
$n = 37$

$s_{37} = \dfrac{(22 + 94) \cdot 37}{2} = 2146$

29. $a_1 = 16, a_n = 92, d = 4$
$92 = 16 + (n - 1)4$
$n = 20$

$s_{20} = \dfrac{(16 + 92) \cdot 20}{2} = 1080$

31. $a_1 = (1)\$0.10,$
     $a_2 = (2)\$0.10, \ldots, a_{31} = (31)\$0.10$

$$s_{31} = \frac{(0.10 + 3.10) \cdot 31}{2}$$

$$= \$49.60$$

33. $a_1 = 20, a_2 = 19, \ldots, a_{20} = 1$

$$s_{20} = \frac{(20 + 1) \cdot 20}{2} = 210$$

35. Let a_1, $a_1 + d$, $a_1 + 2d$ be the numbers. Then

$$a_1 + a_1 + d + a_1 + 2d = 60$$

$$\frac{a_1 + 2d}{a_1} = 7$$

or

$$3a_1 + 3d = 60$$
$$6a_1 - 2d = 0$$

Solving this system, we get $a_1 = 5$, $d = 15$. Thus the three numbers are 5, 20, 35.

37. Let a_1, $a_2 = a_1 + d$, $a_3 = a_1 + 2d$ be the numbers. Then

$$a_1 + a_1 + d + a_1 + 2d = 33$$
$$a_1^2 = (a_1 + 2d)^2 = 274$$

or

$$a_1 + d = 11$$
$$a_1^2 = (a_1 + 2d)^2 = 274$$

From the first equation it follows that $a_1 = 11 - d$. Substituting into the second, we obtain

$$(11 - d)^2 + (11 + d)^2 = 274$$
$$2d^2 = 274 - 242 = 32$$
$$d = \pm 4$$

If $d = 4$, we obtain $a_1 = 7$, $a_2 = 11$, $a_3 = 15$.

If $d = -4$, we obtain $a_1 = 15$, $a_2 = 11$, $a_3 = 7$.

Thus, the numbers are 7, 11, 15.

39. $a_1 = 1$, $a_n = 2n - 1$

$$s_n = \frac{(\cancel{1} + 2n - \cancel{1})n}{2} = n^2$$

Exercises 12.5

1. $a_1 = 2, a_2 = 4, a_3 = 8$

$$r = \frac{a_2}{a_1} = 2$$

$$a_4 = 2a_3 = 16, a_5 = 2a_4 = 32$$

3. $a_1 = 1, a_2 = -1/2, a_3 = 1/4$

$$r = \frac{a_2}{a_1} = -1/2$$

$$a_4 = -1/8, a_5 = 1/16$$

5. $a_1 = 1/2, a_2 = 1/3, a_3 = 2/9$

$$r = \frac{a_2}{a_1} = 2/3$$

$$a_4 = 4/27, a_5 = 8/81$$

7. $\log 3, \log 9, \log 81, \ldots$

$$r = \frac{\log 9}{\log 3} = \frac{\log 3^2}{\log 3}$$

$$= \frac{2 \cancel{\log 3}}{\cancel{\log 3}} = 2$$

$$a_n = (\log 3)2^{n-1}, n \geq 1$$

9. $\log 16, \log 4, \log 2, \ldots$

$$r = \frac{\log 4}{\log 16} = \frac{\log 4}{\log 4^2}$$

$$= \frac{\log 4}{2 \log 4} = \frac{1}{2}$$

$$a_n = \log 16 \left(\frac{1}{2}\right)^{n-1}, \; n \geq 1$$

11. $a_1 = 5, \; r = 4$
$a_2 = a_1 r = 20$
$a_3 = a_1 r^2 = 80$
$a_4 = a_1 r^3 = 320$

13. $a_1 = 3/8, \; a_2 = 1/4$

$$r = \frac{1/4}{3/8} = 2/3$$

$$a_5 = a_1 r^4$$

$$= \left(\frac{3}{8}\right)\left(\frac{2}{3}\right)^4$$

$$= \frac{2}{27}$$

15. $a_3 = 2, \; a_6 = -1/4$
$2 = a_1 r^2$
$-1/4 = a_1 r^5$

$$-8 = \frac{1}{r^3}, \text{ so } r = -1/2$$

$$2 = a_1 \left(-\frac{1}{2}\right)^2$$

$$a_1 = 8$$

17. $a_1 = 6, \; r = -2$

$$s^{10} = \frac{a_1(1 - r^{10})}{1 - r}$$

$$= \frac{6(1 - (-2)^{10})}{1 - (-2)}$$

$$= \frac{6(1 - 1024)}{3}$$

$$= -2046$$

19. The terms in the sum

$$\sum_{k=1}^{12} (-2)^{k-1} \text{ form a geometric}$$

sequence whose ratio is $r = -2$. since $a_1 = (-2)^{1-1} = 1$, it follows that

$$s_{12} = \frac{a_1(1 - r^{12})}{1 - r}$$

$$= \frac{1 - (-2)^{12}}{1 - (-2)}$$

$$= -\frac{4095}{3} = -1365$$

21. $$\sum_{k=1}^{10} 3^{k-1} = \frac{1(1 - 3^{10})}{1 - 3}$$

$$= \frac{-59048}{-2}$$

$$= 29524$$

23. $1 - \frac{1}{2} + \frac{1}{4} - \frac{1}{8} + \ldots$

$a_1 = 1, \; r = -1/2$

$$s = \frac{a_1}{1 - r} = \frac{1}{1 - (-1/2)} = \frac{1}{3/2}$$

$$= 2/3$$

25. $3 + \frac{3}{4} + \frac{3}{16} + \frac{3}{64} + \ldots$

$a_1 = 3, \; r = 1/4$

$$s = \frac{3}{1 - 1/4} = \frac{3}{3/4} = 4$$

27. $27 + 9 + 3 + 1 + \ldots$
$a_1 = 27, \; r = 1/3$

$$S = \frac{27}{1 - (1/3)} = \frac{27}{2/3}$$

$$= \frac{81}{2} = 40.5$$

29. $\displaystyle\sum_{n=1}^{\infty} \left(\frac{5}{6}\right)^{n-1} = \frac{1}{1-(5/6)} = 6$

31. $\displaystyle\sum_{n=1}^{\infty} \left(-\frac{4}{5}\right)^{n} = \frac{(-4/5)}{1-(-4/5)} = \frac{-4/5}{9/5}$

$= -4/9$

33. $a_1 = 18/100$, $r = 1/100$

$s = \dfrac{a_1}{1-r} = \dfrac{18/100}{1-(1/100)}$

$= \dfrac{18/100}{99/100} = 18/99 = 2/11$

35. $a_1 = 63/100$, $r = 1/100$

$s = \dfrac{63/100}{1-(1/100)} = \dfrac{63/100}{99/100}$

$= 63/99 = 7/11$

37. $0.3\overline{18} = 0.3 + 0.018$
$+ 0.00018 + 0.0000018 + \ldots$

The first term of the infinite series
$0.018 + 0.00018 + 0.0000018 + \ldots$
is $a_1 = 0.018$ and the ratio is
$r = 0.01$. Hence, its sum is

$\dfrac{0.018}{1-0.01} = \dfrac{0.018}{0.99} = \dfrac{18}{990}$

$= \dfrac{1}{55}.$

Thus

$0.3\overline{18} = 0.3 + \dfrac{1}{55}.$

$= \dfrac{3}{10} + \dfrac{1}{55} = \dfrac{35}{110} = \dfrac{7}{22}$

39. $0.12\overline{42} = 0.12 + 0.0042$
$+ 0.000042 + \ldots$

$0.00\overline{42} = \dfrac{0.0042}{1-0.01} = \dfrac{0.0042}{0.99}$

$= \dfrac{42}{9900} = \dfrac{7}{1650}$

Thus

$0.12\overline{42} = \dfrac{12}{100} + \dfrac{7}{1650}$

$= \dfrac{396 + 14}{3300} = \dfrac{410}{3300}$

$= \dfrac{41}{330}$

41. Let d be the total distance traveled by the tennis ball. Then

$d = 6 + 2(6) \displaystyle\sum_{n=1}^{\infty} r^n$

with $r = 0.8$. Thus

$d = 6 + 12 \cdot \dfrac{0.8}{1-0.8}$

$= 6 + 12 \cdot \dfrac{0.8}{0.2}$

$= 6 + 12 \cdot 4$

$= 54$ ft

43. $a_1 = 200$, $r = 1.2$
$a_n = 200(1.2)^{n-1}$, $n \geq 1$
$a_4 = 200(1.2)^3 \simeq 346$

45. $a_0 = 100$, $r = 1.30$
$a_n = 100(1.30)^n$
$a_{10} = 100(1.30)^{10} \simeq 1379$

47. The total number of revolutions is

$$1200 + 1200(1/3)$$
$$+ 1200(1/3)^2 + \ldots$$
$$= 1200[1 + (1/3) + (1/3)^2 + \ldots]$$

$$= 1200[\frac{1}{1 - (1/3)}]$$

$$= 1200[\frac{1}{2/3}]$$

$$= 1200(\frac{3}{2})$$

$$= 1800 \text{ revolutions}$$

49. The total amount of money spent as a result of the convention is

$$T = 200,000 + 200,000(3/4)$$
$$+ 200,000(3/4)^2 + \ldots$$

$$= 200,000[1 + (3/4)$$
$$+ (3/4)^2 + \ldots]$$

$$= 200,000[\frac{1}{1 - (3/4)}]$$

$$= 200,000[\frac{1}{1/4}]$$

$$= \$800,000$$

Exercises 12.6

1. $P(6,2) = 6!/(6 - 2)! = 6!/4! = 30$

3. $P(8,3) = 8!/5! = 336$

5. $P(6,6) = \frac{6!}{0!} = 720$

7. $P(40,4) = 40!/36! = 2193360$

9. $P(80,75) = 2,884,801,920$

11. $C(5,2) = \frac{5!}{2!3!} = 10$

13. $C(9,3) = \frac{9!}{3!6!} = 84$

15. $C(10,6) = 10!/6!4! = 210$

17. $C(40,2) = 40!/38!2! = 780$

19. $C(52,13) \approx 6.3501356 \times 10^{11}$

21. $\Gamma(10,3,3,4) = 10!/3!3!4!$
$$= 4200$$

23. $\Gamma(52,5,5,5,5,32)$
$$\approx 1.4782628 \times 10^{24}$$

25. $(10)(9) = 90$

27. $(8)(7)(6) = 336$

29. $(10)(9)(8) = 720$

31. $5! = 120$

33. $P(4,3) = 24$
$P(4,4) = 24$

35. $P(6,3) = 120$

37. $C(12,5) = 792$

39. $C(52,7) = 133,784,560$

41. $C(8,2) = 28$

43. $\Gamma(10,3,3,4) = 4200$

45. $\Gamma(52,13,13,13,13)$
$$\approx 5.3644738 \times 10^{28}$$

47. $\Gamma(11,4,4,2,1) = 34650$

49. $C(14,4) = 1001$
$C(8,2) \, C(6,2) = 420$

Exercises 12.7

1. E = {1,3,5},
    S = {1,2,3,4,5,6}
    P = n(E)/n(S) = 3/6 = 1/2

3. 2/3

5. 1

7. E = {(1,5),(2,4),(3,3),(4,2)
        (5,1)};
    n(S) = 36
    P = n(E)/n(S) = 5/36

9. 21/36 = 7/12

11. 15/36 = 5/12

13. 0

15. E = {(H,H)},
    S = {(H,H),(H,T),(T,H),(T,T)}
    P = n(E)/n(S) = 1/4

17. Sample space

    (1,2) (1,3) (1,4) (1,5)
    (2,1) (2,3) (2,4) (2,5)
    (3,1) (3,2) (3,4) (3,5)
    (4,1) (4,2) (4,3) (4,5)
    (5,1) (5,2) (5,3) (5,4)

    P = n(E)/n(S) = 12/20 = 3/5

19. 2/20 = 1/10

21. Sample space

    (T,T,T) (T,T,F) (T,F,T) (T,F,F)
    (F,T,T) (F,T,F) (F,F,T) (F,F,F)
    P = n(E)/n(S) = 1/8

23. 4/8 = 1/2

25. $n(S) = 2^5 = 32$

    a) $n(E) = \binom{5}{5} = 1$, P = 1/32

    b) $n(E) = \binom{5}{3} = 10$,

       P = 10/32 = 5/16

27. A deck of 52 playing cards has 4
    suits (diamonds, hearts, spades,
    and clubs) each consisting of 13
    cards.

    $n(S) = \binom{52}{5}$, $n(E) = 4\binom{13}{5}$

    $P = 4\binom{13}{5}/\binom{52}{5} \approx 1.98 \times 10^{-3}$

29. $\binom{4}{3}\binom{4}{2}\binom{44}{0}/\binom{52}{5} \approx 9.23 \times 10^{-6}$

31. $(4)(9)/\binom{52}{5} \approx 1.385 \times 10^{-5}$

33. a) $\dfrac{1}{13} + \dfrac{1}{13} = \dfrac{2}{13}$

    b) $\dfrac{1}{4} + \dfrac{1}{4} = \dfrac{1}{2}$

35. a) $\dfrac{6}{4+6+8} = \dfrac{6}{18} = \dfrac{1}{3}$

    b) $\dfrac{4+8}{4+6+8} = \dfrac{12}{18} = \dfrac{2}{3}$

37. $1/\binom{8}{2} = 1/28$

39. 1/3

41. (a) $p = 5/8$

   $p' = 1 - p = 1 - 5/8 = 3/8$

   $$\frac{p}{p'} = \frac{5/8}{3/8} = \frac{5}{3}$$

   5 to 3

   (b) $p = 0.3 = 3/10$

   $p' = 0.7 = 7/10$

   $$\frac{p}{p'} = \frac{3/10}{7/10} = \frac{3}{7}$$

   3 to 7

43. (a) $$\frac{p}{1 - p} = \frac{3}{2}$$

   $2p = 3(1 - p)$
   $5p = 3$

   $$p = \frac{3}{5} = 0.6$$

   (b) $$\frac{p}{1 - p} = \frac{3}{5}$$

   $5p = 3(1 - p)$
   $8p = 3$

   $$p = \frac{3}{8} = 0.375$$

45. (a) $$p = \frac{4}{6} = \frac{2}{3}$$

   $$p' = 1 - p = \frac{1}{3}$$

   Odds in favor: 2 to 1

   (b) There are 13 diamond cards in a standard 52-card deck. The probability of drawing a diamond is

   $$p = 13/52 = 1/4$$

   Thus, $p' = 1 - p = 3/4$ and

$$\frac{p}{p'} = \frac{1/4}{3/4} = \frac{1}{3}$$

Odds in favor: 1 to 3

47. (a) The probability of selecting a defective battery is

   $$p = 3/10.$$

   (b) The probability of selecting a good battery is

   $$p = 7/10.$$

   Odds in favor: 7 to 3.

49. The probability of selecting a bottle of white Burgundy is

   $$p = 18/55.$$

   Since

   $$1 - p = 37/55,$$

   it follows that the odds are 18 to 37.

Review Exercises - Chapter 12

1. 816

3. $\dfrac{1}{32}a^5 + \dfrac{5}{8}a^4b + 5a^3b^2$

   $+ 20a^2b^3 + 40ab^4 + 32b^5$

5. $\binom{7}{5}(2u^2)(-b^5) = -84u^2b^5$

7. $\binom{6}{3}(3x^3)(-4)^3 = -34560x^3$

9. Fifth term:

   $\binom{8}{4}(a^{1/2})^4(b^{1/2})^4 = 70a^2b^2$

11. $(1.02)^6 = (1 + 0.02)^6 \simeq 1.1261624$

13. $9/2, 4, 7/2, 3, 5/2$

15. $1, -1/2, 1/6, -1/24, 1/120$

17. $0, 1/3, -1/2, 3/5, -2/3$

19. $\begin{aligned} a_n &= s_n - s_{n-1} \\ &= (4^n + 1) - (4^{n-1} + 1) \\ &= 4^n - 4^{n-1} \\ &= 4^{n-1}(4 - 1) \\ &= 3(4^{n-1}) \end{aligned}$

21. $n(n + 1)$

23. $2.1, 2.71\overline{6}, 2.6466769, 2.6457515$

25. $d = -3; \ 1, -2$

27. $d = -\ln 3; \ \ln(1/9), \ \ln(1/27)$

29. $d = 5/2, \ a_{10} = 47/2$

31. $11, 528$

33. $r = 5/6; \ 25/72, 125/432$

35. $r = 1.5; \ 16.875, 25.3125$

37. $r = 2/3, \ a_2 = 2/3$

39. $\displaystyle \sum_{n=1}^{\infty} 3\left(-\frac{1}{3}\right)^{n-1} = 3\sum_{n=1}^{\infty} \left(-\frac{1}{3}\right)^{n-1}$

$\displaystyle = 3 \cdot \frac{1}{1 - (-1/3)} = 9/4$

41. $0.\overline{27} = \dfrac{0.27}{1 - 0.01} = \dfrac{27}{99} = \dfrac{3}{11}$

43. $0.0\overline{24} = \dfrac{0.024}{1 - 0.01} = \dfrac{24}{990} = \dfrac{4}{165}$

45. $\begin{aligned} 4 + 8 + &\ldots + 4k + 4(k + 1) \\ &= 2k(k + 1) + 4(k + 1) \\ &= 2(k + 1)(k + 2) \end{aligned}$

47. $\begin{aligned} a(b_1 + &\ldots + b_k + b_{k+1}) \\ &= a(b_1 + \ldots + b_k) + ab_{k+1} \end{aligned}$

$= ab_1 + \ldots + ab_k + ab_{k+1}$

49. $\begin{aligned} (1 + x)^{k+1} &= (1 + x)^k(1 + x) \\ &\geq (1 + kx)(1 + x) \\ &= 1 + (k + 1)x + kx^2 \\ &\geq 1 + (k + 1)x \end{aligned}$

51. $\begin{aligned} (a + b)^{k+1} &= (a + b)^k(a + b) \\ &> (a^k + b^k)(a + b) \\ &= a^{k+1} + b^{k+1} + a^k b + ab^k \\ &> a^{k+1} + b^{k+1} \end{aligned}$

53. $2^n = (1 + 1)^n$

$= \binom{n}{0}1^n + \binom{n}{1}1^{n-1}(1) + \ldots$

$+ \binom{n}{n-1}(1)1^{n-1} + \binom{n}{n}1^n$

$= \binom{n}{0} + \binom{n}{1} + \ldots + \binom{n}{n-1} + \binom{n}{n}$

55. $a^n = 16(2n - 1), \ s_n = 16n^2$

$\begin{aligned} a_{11} &= 16(2 \cdot 11 - 1) \\ &= 16(21) \\ &= 336 \text{ ft} \end{aligned}$

$\begin{aligned} s_{11} &= 16(11)^2 \\ &= 1936 \text{ ft} \end{aligned}$

57. $a_1 = 12, \ d = -1/8, \ n = 11$

$\begin{aligned} a_{11} &= 12 + (11 - 1)(-1/8) \\ &= 43/4 \text{ inches} \end{aligned}$

$s_{11} = \dfrac{(12 + 43/4)11}{2}$

$= 1001/8 \text{ inches}$

59. $a_1 = 2, \ r = 0.8$

$s = \dfrac{2}{1 - 0.8} = \dfrac{2}{0.2}$

$= 10 \text{ meters}$

61. a) 336
     b) 151,200
     c) 252
     d) 495
     e) 280
     f) 1260

63. $P(4,4) = 4! = 24$

65. $\binom{8}{2} = 28$

67. $2\binom{8}{2} = 56$

69. $\binom{18}{12} = 18564$

71. $\dfrac{5}{36} + \dfrac{4}{36} + \dfrac{3}{36} + \dfrac{2}{36} + \dfrac{1}{36}$

   $= \dfrac{15}{36} = \dfrac{5}{12}$

73. $n(S) = 15;\ n(E) = 3$
     $P = 3/15 = 1/5$

75. $P = 1/\binom{20}{2} = 1/380$

77. $\binom{13}{3}\binom{13}{2}/\binom{52}{5} \simeq 0.0086$

79. $\binom{5}{3}\binom{5}{2}/\binom{10}{5} \simeq 0.3968$

Chapter 12 Test

In Problems 1 - 5, use mathematical induction to prove each statement.

1. $1 + 3 + 5 + \ldots + (2n - 3) = (n - 1)^2$, $n > 1$.

2. $\dfrac{1}{1 \cdot 2} + \dfrac{1}{2 \cdot 3} + \ldots + \dfrac{1}{(n - 1)n} = \dfrac{n - 1}{n}$, $n > 1$

3. $1 + 5 + 9 + \ldots + (4n - 3) = (2n - 1)n$, $n \geq 1$.

4. $2^{2n} - 1$ is divisible by 3 for all $n \geq 1$.

5. $n^2 > 2n - 1$ for $n > 1$.

6. Evaluate $\binom{14}{10}$.

7. Using the Binomial Theorem expand $(a - b)^6$.

8. Find the third term in the expansion of $(2x + y)^8$.

9. Find the middle term in the expansion of $(\sqrt{x} - \sqrt{3})^4$.

10. Find the two middle terms in the expansion of $(3x - 2y)^5$.

11. Write the first six terms of the sequence whose general term is

    $$a_n = \frac{(-1)^{n-1}n}{n + 1}.$$

12. Write the first four terms of the sequence defined by the following recurrence relation $a_1 = 5$, $a_k = k/2a_{k-1}$.

13. Let $s_n = 3n(n + 3)/2$ be the nth partial sum of a series. Find the general term a_n of the sequence which originates the series.

14. Use Newton's method to find four decimal approximations of $\sqrt{7}$. Take as initial guess $x_1 = 2$.

15. Find the common difference and determine the next three terms of the arithmetic sequence $\ln 3$, $\ln 9$, $\ln 27$, $\ldots$.

16. The fourth and seventh terms of an arithmetic sequence are 4 and -2 respectively. Find the eleventh term.

17. Let $a_1 = 58$ and $d = -4$ be the first term and the common difference of an arithmetic sequence. Find s_{17}.

18. Find the sum of all two-digit odd numbers.

19. What is the ratio of the geometric sequence 27/1000, 9/100, 3/10, ... ? Find the next three terms.

20. The first and fourth terms of a geometric sequence are 8 and -1/8. Find the sixth term.

21. Write the repeating decimal $0.\overline{24}$ as a fraction in lowest terms.

22. Compute the infinite sum $\sum\limits_{n=0}^{\infty} (2/3^n)$.

23. Find C(18,15).

24. Find Γ(12,6,4,2).

25. Find the number of distinguishable permutations obtained with the letters of the word BARBARIC.

26. You are dealt five cards from a standard 52-card deck. What is the probability that you'll receive four spades and one heart?

27. A school committee of four is selected at random from a group of 5 boys and 3 girls. What is the probability that 3 boys and 1 girl will form a committee?

28. Assuming that all birthdays are equally likely, what is the probability that a person's birthday will fall in April?

29. From a wine cellar where there are 15 bottles of red Burgundy, 18 of white Burgundy, 10 of red Bordeaux, and 12 of white Bordaux, a bottle is selected at random. What is the probability of selecting a white Burgundy? A red wine?

30. In how many ways can 9 people be divided into 3 comittees of 4, 3, and 2 people with no person in more than 1 committee?

Answers To Test Problems

Chapter 1

1. $2^3 \times 3 \times 7$ 2. 0.53125   3. 7/30   4. $0.\overline{79}$ 5. 35/148
6. a) rational, b) irrational, c) rational, d) irrational
7. $x^3/3y^7$ 8. $9b^8/4a^4$ 9. $\approx 1.89 \times 10^{-5}$ 10. 6.94×10^{-6}
11. $6x^3 + 10x^2 + 8$ 12. $x - 4a + 5a^2x^2$ 13. $2(5a + 6)(5a - 6)$
14. $(x + 2)(x + 1)(x - 1)$ 15. $(25x^2 + 16)(5x + 4)(5x - 4)$ 16. $1/(x + 3)$
17. y^3 18. $-(x + 6)/(x + 2)(x + 3)$ 19. $xy/(y - x)$ 20. 0
21. $(19 + 6\sqrt{2})/17$ 22. $1/(\sqrt{x + h} + \sqrt{x})$ 23. $a^{3/2}b^{5/2}$
24. $4b^3/5a^3$ 25. $6\sqrt{4^2 \cdot 3^3}$ 26. $(4x + 3)/\sqrt{x + 1}$
27. Note that $(2\sqrt{3} - \sqrt{5})^2 = 17 - 4\sqrt{15}$ 28. $34 - 14i$
29. $(12/37) - (2/37)i$ 30. $x = 3$

Chapter 2

1. $(1 + 2\sqrt{5})/3$ 2. $7/5 < \sqrt{2} < 17/12 < 3/2$ 3. 114, 116, 118, 120
4. $x > 10$ 5. 9, -9   6. 4, 5   7. 12, 13 or -13, -12
8. $x < -4$ or $x > 5$ 9. 18   10. 1210 pounds   11. 5, -3
12. $x > 2$ 13. -1, 12   14. -1   15. 1 inch
16. $-1 < x < 0$ or $0 < x < 1$ 17. 12/5
18. 50 mg of caffeine, 150 mg of phenacetin, 300 mg of acetylsalicylic acid
19. $-3x^3$ 20. $x < -2/3$ or $x > 2$ 21. 9, 3   22. 0, 6   23. 8 seconds
24. $h = 3V/\pi(r^2 + rR + R^2)$ 25. 4 hours   26. 3, 3-$\sqrt{2}$, 6-$\sqrt{2}$
27. 125 mi/hr to 150 mi/hr   28. $(3 \pm i\sqrt{127})/4$ 29. $-3/2 < x \leq 0$ or $1/3 \leq x < 1$
30. $t > 55$

Chapter 3

1. $(-2, -2)$ 2. 5, -17 3. -2
4. $d(A, B) = d(C, D) = \sqrt{52}$, $d(A, D) = d(B, C) = \sqrt{34}$ 5. $a = -6$, $b = 4$
6. Yes, $d(A, B) = \sqrt{25}$, $d(B,C) = \sqrt{25}$, $d(A, C) = \sqrt{50}$ 7. 4 8. 7
9. $y = x/3 + 17/3$ 10. $2x - 5y + 30 = 0$ 11. 5/6, -1/2; neither
12. -12/5 13. $y = (-3/2)x - 8$ 14. $(-3/2, -1)$, -8/5, $10x - 16y - 1 = 0$
15. $9X + 4Y + 2 = 0$ 16. $(x + 4)^2 + (y - 2)^2 = 65$
17. $(x - 4)^2 + (y + 1/2)^2 = 29/4$ 18. Symmetric with respect to the line $x = 3$
19. Circle with center $(2, -1)$ and radius 6 20. $(-1, 3)$, $(1, -1)$
21. $k > -1/3$ 22. $-6 < b < 6$ 23. $V(1, 3)$, $F(3/2, 3)$, $y = 3$, $x = 1/2$
24. $(x - 3)^2 = -8y$ 25. $C(-1, 0)$, $a = 2$, $b = 1$ 26. $x^2/25 + y^2/9 = 1$
27. $(x - 2)^2/25 + (y + 1)^2/16 = 1$ 28. $x^2/8 - y^2/1 = 1$
29. $(y - 1)^2/9 - (x - 2)^2/16 = 1$
30. $(x - 1)^2 - (y + 1)^2/4 = 1$, center: $(1, -1)$, vertices: $(0, -1)$, $(2, -1)$

Chapter 4

1. Domain: $\Re$, range: $y \geq -5$ 2. Domain: $-3 \leq x \leq 3$, range: $0 \leq y \leq \sqrt{18}$
3. Domain: $\Re$, range: $y \geq 0$ 4. Domain: $x < 3/2$, range: $y > 0$
5. -14, 9/4, 6, -6 6. a) 1/3, b) not defined, c) 2, d) 1

7. a) $2\left(a + \dfrac{1}{a}\right)$, b) 0, c) $\left(a + \dfrac{1}{a}\right)^2$, d) $\dfrac{a^2 + 1}{a} + \dfrac{a}{a^2 + 1}$

8. a) $(a - b)/(a + 1)(b + 1)$, b) $a/(a + b)$, c) $a(b + 1)/b(a + 1)$, d) $a/(2a + 1)$
9. a) $(-x^2 + 11x - 25)/x(x - 5)$, $x \neq 0$, $x \neq 5$, b) $1/x$, $x \neq 0$, $x \neq 5$
10. a) $1 + x$, $x \neq 0$, b) $(x + 4)/(x^2 - 4)$, $x \neq 0$, $x \neq \pm 2$ 11. -5
12. $-1/x(x + h)$ 13. $16 + 4h$ 14. $1/2(2 + h)$ 15. $A = h(h - 6)/4$
16. $A = 2d^2/5$ 17. $V = \pi d^2/6$ 18. $J = 1/t + 1/(t - 2)$ 23. odd
24. even 25. odd 26. even 27. Increasing on $(-\infty, 1)$, decreasing on $(1, \infty)$
28. Decreasing on $(-\infty, -2)$, increasing on $(-2, \infty)$ 29. $w = kx^2/yz$, w is doubled
30. $C = ka\ell/m$

Chapter 5

1. $V(1, -32)$, $x = 1$, x-intercepts: -3 and 5, y-intercept: -30, upward
2. $V(-5, 9)$, $x = -5$, x-intercepts: -8 and -2, y-intercept: -16, downward
3. $V(7, 0)$, $x = 7$, x-intercept: 7(parabola tangent to the x-axis), y-intercept:
   -98, downward 4. $V(0, 5)$, $x = 0$, no x-intercepts, y-intercept: 5, upward
5. $(5, -9)$, $x = 5$ 6. $(-3, -3)$, $x = -3$ 7. $k < -4$ or $k > 0$ 8. $\pm 4\sqrt{3}$
9. $0 < a < 9$ 10. 2500 units, $37,500 11. -32 and 32
12. a) even, b) x-intercepts: $\pm\sqrt{2}$, y-intercept: 1, c) $c(x) > 0$ on $(-\sqrt{2}, \sqrt{2})$,
    $c(x) < 0$ on $(-\infty, -\sqrt{2})$ or $(\sqrt{2}, \infty)$ 13. Symmetric about the point $(-2, 4)$
14. Symmetric about the line $x = 5$ 15. 7 ft by 14 ft
16. Domain: $x \neq 3/2$, V.A.: $x = 3/2$, H.A.: $y = 0$, y-intercept: -5/3
17. Domain: $x \neq -6$, V.A.: $x = -6$, H.A.: $y = 2$, x-intercept: -5/2,
    y-intercept: 5/6

18. Domain: $x \neq 3/2$ and $x \neq 1$, V.A.: $x = 3/2$ and $x = 1$,
    H.A.: $y = 0$, x-intercept: $1/2$, y-intercept: $-2/3$
19. Domain: $x \neq 1/2$,
    V.A.: $x = 1/2$, H.A.: $y = 1/4$, x-intercepts: ± 1, y-intercept: -1
22. a) $0 \leq x \leq 200$, b) $\$2.55$, $\$2$, $\$1.275$, $\$0.85$
23. $2\sqrt{x - 4} + 1$, $\sqrt{2x - 3}$, $\sqrt{\sqrt{x - 4} - 4}$ 24. 3, $1/5$, 1
25. $f^{-1}(x) = -\sqrt{3 - x}/2$, $x \leq 3$ 28. $0 \leq x < 100$ 29. $\$80,000$, $\$240000$, $\$720,000$.
    No, as x approaches 100 the cost increases without bound
30. $(R \circ q)(n) = 600n - 6n^2$

Chapter 6

1. $1/36$, 36, $1/6\sqrt{6}$, $6^3\sqrt{6}$ 2. $17/8$, 3, 4, 34
3. a) 0.497, b) 1.284, c) 0.472, d) 0.091   4.   3.29, 3.317, 3.322; 3.3219971
5. 5, $25/8$, 35   6. $A = -5$, $B = 2$; $-5/8$
7. $N(t) = 1600 \cdot 2^{t/4}$, a) 6400, b) $6400\sqrt{2} \approx 9051$
8. a) the initial amount of Carbon 14, b) 5600 years, c) 10 mg
9. $77°F$ 10.   9 pounds per square inch, 3.3 miles   11.   491,238, 34.7 years
12. 1992   13.   a) $\log_6 1296 = 4$, b) $\log_4(1/1024) = -5$, c) $\log_3 a = 8$, d) $\log_x 15 = -3$
14. a) $343 = 7^3$, b) $10 = 100^{1/2}$, c) $x = 4^{-2}$, d) $18 = 5^y$ 15. $3(1 - \log_2 5)/2$
16. $\log_4(1/10)$ 17. $\log[(3x + 1)^{3/4}(2x + 7)^{1/2}/(x + 4)]$
18. $\ln 8 + 2 \ln p + 5 \ln q - 4 \ln r$ 19.   a) 0.8669, b) -1.6635
20. a) 3.84×10^3, b) 7.46×10^{-2} 21.   105 decibels   22. $I_1 = 10 I_2$
23. a) 3.32, b) 5.16   24.   1.1   25. $5/3$ 26.   5   27. $\log 24/\log(16/9)$
28. No solution   29.   0, 999   30. $\log_2 5$

Chapter 7

1. a) $60°$, b) $150°$ 2.   a) $5\pi/6$, b) $3\pi/4$ 3.   a) second, b) third
4. a) $-11\pi/6$, b) $5\pi/6$ 5.   a) $220°$, b) $-70°$ 6.   4.5 radians
7. a) $\pi/6$ radians $= 30°$, b) $\pi/18$ radians $= 10°$ 8.   a) $114.59°$, b) $32.73°$
9. a) $11\pi/10$ rad/s, b) $11\pi/2$ rad, c) $33\pi/5$ in/s   10.   1885 mi
11. $\sin \beta = 20/29$, $\cos \beta = 21/29$, $\tan \beta = 20/21$, $\cot \beta = 21/20$, $\sec \beta = 29/21$,
    $\csc \beta = 29/20$
12. $\sin \alpha = 2\sqrt{2}/3$, $\cos \alpha = 1/3$, $\tan \alpha = 2\sqrt{2}$, $\cot \alpha = 1/2\sqrt{2}$,
    $\sec \alpha = 3$, $\csc \alpha = 3/2\sqrt{2}$
13. $\sin \alpha = \sqrt{55}/8$, $\tan \alpha = \sqrt{55}/3$, $\cot \alpha = 3/\sqrt{55}$,
    $\sec \alpha = 8/3$, $\csc \alpha = 8/\sqrt{55}$
14. $\sin \beta = 1/4$, $\cos \beta = \sqrt{15}/4$, $\tan \beta = 1/\sqrt{15}$,
    $\cot \beta = \sqrt{15}$, $\sec \beta = 4/\sqrt{15}$ 15. $\sin \alpha = 0.40$, $\cos \alpha = 0.92$
16. $\sin \alpha = 3/\sqrt{13}$, $\cos \alpha = 2/\sqrt{13}$ 17. $6\sqrt{2}$ inches   18.   6 ft, $6\sqrt{3}$ ft
19. a) 0.7254, b) 0.3772   20.   a) 0.9033, b) 1.1044
21. $\beta = 60°$, $b = 8\sqrt{3}$, $c = 16$ 22. $\alpha = 45°$, $b = 6$, $c = 6\sqrt{2}$
23. $\beta = 49°50'$, $a \approx 2.9$, $c \approx 4.4$ 24. $\beta \approx 74°30'$, $b \approx 20.2$, $c \approx 21$ 25. $8\sqrt{2}$ m
26. 38.04 ft^2 27.   48.56 ft $\approx$ 49 ft   28.   357 ft   29.   112.48 m   30.   0.25 miles

Chapter 8

1. $(-\sqrt{3}/2,\ 1/2)$, $\sin\theta = 1/2$, $\cos\theta = -\sqrt{3}/2$, $\tan\theta = -1/\sqrt{3}$, $\cot\theta = -\sqrt{3}$, $\sec\theta = -2/\sqrt{3}$, $\csc\theta = 2$ 2. $(-1/2),\ -\sqrt{3}/2)$, $\sin\alpha = -\sqrt{3}/2$, $\cos\alpha = -1/2$, $\tan\alpha = \sqrt{3}$, $\cot\alpha = 1/\sqrt{3}$, $\sec\alpha = -2$, $\csc\alpha = -2/\sqrt{3}$

3. $\tan 1/2/\sqrt{2}$, $\cos\alpha = 2\sqrt{2}$, $\sec\alpha = 3/2\sqrt{2}$, $\csc\alpha = 3$

4. $\sin\beta = 1/4$, $\cos\beta = \sqrt{15}/4$, $\tan\beta = 1/\sqrt{15}$, $\cot\beta = \sqrt{15}$, $\sec\beta = 4/\sqrt{15}$

5. $\sin\alpha = 0.8799$, $\cos\alpha = -0.4752$, $\tan\alpha = -1.8517$, $\cot\alpha = -0.5400$, $\csc\alpha = 1.1365$

6. $\sin\alpha = -3/\sqrt{10}$, $\cos\alpha = 1/\sqrt{10}$, $\tan\alpha = -3$, $\cot\alpha = -1/3$, $\sec\alpha = \sqrt{10}$, $\csc\alpha = -\sqrt{10}/3$

7. $\sin\alpha = 4/5$, $\cos\alpha = -3/5$, $\tan\alpha = -4/3$, $\cot\alpha = -3/4$, $\sec\alpha = -5/3$, $\csc\alpha = 5/4$

8. $\sin\alpha = -1/\sqrt{5}$, $\cos\alpha = 2/\sqrt{5}$, $\tan\alpha = -1/2$, $\cot\alpha = -2$, $\sec\alpha = \sqrt{5}/2$, $\csc\alpha = -\sqrt{5}$

9. $\sin\alpha = 4/\sqrt{17}$, $\cos\alpha = 1/\sqrt{17}$, $\tan\alpha = 4$, $\cot\alpha = 1/4$, $\sec\alpha = \sqrt{17}$, $\csc\alpha = \sqrt{17}/4$

10. $(-3/5,\ 4/5)$;   cosine and sine of the angle in standard position whose terminal side contains the point $(-6,-8)$ 11.   30°

12. $\pi/3$ 13. $-\sqrt{3}/2$ 14. -1 15.   49°40′, $\cos 130°20' = -\cos 49°40' = -0.6472$

16. Amplitude: 4, period: $2\pi/3$ 17.   Amplitude: 3, period: 8π

18. Amplitude: 2, period: $\pi/2$, phase shift: $\pi/4$

19. Amplitude: 5, period: 4, phase shift: $\pi/2$ 22.   even

23. a) 3 cm, b) angular frequency: 5, frequency: $5/2\pi$ seconds, c) $k = 6.250 \times 10^3$

24. $y = 0.5 \cos 4t$, $T = \pi/2$ seconds $\simeq 1.6$ seconds   25.   9.87 m/s^2

26. 5400, 4200, January 1, 1985   27. $2/\sqrt{5}$ 28. $1/\sqrt{5}$

29. $x = (1/2)\ \text{arc}\ \cos\ (y/5)$, $-5 \le y \le 5$ 30. $x = \pi + \text{arc}\ \sin\ (y/3)$, $-3 \le y \le 3$

Chapter 9

3. a) $(-\sqrt{2} - \sqrt{6})/4$, b) $(\sqrt{2} - \sqrt{6})/4$ 4. $\sin 59°$ 5. $\cos 3a - \cos 7a$

6. $(2 - \sqrt{63})/10$, $-(\sqrt{6} + \sqrt{21})/10$, third quadrant

8. $\sin 2\alpha = -24/25$, $\cos 2\alpha = 7/25$, $\tan 2 = -24/7$

9. $\sin(\alpha/2) = 3/\sqrt{13}$, $\cos(\alpha/2) = 2/\sqrt{13}$, $\tan(\alpha/2) = 3/2$ 10. $\sqrt{2 - \sqrt{3}/2}$

11. 0.9102   13. $\pi/2$, $3\pi/2$ 14.   270°   15. $\pi/3$, $2\pi/3$ 16.   0°, 30°, 150°, 180°

17. $\gamma = 100°$, $a = 3.52$, $c = 6.94$ 18. $\gamma = 81.1°$, $a = 10.67$, $b = 13.15$

19. 28.1°, 61.9°, 90°   20.   5.04 cm   21.   271 ft, 302 ft   22. $36\sqrt{2}$ mi $\simeq 51$ mi

23. $\sqrt{28}$ in $\simeq 5.3$ in   24.   8.8km

25. $3(\cos\dfrac{\pi}{2} + i\sin\dfrac{\pi}{2})$ 26. $30(\cos 162° + i\sin 162°)$ 27. $-8 + 8\sqrt{3}i$

28. $\dfrac{\sqrt{3}}{3} - \dfrac{i}{3}$ 29.   5, $5(\cos\dfrac{2\pi}{3} + i\sin\dfrac{2\pi}{3})$, $5(\cos\dfrac{4\pi}{3} + i\sin\dfrac{4\pi}{3})$

Chapter 10

1. $(3,2)$ 2. $(1/2,-2)$ 3. $(-1/3,1)$ 4. 400 km/h, 50 km/h 5. 240 liters
6. $(3,1,-2)$ 7. $(2,-2,2)$ 8. $(2,0,1)$ 9. $(-z\ -z,\ z)$, z any number
10. 20 nickels, 15 dimes, 35 quarters

11. $\begin{bmatrix} 0 & 1 \\ 2 & -2 \\ 5 & -4 \end{bmatrix}, \begin{bmatrix} 5 & -8 \\ 4 & -4 \\ 0 & -3 \end{bmatrix}$ 12. $AB = BA = \begin{bmatrix} 1 & 3 & 6 \\ 0 & 1 & 3 \\ 0 & 0 & 1 \end{bmatrix}$ 13. $\begin{bmatrix} 1 & 0 \\ -2 & 1 \end{bmatrix}$

14. $\begin{bmatrix} -1/3 & 0 & 2/3 \\ 0 & 1 & 0 \\ 2/3 & 0 & -1/3 \end{bmatrix}$ 15. $(-1,2,1)$ 16. $\begin{bmatrix} 2 & 0 \\ 1 & 0 \end{bmatrix}$, $A_{32} = (-1)^{3+2}\det M_{32} = 0$

17. 13 19. $(-3,-2,2)$ 20. $(3,2,-1)$ 21. $(0,0)$ or $(-2,4)$
22. $(1,3/2)$ or $(-1,-3/2)$ 23. $(-1,3)$ or $(1,3)$ 24. -12 and 20
25. 6 by 8 meters 26. $(-2,3)$, yes; $(4,-2)$, no 27. Maximum $L = 3$ at $(1,0)$,
    minimum $L = -11$ at $(-3,2)$ 28. $L = 9$ at $(5,0)$ 29. $840
30. 2 kg of A, 3 kg of B

Chapter 11

1. a) 0, b) 9, c) 0, d) 27/8 2. $A = 1$, $B = 4$, $C = 6$ 3. $3x^2 + 12x + 47$, 194

4. 6 5. $2x^2 + 2x/3 - 4/3$, $-4x/3 - 7/3$ 6. $4x + \dfrac{-6x + 3}{x^2 + 1}$ 7. -30
8. $3x^2 - 13x + 39$, -112 9. $x^2 - 3x + 4$, 0 10. 49 11. 15 12. -2 13. $4 + i$
14. Yes, other factors: $x - 3$, $x - 1$, $x + 1$ 15. $2 - i$, 1 16. -1, 3
17. $2x^3 - 4x^2 - 6x$ 18. $x^4 + 2x^3 - 2x^2 + 8$ 19. -3, -4, $1/2$ 20. 1, -2, i, $-i$
23. $(x - 4)(x + i)(x - i)$ 24. $(x + 2)(x - 1 - 3i)(x - 1 - 3i)$

25. $A = 2$, $B = -1$ 26. $A = 1$, $B = -3$, $C = 1$ 27. $\dfrac{1}{x + 5} + \dfrac{1}{x - 2}$

28. $\dfrac{2}{x} - \dfrac{1}{x^2} + \dfrac{3}{x + 1}$ 29. $\dfrac{5}{x} - \dfrac{2x}{x^2 + 1}$ 30. $\dfrac{3x}{2x^2 + 1} - \dfrac{1}{(2x^2 + 1)^2}$

Chapter 12

In Problems 1 - 5, we only show how the statement $P(k + 1)$ can be deduced from the statement $P(k)$.

1. $1 + 3 + 5 + \ldots + (2k - 3) + (2k - 1) = (k - 1)^2 + 2k - 1$

$$= k^2 - 2k + 1 + 2k - 1$$

$$= k^2$$

2. $\dfrac{1}{1 \cdot 2} + \dfrac{1}{2 \cdot 3} + \ldots + \dfrac{1}{(k-1)k} + \dfrac{1}{k(k+1)} = \dfrac{k-1}{k} + \dfrac{1}{k(k+1)}$

$$= \dfrac{(k-1)(k+1) + 1}{k(k+1)}$$

$$= \dfrac{k^2 - 1 + 1}{k(k+1)}$$

$$= \dfrac{k}{k+1}$$

3. $1 + 5 + \ldots + (4k - 3) + (4k + 1) = (2k - 1)k + (4k + 1)$

$$= 2k^2 - k + 4k + 1$$

$$= 2k^2 + 3k + 1$$

$$= (2k + 1)(k + 1)$$

4. $2^{2(k+1)} - 1 = 2^{2k} \cdot 2^2 - 1$

$$= 2^{2k} \cdot 2^2 - 2^2 + 2^2 - 1$$

$$= 2^2(2^{2k} - 1) + (2^2 - 1)$$

$$= 2^2(2^{2k} - 1) + 3$$

5. $(k + 1)^2 = k^2 + 2k + 1$

$$> 2k - 1 + 2k + 1$$

$$= 2k + 2k$$

$$= 2k + 2 + 2k - 2$$

$$= 2(k + 1) + 2(k - 1)$$

$$[2(k - 1) > -1 \text{ for all } n > 1]$$

$$> 2(k + 1) - 1$$

6. 1001 7. $a^6 - 6a^5b + 15a^4b^2 - 20a^3b^3 + 15a^2b^4 - 6ab^5 + b^6$ 8. $1792x^6y^2$

9. $18x$ 10. $1080x^3y^2 - 720x^2y^3$ 11. 1/2, -2/3, 3/4, -4/5, 5/6, -6/7

12. 5, 1/5, 15/2, 4/15 13. $3(n + 1)$ 14. $x_1 = 2$, $x_2 = 2.75$, $x_3 = 2.6477273$, $x_4 = 2.645720$ 15. $\ln 3$; $\ln 81$, $\ln 243$, $\ln 729$ 16. -10 17. 442 18. 2475

19. 10/3; 1, 10/3, 100/9 20. -1/128 21. 8/33 22. 3 23. 816 24. 13860

25. 5040 26. $13\binom{13}{4}/\binom{52}{5} \approx 0.0036$ 27. $3\binom{5}{3}/\binom{8}{4} = 3/7$ 28. $30/365 = 0.082$

29. 18/55, 5/11 30. 1260